NEW YORK REVIEW BOOKS
CLASSICS

HEAVEN'S BREA⊺

LYALL WATSON (né Malcolm Lyall-Watson; 1939–2008) was born in Johannesburg, South Africa, the oldest of three sons of a Scottish architect father and a radiologist mother who was descended from the first Dutch governor of the Cape Colony. Raised in part on his grandparents' farm, he attended boarding school in Cape Town and at the age of fifteen matriculated at the University of Witwatersrand, where he earned degrees in botany and zoology. After receiving a doctorate in ethology at the University of London, he embarked on a wide-ranging career that included writing and producing BBC nature documentaries, leading scientific expeditions to the Amazon, serving as director of the Johannesburg Zoo, presenting sumo wrestling tournaments on British television, and writing more than two dozen books, including the best-sellers *Super-nature: A Natural History of the Supernatural* (1973) and *The Romeo Error* (1974). In the late 1970s, he was appointed the informal Commissioner for Whales for the Seychelles, where he was instrumental in creating the Indian Ocean Whale Sanctuary. Watson spent time in the United States and England, but lived his final years in West Cork, Ireland.

NICK HUNT is the author of *Where the Wild Winds Are: Walking Europe's Winds from the Pennines to Provence*, which was a finalist for the Stanford Dolman Travel Book of the Year. His other books include *Walking the Woods and the Water* and *The Parakeeting of London*. He also writes fiction, and works as an editor for the Dark Mountain Project. He lives in Bristol, England.

HEAVEN'S BREATH
A Natural History of the Wind

LYALL WATSON

Introduction by
NICK HUNT

NEW YORK REVIEW BOOKS

New York

THIS IS A NEW YORK REVIEW BOOK
PUBLISHED BY THE NEW YORK REVIEW OF BOOKS
435 Hudson Street, New York, NY 10014
www.nyrb.com

Library of Congress Cataloging-in-Publication Data
Names: Watson, Lyall, author.
Title: Heaven's breath : a natural history of the wind / by Lyall Watson.
Description: [2019 edition]. | New York : New York Review Books, 2019. |
  Series: New York Review Books classics | Originally published: New York :
  Morrow, 1984. | Includes bibliographical references and index.
Identifiers: LCCN 2019012599| ISBN 9781681373690 (alk. paper) | ISBN
  9781681373706 (epub)
Subjects: LCSH: Winds. | Atmospheric circulation. | Ecology.
Classification: LCC QC931 .W29 2019 | DDC 551.5/18—dc23
LC record available at https://lccn.loc.gov/2019012599

ISBN 978-1-68137-369-0
Available as an electronic book; ISBN 978-1-68137-370-6

Printed in the United States of America on acid-free paper.
10 9 8 7 6 5 4 3 2

CONTENTS

INTRODUCTION

Sailing back from the Trojan War, Odysseus and his men land on the island of Aeolia, the domain of Aeolus, the Keeper of the Winds. Aeolus gives them an oxhide sack that contains the captive winds of the world—apart from a gentle westerly to waft them safely home—with strict instructions not to open it. Believing there is treasure inside, the greedy crew do just that. The opening of the sack unleashes a chaotic gale that hurls their ship upon the waves, out across the wine-dark sea, blowing Odysseus madly off course. His adventures last ten years.

The book in your hands is that oxhide sack. All the winds of the world are inside. If you open it, you will be blown to places you never expected.

Lyall Watson is a modern-day Keeper of the Winds. His natural history of the great, invisible forces that shape our planet—from the sand dunes of the Sahara to the serotonin inside our brains—twists and turns, uplifts and surprises like the subject it describes. "Wind is defined as air in motion," he tells us early on, then explodes this apparently simple statement in every conceivable way. The reader is propelled back through recorded history into deep time, from the formation of the universe, through physics and mythology, biology and psychology, religion and sociology, in and out of an extraordinary diversity of cultures. We are blown from the macro to the micro, often in the space of a single page: from the planetary scale of the solar system to the infinitesimal particles that flow into us with each breath.

For wind, as *Heaven's Breath* reveals, is much more than moving air. It is one of the great circulatory systems of our planet, connecting—like the book itself—far-flung climates and cultures. It ensures the spread of life, distributing pollen, seeds, and spores over vast distances, accompanied by a tidal soup of floating microorganisms. The atmosphere above our heads, far from being empty space, surges with windborne life, from thermal-surfing spiderlings ("arachnauts," in Watson's playful term) to the "living rain" of tens-of-millions-strong butterfly migrations.

Wind also shapes the Earth, sculpting the land beneath it. The sand dunes of the world's great deserts are the most obvious example, pure creations of the air that mimic the ocean's rolling waves, and abrasive airborne particles scrape and erode like sandpaper, creating fantastical rock formations millennia in the making. Godlike, wind gives with one hand and takes away with the other: the loess landscapes of China are composed of aeolian silt deep enough to carve entire towns from, while in the dust bowl of the 1930s, topsoil from the American Great Plains was whipped away in vast clouds, skinning "windhollows" in the land up to fifty meters deep.

The influence of wind, however, is not only physical. With a polymath's panache, Watson plunges from natural history deep into the spiritual, for wind has always blown inside the human imagination. From the earliest times, this mysterious power—invisible yet tangible, existing nowhere yet everywhere—has been central to the origin myths of many cultures. In the Judeo-Christian story God breathes life, in the form of a gust of air, into lifeless clay, and numerous other religions place wind at the beginning—and sometimes at the end—of all things. Wind gods and spirits feature in belief systems from the Maori to the Mayan, often linked not only with creation but with pro-creation. Of the ancient Greek wind deities—collectively called the Anemoi, which derives from *anima*, "soul"—Zephyrus, the god of the warm west wind, was associated with fertilization, while Boreas, the god of the frozen north, was believed to be so lusty that mares grazing with northerly-facing hindquarters could be impregnated; an example of aeolian immaculate conception.

One of the greatest delights of *Heaven's Breath* lies in the sinuous journey it takes through the world's multivarious cultures, exploring the ways in which people have attempted not only to understand wind—and to translate its invisible language into portents and pre-dictions—but to defend their communities from its more destructive urges. Watson relates how a New Guinea tribe fixed spears to the roofs of their houses "in order to pierce the wind's belly," and how, in South Africa, a Xhosa priest-diviner would climb a hill to spit potion into the eye of the wind. In other cultures, malevolent breezes were assailed with rocks, seaweed whips, gongs, lances, flaming torches, urine, and even (in Scotland) left shoes. These different tactics are all part of the same "ancient and common code" that Watson returns to time and again, drawing connections between scattered points that seem, at first glance, unrelated. Wind, unconstrained by borders, is the unifying thread between them.

All of this is a reflection of the breadth of Watson's fascinations. A self-declared "scientific nomad," he had degrees in botany, zoology, marine biology, chemistry, geology, ecology, and anthropology—plus a doctorate in ethology, the study of animal behavior—and wrote books on subjects as diverse as the unconscious, the supernatural, elephants, water, inanimate objects, and sumo wrestling. Interested in everything, he had a truly Gaian outlook, regarding the world not as a lump of insensible rock but as a living thing: "An organism in its own right, growing, wrapping itself in a moist and luminous membrane of air." In following the paths of the wind, his restless curiosity had free reign.

Heaven's Breath is not a book to read if you want facts in a hurry. It does not take the reader in a straight line from one point to the next. Although its chapters are divided into five broad themes, or breaths—dealing with wind's relationship to Earth, time, life, body, and mind—the narratives constantly interweave and overlap. Watson delights in the meander, which makes wind his perfect subject. The thing that keeps this informative flood from spilling into breathlessness is the controlled pace of the prose and the precision of the language. This is especially evident in the sections dealing with the science of wind: meteorology is a complex subject, and describing something that cannot be seen—as many artists have discovered—is a particularly daunting task. But Watson has a gift for fixing the invisible in concrete images. Warm air lifts "in great buoyant bubbles, exactly like the vapour that rises from the bottom of a kettle of gently-boiling water." A cold front "edges forward beneath the warmer air in its path, prying it up and rolling it back like a toe tucked under carpet." And the sun, as it formed and heated, "spread radiation like a firehose." Such images have an anchoring effect, keeping the reader's feet on the ground and preventing us from being blown off course.

Myth and science are drawn together in the field of biometeorology, the study of how atmospheric conditions affect living organisms. Farmers have long observed that animals can grow skittish in wind, and teachers know that windy days create anarchy in the playground. Stories of "ill winds" exist in every part of the world, documented through generations of folk wisdom and weather lore, blamed for everything from minor misfortune to national calamity. Easterly winds are often associated with bad luck. The mistral in the south of France is known as "the wind of madness." In parts of Europe, the blowing of certain winds was historically considered a mitigating factor in crime cases; if you could prove that the wind made you do it, you might receive leniency in court. Goethe, Voltaire, Ruskin, Schiller, and Nietzsche—to name but a few weather-sensitive souls—have all drawn links between ill winds and violent changes in mood.

I have a particular reason to believe in such accounts. In 2016, I walked the routes of several European winds to explore the effects they have on landscapes, peoples, and cultures. I was frozen by the helm wind in northern England, knocked off my feet by the bora (named after Boreas, the horse-ravishing god of the north), enervated by the sirocco, and battered by the mistral. But it wasn't until I reached Switzerland that I truly understood wind's power. In valley after valley, I had heard claims that the föhn—the warm, desiccating wind that melts the snow from the mountain peaks—causes headaches, insomnia, irritability, anxiety, and a host of other ailments. It is even rumored to cause a spike in the suicide rate. These stories sounded like old wives' tales and I took them with a pinch of salt, until, in the Haslital valley, the föhn began to blow. After a day and a night of its relentless blast— like the slipstream of an invisible train thundering constantly around me—I found myself so exhausted and depressed that I could hardly move. Despairing, almost paranoid, I fled to another valley, where, out of range of the wind, I felt completely normal again. I had fallen victim to *Föhnkrankheit*, the notorious "föhn sickness."

Heaven's Breath helped me understand what I had experienced. Studies have linked föhn-type winds—characterized by a downslope flow of warm, dry air—to an increase in positively charged ions in the atmosphere. Evidence suggests that positive ions play havoc with serotonin levels, causing dramatic mood swings; pressure fluctuations and alarm reactions might also play their parts. The baleful effects of such winds are known from Israel (in the form of the sharav) to Los Angeles, where the notorious Santa Ana is alleged to drive up homicides. Not for nothing is it also called the Devil Wind.

On the subject of named winds—which blow from specific directions at specific times of year—one of the final delights of the book is the aeolian dictionary that leads us from the aajej (a Moroccan whirlwind) to the zonda (a westerly in the Andes). On the way we meet the xlokk, the southerly burster, and the bad-i-sad-o-bist-roz, the "Wind of One Hundred and Twenty Days" that rages in Iran and Afghanistan. It is hard not to be carried away by fairy-tale names such as these. Whether baleful or beneficial, avatars of creation or destruction, these great forces are introduced as a cast of characters, all with distinctive moods and personalities.

It is a fitting tribute to the spirits in whose midst we dwell. As Watson reminds us, the words for wind, spirit, and breath are the same in many languages, from the Hebrew and the Arabic to the Sioux and the Dakota. In the *anima* of the ancient Greeks we even catch a glimpse of ourselves: we are breathing *animals*, brought to life by the air flowing through us.

The oxhide sack in your hands, then, is not just full of wind. It is full of life.

—Nick Hunt

HEAVEN'S BREATH

ON THE WIND

If the trend toward specialisation in science is defined as knowing more and more about less and less, then this is its logical conclusion.

Everything you always wanted to know about nothing.

It began as an essay on experience of the ineffable, but grew, as wind will, to have a life of its own. As it gathered strength, drawing on surprising resources, it became apparent that wind is far from hollow. It is the most vital of metaphors.

Part of this vigour is revealed in language. In Arabic, the wind is *ruh*, but the same word also means 'breath' and 'spirit'. While in Hebrew, *ruach* enlarges the sphere of influence to include concepts of creation and divinity. And the Greek *pneuma*, or the Latin *animus* are redolent, not just of air, but of the very stuff of the soul.

Without wind, most of Earth would be uninhabitable. The tropics would grow so unbearably hot that nothing could live there, and the rest of the planet would freeze. Moisture, if any existed, would be confined to the oceans, and all but the fringe of the great continents along a narrow temperate belt, would be desert. There would be no erosion, no soil, and for any community that managed to evolve despite these rigours, no relief from suffocation by their own waste products.

But with the wind, Earth comes truly alive. Winds provide the circulatory and nervous systems of the planet, sharing out energy and information, distributing both warmth and awareness, making something out of nothing.

All wind's properties are borrowed. Our knowledge of it comes at secondhand, but it comes strongly. And this combination of a force that cannot be apprehended, but nevertheless has an undeniable existence, was our first experience of the spiritual. A crack in the cosmos that widened to let the tide of consciousness flow through.

We are the fruits of the wind – and have been seeded, irrigated and cultivated by its craft.

This is a natural history of that process. It is an acknowledgment, and a celebration, of an awesome debt.

LYALL WATSON
Kyoto, Japan; 1983

Measurements:

In accordance with international meteorological practice, all measures of length, speed, distance, area, volume, weight, pressure and temperature used in the text are metric.

'Tons' are metric tons of 1000 kilograms.

Where 'billion' occurs, it means one thousand million.

To assist those more familiar with 'customary' units, a simple conversion table is included as an Appendix at the end of the Bibliography.

PART ONE
Wind and Earth

It all began with a big bang. With an awesome explosion which occurred simultaneously everywhere, 'filling all space from the beginning, with every particle of matter rushing apart from every other particle'.[514]

It was a strangely gentle explosion. As though the cosmos, in the throes of a great idea, had taken a deep breath some twenty thousand million years ago and then, as the consequence of inspiration became apparent, let it all out again in a long and heartfelt sigh.

It seems right somehow, consistent with manifest imperfection, that our universe should have been conceived in disquiet. And appropriate too when the current exhalation ends, perhaps after another ten thousand million years, that the cosmos will pause and then inhale once more. Whereupon the whole extraordinary process will begin to run backwards, giving way to an accelerating contraction that can only end when everything approaches a new state of infinite density; when there will be another breath, another and possibly more orderly, grand expansion.

Everything points to a cosmos which behaves like that, one that evolves and grows and changes.

All astronomers agree that the universe is at present expanding, that most other galaxies in sight are rushing away from us, and from each other, at speeds of many millions of kilometres per hour. And some feel that this expansion cannot go on indefinitely, that gravity will eventually slow everything down and pull the pieces back together again.

They suspect that such oscillations cannot persist, that the universe must run down in much the same way as a bouncing ball eventually comes to a stop. But whatever their theoretical bias, everyone seems to see the cosmos as a dynamic system with a natural history of its own. An ecology in which our Earth is one of the creatures.

This concept of an organic universe, of a cosmos that lives and breathes, is an important one. It carries the seeds of a new and more tuneful understanding.

One rich with the eloquent cadence of the wind.

1. THE PHYSICS OF WIND

There are worlds without wind.

Breathless ones that shine with a hard, diamond-bright, reflected light. Ones like Mercury and the Moon that are too hot or too cold, too small or too old to sustain an atmosphere.

All the rest return light that has been softened and coloured by passage through various layers of vapour. Mars looks red and Venus carries a pale concealing cloak. Jupiter is a glowing gold. Saturn is set in iridescent rings of ice and Earth, we now know, is wrapped in an exuberant blue swirl. It has a halo of its own, an aura.

The scientific jury has not yet returned a verdict on the origin of the solar system, but the best evidence suggests that our Sun condensed out of an extensive cloud of interstellar dust. The cloud gradually collapsed until it became so hot and so dense in the centre that nuclear reactions began to take place there. The energy of these was sufficient to boil away all the lighter molecules nearby, but heavier silicates and iron settled into clumps or planetismals arranged at progressively greater distances along a spinning disc of matter. And in time these coalesced into planets.

At first the planets were cold. But as each contracted under the influence of its own growing gravity, the centre heated and melted to form a glowing core, which in some cases cooled sufficiently to surround itself with a thin solid crust. The skin of a new world.

Space Winds

In the beginning, each orbiting body carried its own envelope of hydrogen and helium, which together still form 99 per cent of all matter in the cosmos. But as our Sun heated up, it spread radiation like a firehose, sending waves of destructive ultraviolet out to scour the surface of all the young planets, stripping them of their light gas atmospheres.

The more distant outer ones, particularly Jupiter and Saturn, remain relatively unscathed. They have kept most of their original hydrogen-helium envelopes and still carry deep atmospheres around relatively small, possibly solid, cores.

But the inner planets are entirely different. They have shallower atmospheres made up of rare and dense gases that are largely homegrown. The thin cool crusts have ruptured and split to allow the planets to breathe and to wrap themselves securely in their own airy cocoons.

In some cases, even these secondary atmospheres have been lost. Mercury, our innermost planet, is so close to the Sun that surface temperatures at the equator reach 430° Centigrade at noon. All molecules everywhere are excited by heat, moving faster at higher temperatures, and those on a world whose surface gravity is low – Mercury's is only 40 per cent as strong as ours – soon reach escape velocity. So whatever atmosphere Mercury once had, all or most of it, has long since boiled away into space.

The other inner planets have been more fortunate. The size and position of Venus are so much like our own that she has long been seen as a sister planet, and considered the next most likely to shelter life in our solar system. Although Venus intercepts twice as much sunlight as Earth does, it was hoped that the constant cover of pale yellow cloud might reflect most of this energy; leaving the surface relatively cool. But a series of Russian and American space probes launched between 1975 and 1979, have completely destroyed that illusion. Surface temperatures on Venus are high enough to melt lead and the thick clouds are saturated with concentrated sulphuric acid, producing searing hot rains of the most corrosive fluid in the solar system.

The Venusian atmosphere is largely carbon dioxide and ninety times heavier than ours, which means that pressure at the surface is equivalent to that experienced by a body a thousand metres down on our ocean floor. Craft which have reached the Venusian surface and continued, however briefly, to relay information, record wind speeds no greater than five kilometres an hour. But ultraviolet photographs of the opaque cloud tops suggest that the entire upper atmosphere flows around the plant at over 350 kilometres an hour in a swathe of high-speed bands of wind, each one as wide as an ocean.

Mars is easier to see and to live on. It is a small planet with a radius only half our own, and a low surface gravity, similar to Mercury's. Most of its atmosphere has therefore escaped, but the equally low surface temperature, which averages −42° Centigrade, has left it with just enough gas to produce an atmospheric pressure 160 times weaker than ours. And yet, despite this shortcoming, every year just as Mars get closest to the Sun, violent winds pick up the fine surface sand of the planet and keep enough of it in suspension to reduce visibility to nil.

This seasonal storm is visible from here. It starts as a shining white trail which extends for thousands of kilometres across the southern hemisphere of the planet. After a few days (Martian days are roughly the same length as our own), the scar turns reddish and moves in a westerly direction, widening and spreading as it goes. Before long it circles the planet and then fills until it envelops the entire hemisphere. While *Mariner 9* was in Martian orbit in 1972, the annual southern storm spilled over into the northern hemisphere and obscured the planet completely for several weeks on end.

To carry sand to such heights and over such distances in the rarefied Martian atmosphere, the winds must blow at well over 150 kilometres an hour, and continue without ceasing for months, in ways that make even the worst of our tropical storms seem like 'sweet-breathing zephyrs'.[457]

Conditions are, predictably, even more extreme on the outer planets.

Jupiter is the giant of the solar system. Its mass is more than twice that of all the other planets put together, but very little of this is solid. It is, in effect, an enormous drop of liquid hydrogen with a tiny nucleus of iron. All atmosphere and hardly any planet. Which is probably just as well. Surface pressure, if there is such a thing as solid surface, would be sixty million times higher than one Earth atmosphere. High enough to crush a closed steel cylinder block like an eggshell.

Like Venus, all we ever see of Jupiter are the uppermost layers of cloud, but these are fascinating. They sweep around the great golden globe at enormous speed, producing alternating belts and zones of darker and lighter hues. The planet itself spins on its axis in less than ten Earth hours, giving top-like stability and cohesion to what would otherwise be dynamic chaos. The result is a dense, cold mass of gas, fuelled by the planet's own internal energy, that has sorted itself out into an orderly pattern of fast-moving jet streams and prevailing winds.

There are many minor curls and eddies within the latitudinal zones, but the only major blemish on the face of the planet is the Great Red Spot in the southern hemisphere. This flares and fades with time, leaving a persistent hollow even when it is no longer clearly visible, like the vortex that wobbles above the plug-hole in a bath. Suspicion grows that it must be a vast cyclonic storm, with an eye over thirty thousand kilometres wide, that has raged throughout the 300 years since its discovery in 1664. And might, if current models of the Jovian atmosphere are correct, go blazing on for 300 thousand more.

Things are a little better on slightly smaller Saturn. Visual examination of the planet itself is complicated by the fact that it is surrounded by a bright halo of ice, by an orbiting snowdrift of giant fifteen-centimetre flakes arranged into the famous 'rings'. The visible surface is of pale gold-coloured clouds layered into equatorial and temperate belts by rapid rotation. There are also short-lived white spots, the longest lasting eighteen of our months, that seem to be storms breaking through the otherwise smooth surface of ammonia cirrus clouds. Voyager spacecraft passing the planet in 1981 heard static from what appeared to be a massive thunderstorm sixty thousand kilometres wide, raging round the equator, with wind speeds of over 1500 kilometres per hour. Conditions below the atmosphere are, it seems, essentially similar to Jupiter's, but appropriately muted by the differences in size and density.

We know little about the remaining planets. Uranus and Neptune appear as small, bluish-green discs without conspicuous surface features. They have shallow, cold, remarkably clear atmospheres of hydrogen over what seem to be solid cores. Nothing is known of their winds. And Pluto seems to be too small to have any kind of atmosphere at all.

Apart from the planets, there are thirty-four other natural satellites in the solar system. Earth has one, Mars two, Jupiter fourteen, Saturn ten, Uranus five and Neptune two moons apiece. Of these, Jupiter's Ganymede and Callisto, and Saturn's Titan, are each larger than the planet Mercury and ought to be considered as worlds in their own right. Titan is almost the size of Mars and has an extensive atmosphere with a pressure higher than the Martian one. Ganymede and Callisto, and even Io, which is slightly larger than our Moon, seem also to have atmospheres. These are sustained perhaps, despite their small size, by constant volcanic activity that vents new gases as the old ones reach escape velocity. But nothing is yet known about possible winds and weather on these fledgling worlds.

All the remaining odds and ends in the solar system, including the asteroids and our Moon, are too small to produce enough gravity to keep a protective envelope of gases wrapped around themselves. And, as a result, all lack the exuberance we take for granted as Earth's right.

None have membranes that are able to catch and hold energy, banking it against lean times.

None can edit the Sun or provide shelter from the showers of meteorites that hammer on the outer edges of the atmosphere, 'like the random noise of rain on the roof at night'.[76]

None know radiance that scatters the bright white light of space and tempers it into rainbows, sunsets and twilights, softening the hard edges of things and allowing shapes and colours the freedom to flow one into another.

None feel the fluctuations in pressure that set particles in motion, spreading agitation in trains of waves, carrying news and information in the form of sounds.

None provide the cushion of relative stability that makes life possible, or the spur of infinite variety that requires continual adjustment and results in evolution and creativity.

They are, like Tennyson's Maud, 'faultily faultless, icily regular, splendidly null'.

An atmosphere is the most vital prerequisite for life and mind. Without it, worlds are hard-pounded deserts, dry as old bones, where alien feet fall soundlessly and even the dust is dead. With it, everything becomes possible. Orbiting rockpiles organise themselves, taking on the self-contained look of living creatures, giving and receiving information, learning the essential skills involved in managing the sun.

Earth is such a creature. An organism in its own right, growing, wrapping itself in a moist and luminous membrane of air. And within this semi-transparent envelope, fed by a network of capillary breezes, carrying energy and information from one extremity to the other, are the great arteries of the prevailing winds.

There is more than a hint of something in the air.

There is something about the air itself that is full of surprises, suggesting that it could never have arisen by accident nor persisted by chance.

In the eyes of a chemist, air is a mixture of gases. But it is a mixture so curious and incompatible, so inherently unstable, that it begins to look unreasonable. Or, what is even more exciting, to look as though it had a reason for being.

Left to themselves, things everywhere tend to become more and more disorderly, running down toward randomness, which seems to be the final and natural state of matter. This is one of the laws of the universe and, in the midst of prevailing chaos, any kind of order has to be seen to be statistically unlikely and regarded with suspicion.

Nothing is more suspicious, more improbable, than life itself. It is an unnatural state, possessed of the outlaw qualities of regularity and organisation which it shares with, and may even have borrowed from, certain crystal structures. And which it may, in turn, have imparted to the atmosphere.

Earth's aura, our air, is unusual. It turns out on examination to be unlike any other atmosphere we know of. It has strange properties that make it necessary to include the air we breathe amongst Earth's growing collection of rare and unreasonable things.

It would be doing unnecessary violence to the laws of chance to suggest that so many singularities came by coincidence to the third planet of a minor star. So, in the face of such wild improbability, it becomes necessary to look for connections.

Some biologists suspect that life may have borrowed the maverick capacity for replication from substances like the crystalline clays, whose regular geometric forms arise spontaneously, but go on reproducing themselves in a stable and organised way. Now, perhaps it becomes necessary to extend the chain of revolutionary circumstance and suggest that air, far from being a random catalogue of stray gases, might in fact be a biological ensemble.

Something created by, and maintained by, the biosphere for its own benefit.

Look at the facts. Our Sun, being a typical star of the main sequence, has evolved in the usual way and now gives out up to three times as much radiant energy as it did when Earth first was formed some five thousand million years ago. But Earth is *not* now three times as hot as it was then.

Our planet, like the others, once had a molten core in which lighter substances floated to the top, forming a solid crust. In the beginning, there was nothing outside this. The rays of the Sun swept the surface clean and Earth was 'without form and void'.

Then it belched. The hot breath of the young planet bubbled up through cracks and fissures in the crust, breaking out in the form of geysers, fumaroles and explosive volcanic eruptions. Methane, ammonia and water vapour – and possibly large quantities of carbon dioxide – were released from combination with hot rocks and thrown almost at random into a mechanical mixture that formed the first atmospheric blanket. This acted like the glass in a greenhouse, cutting down on the amount of heat radiated back from Earth's surface and keeping it reasonably warm, probably somewhere between 15° and 30° Centigrade. Warm enough for life to begin.

Once living things appeared on Earth, they started to feed on the atmosphere, nibbling away at the blanket that protected them. By rights, the planet should soon have become icebound and lifeless, stripped bare as the guardian gases boiled away under increased radiation from an ever more active Sun. But neither of these things happened.

Somehow, despite drastic changes in the composition of the early atmosphere, and a large increase in the mean solar flux, the temperature of Earth's surface has been kept within astonishingly narrow limits for many hundreds of millions of years. And these limits are precisely the ones that life requires.

Before there was a significant amount of oxygen in the air, it seems that it was ammonia, which also has heat-absorbing and retaining properties, which helped to keep the early Earth warm. But this gas on its own would not have been sufficient to ensure temperature stability. The failure of one year's crop of ammonia, whether seeping from Earth or produced by living things, would have been enough to lead to a self-perpetuating temperature decline and the extinction of life.

Later, when green plants practising photosynthesis had com-

19

pleted their task of oxidising the atmosphere, carbon dioxide became the main heat-controlling gas. But until this development took place about 2000 million years ago, it seems that Earth's thermostat was at least partly under the control of vast mats of pioneer algae which, by changing their colour from light to dark and back, were able to alter the amount of heat being radiated from large areas of Earth's surface. And, right from the start, equable distribution of all radiant energy was ensured by the circulation of heat through the atmosphere by the wind. Without such movement, half the globe would have become progressively hotter and the other half progressively colder, with life possible only along a narrow band between the two extremes at a latitude of about thirty-eight degrees north and south of a boiling equator.

For 3500 million years, most of the surface of Earth has maintained a temperature which was both favourable for life and astonishingly constant, despite the fact that our planet turns before an uncontrolled radiant heater with an erratic output. It is not easy to identify the precise mechanism responsible for this delicate thermostatic control, but it becomes difficult to deny that there is control, that air temperature and perhaps even the entire climate of our planet are being manipulated.

The composition of the atmosphere has been equally serendipitous and remains largely mysterious. Almost everything about it violates the laws of chemistry. For a start, it is a highly combustible mixture and there should by now be no free nitrogen in it. Nitrogen normally reacts with oxygen, and all or most of both gases should long since have ended up in the ocean in the form of stable nitrate ions. But the air we breathe remains stubbornly uncombined, with concentrations of oxygen and nitrogen at a steady and separate 21 and 78 per cent respectively.

Nor are either of these proportions arbitrary. If the abundance of oxygen was any greater, all life would be at risk. The probability of a forest fire increases by 70 per cent for each 1 per cent rise in oxygen concentration above the present level. If it formed 25 per cent of the air, oxygen would leave all land vegetation in raging conflagration. Everything from the driest arctic tundra to the wettest tropical rain forest would burst into flame. And if the level of nitrogen, which is largely responsible for air pressure, were to fall to 75 parts per hundred, nothing could prevent the onset of global and possibly permanent glaciation.[295]

The presence of methane in the contemporary atmosphere is equally problematic. It is toxic and highly unstable, combining readily with oxygen and vanishing almost as fast as it is formed. But something like 5 per cent of all the photosynthetic energy of the entire biosphere is devoted to producing methane, mostly by fermentation in the fetid muds of marshes and wet lands. It could be regarded simply as a waste product, part at least of it comes from the farts of ruminant animals, but the fact is that without methane, oxygen concentration would rise by a dangerous 1 per cent in every twelve thousand years.[318]

Another significant portion of Earth's energy budget is devoted to adding ammonia to the atmosphere. It is no longer needed to help keep the planet warm, but the biosphere still produces about 1000 million tons of it each year. Why? Well, one of the consequences of having nitrogen and oxygen in the air, is their tendency to combine under certain circumstances to produce powerful corrosives. Every thunderstorm creates tons of nitric acid and were it not for the neutralising effect of ammonia, the soils all over Earth would soon become intolerably sour and hostile.

Everything points to the same conclusion. The atmosphere cannot be just a fortunate one-off emanation from some ancient rocks. Life does not merely borrow gases from the environment and return them unchanged. Our air begins to look more and more like an artefact, like something made by, and maintained by, living things for their own ends.

In other words, life defines the conditions necessary for its own survival and somehow makes sure that they stay there.

The biosphere on its own, however, contains only part of the necessary machinery. A solid ball of living matter isolated in space could feed on itself for a while, but could never achieve homeostasis without a proper substrate providing not only raw materials, but essential stability.

The land and water surfaces of Earth, together with living matter and air, between them form a giant complex which is far from being passive. It deserves to be regarded as a single organism in its own right, a living creature, the largest one in the solar system.

James Lovelock, who is amongst the last of the great independent natural philosophers, has also been the first to quantify an old belief in the existence and reality of Mother Earth. He has recommended that this giant creature be called Gaia after the Greek Earth goddess.[294]

Gaia is an entity which, like all other symbiotic associations in biology, has collective properties which are far greater than the sum total of its parts. And one of these is the capacity to contrive a unique and dynamic atmospheric security blanket that keeps the system intact.

The air, our aura, serves the same function as the fur on a fox or the shell of a snail, but it is more sensitive and more responsive. It is a strange and beautiful anomaly, not itself alive in the same sense that we would normally define living things, but very much more than an insubstantial gaseous veil or a random mixture of gases.

We shall see why.

Earth Winds

There is a surprising quantity of air on Earth. At any one time, we are wrapped in 5600 million million tons of it.

If, in some universal dispensation, all the material things of this world were to be divided equally among all the human inhabitants, each person would receive as their share a million tons of air. That is enough to fill a thousand buildings the size of the Empire State, 3000 Astrodomes, or fifty thousand Albert Halls.

Each of us breathes about ten million times a year, using something like five million litres. At this rate, an allotment of a million tons would last each person for over 160 thousand years. The first Neanderthal man, if he were still alive today, would by now be only halfway through his own personal share.

These are comforting calculations, but they are also grossly misleading. For the beautiful thing about our atmosphere is that it is not a static, closed system – nothing like a body of dead air trapped in a hall. All constituents of the free air, including the oxygen we need to breathe, are being constantly replenished. There is an open dynamic interchange between Earth, biosphere and air that stirs things up and keeps Gaia herself alive and breathing, whistling up the wind.

Wind is defined as air in motion.

Usually a large quantity of air moving, for preference, sideways rather than up and down. There are vertical movements of air, which are vital in shaping climate and in moving things, but most motions are lateral, taking place within distinct layers of the atmosphere.

The layers themselves are a surprise. Common sense suggests that the atmosphere ought to be featureless, receding smoothly upward and merging imperceptibly with outer space. Early astronomers were sure that it was so. But the first high-altitude balloonists of the nineteenth century bumped right into several unexpected discontinuities.

Air is certainly densest at Earth's surface and, since the pressure at every level must be just enough to support the weight of air above, it is under less pressure and is therefore less dense at greater altitudes. Higher thinner air means that there are fewer molecules per cubic metre and fewer collisions between particles, which consequently move more slowly. As temperature is essentially a measure of the amount of energy in moving matter, it follows that thinner air is also going to be cooler air.

The scientific logic in all this is impeccable and it works in practice, but only up to a point. Air does get cooler the higher you go, dropping by about 1°Centigrade for every 150 metres, until you reach a critical point where the thermometer seems to stick.

On our imperfect sphere, this natural ceiling lies between an altitude of eight kilometres over the poles, where the temperature remains at −60°Centigrade, and fifteen kilometres over the equator, where it usually falls even lower.

Everything below the ceiling, where the rules work normally, is known as the troposphere from the Greek *tropos*, meaning 'a turn'. This is the zone in which the air truly turns, where 75 per cent of the mass of the air and virtually all the water vapour and weather are confined. Its upper limit lies where thunderheads flatten into shining anvils and cascades of crystals whip off the ends of cirrus clouds, forming tenuous mare's tails of ice, too insubstantial even to cast a shadow.

Above the ceiling lies a second zone, more than twice as deep and ending at an altitude of over fifty kilometres, where the temperature remains relatively constant or even tends to increase.

This is the stratosphere – from the Latin *stratos*, meaning 'something flat, in layers', because within this realm the air is actually visibly layered. There are thin shaded bands of blue sky, sharply etched against the darkness of space, hovering over Earth like a succession of haloes.

The stratosphere is highly rarefied, having a density which is less than one thousandth of that of air at sea level. But a significant number of its highly scattered molecules are of oxygen which has been persuaded, under the influence of solar radiation, to pick up an extra atom – converting familiar O_2 to unstable O_3 or ozone.

Between twenty and thirty kilometres above Earth's surface, lies a deep layer of this ozone. It is not uniform and it follows the same profile as the weather ceiling, coming closest to the ground over the poles and reaching its maximum height over the equator. It is a vast yet insubstantial thing. If the entire layer were to be brought down to sea level and subjected to full atmospheric pressure, it would be just three millimetres deep, about the thickness of a beach umbrella. And yet without this flimsy parasol, which effectively filters out most damaging ultraviolet radiation, life on Earth would probably never have evolved, at least in its present form.

It is the absorption of solar energy by this layer that produces the unexpected temperature rise in the stratosphere. Where ozone is most concentrated, temperatures climb above 30° Centigrade, higher than most places at sea level on a hot summer's day.

The relatively high density of this warm layer also gives it some extraordinary reflective powers. The sound of great batteries of cannon being fired at Queen Victoria's funeral in 1901, was clearly heard in parts of Germany, but skipped 500 kilometres of France and Belgium in between.

Beyond the limits of ozone at about sixty kilometres, temperatures once again begin to fall and continue to do so throughout the third great atmospheric belt, the mesosphere. Between eighty and a hundred kilometres above the surface, this too reaches a ceiling at a point where temperatures have been known to fall to $-143°$ Centigrade, the lowest natural levels ever recorded anywhere in Gaia's being.

This icy place is the home of mysterious noctilucent clouds that shine with a silvery-blue glow well into the short summer nights near the polar circles. They first appear just after sunset, materialising out of an otherwise clear sky sixty kilometres above all other weather phenomena, glowing above the twilight arch at the horizon. In the words of geophysicist Louise Young, 'their delicate feathery ripples of foam, look like waves in a phantom sea'.[537]

These strange night lights glowed for several months after a gigantic meteor destroyed thousands of square kilometres of taiga forest along the Tunguska River in Siberia on June 30th, 1908, and suspicion grew that it was dust from the impact that produced the necessary reflection. But it was difficult for scientists to believe that dust from the Earth's surface could rise that high, so when appropriate technology became available, samples were collected by firing rockets into several such clouds that formed over Sweden in the summer of 1962.[454]

Analysis showed water ices surrounding larger particles whose chemistry indicated an extraterrestrial origin, but nothing is yet known about the forces which concentrate such filaments of cosmic dust or throw the fragments of a shooting star together like spindrift on the shores of a cold, high and otherwise invisible sea.

Above the frozen mesosphere lies another 300 kilometres or more of air so thin no part of it reaches a density of more than one millionth of an atmosphere. But this is nevertheless Gaia's first line of defence, the point where Earth is bombarded by a relentless electromagnetic cannonade. Ultraviolet, infrared, radio and X-rays, corpuscular streams from our Sun, similar radiations from other stars, and cosmic rays from interstellar space all buffet the outer edges of the atmosphere. Oxygen and nitrogen molecules are split apart and free electrons gather in electrically charged groups of ions that arrange themselves in belts around Earth.

This is the thermosphere or ionosphere, where temperatures soar to over 2000° Centigrade, but particles are so scarce that not enough of them strike a body in orbit to transfer such awesome heat. There are enough of them, however, to reflect radio short waves back to the surface, making it possible to bounce signals around Earth's curve. Enough too to combine with other charged particles thrown into the edges of our atmosphere by gusts of solar wind. The result of these interactions is an awesome display of pyrotechnics, of wavering, luminous sheets of colour that form and break and move, 'like firelight flickering on the ceiling of the world'.[537]

Such auroras concentrate around the polar circles, appearing a few hours after sunset as pale greenish glows that brighten until phantom curtains of lime green, gold and magenta arch across the sky, as though a gigantic explosion had taken place just over the edge of the world and heaven itself was on fire. Small wonder that Nordic sagas tell of reflections from the golden shields on which Valkyries carry fallen heroes to Valhalla. Or that Inuit of Hudson Bay still speak of the light of lanterns borne by demons searching the universe for lost souls.[289]

We know now that both aurora borealis and australis wax and wane in eleven year cycles in synchrony with spots and flares on the Sun. Earth's influence does not end even 500 kilometres out, where the ionosphere shades into exosphere and particles become too diffuse to measure.

Our aura and the Sun's flow together, bending and shaping each other, interacting in a multitude of ways that we are only just beginning to understand.

Gaia moves and lives and breathes in this highly-charged medium, buffeted by the effects of a nuclear war which rages constantly at the surface of the Sun, destroying four million tons of matter every second.

The solar wind rushes out from this conflagration, producing waves of energy that travel at 300 thousand kilometres per second. Eight minutes and twenty seconds after leaving the Sun, these waves sweep past Earth and a segment of their front is shattered by contact with our atmosphere. Thirty per cent of the radiant fragments are reflected back into space by air particles, by the mirror faces of clouds, by snowfields and the surface of the sea. A further 20 per cent are snatched up and absorbed by our atmospheric gases. Oxygen seizes bits of just the right shade of red, nitrogen has an affinity for violet, and ozone soaks up most of the very short wavelength ultraviolet rays. The rest, approximately half of all solar radiation coming our way, successfully runs the aerial gauntlet and is absorbed by Earth's surface just one thousandth of a second after the wave breaks on our outer shores.[39]

Earth eventually radiates this energy back, but it does so only at very long wavelengths. Long enough to be intercepted by molecules of water in the air, which are too widely spaced to interfere with the shorter solar wavelengths coming in. Moisture therefore acts as a one-way valve, allowing radiation to flow downwards, but inhibiting its return. As a result, the moist lower atmosphere forms a blanket that holds heat close to Earth, letting it escape back to space only gradually.

All parts of the solar shower contribute directly to Gaia's body temperature, but most of our warmth is derived indirectly, secondhand, from radiation re-released by Earth some time after absorbing it.

If, like pit-vipers that hunt their prey by detecting body heat, we had the capacity to use infrared light, there would be no darkness for us. Even on the blackest nights, every object would be visible, lit by a radiance as great as that provided by a full moon, but with a weird difference. There would be no shadows on the ground, because Earth itself would be the principal source of light.

It is always the main source of heat. On a clear sunny day, for instance, beach sand gets too hot to walk on with bare feet, reaching temperatures as high as 60° Centigrade. The temperature of still air around the shoulders of someone sitting on the sand is much less, perhaps 40° Centigrade. And when such a sunbather stands up, their head moves into an air layer that may be at just 30° Centigrade. Within the space of less than two metres, from the soles of the feet to the top of the head of a single human being, the temperature range can be 30° Centigrade. A difference as great as that between average July temperatures at Timbuktu in the Sahara and Reykjavik in Iceland, at both of which places the official temperatures are made at a standard height above the ground.

The reason the Sahara is hotter, is that it is closer to the equator, where incoming solar rays are more nearly perpendicular to the surface and penetrate the atmosphere most easily. In Iceland, rays strike the ground at a glancing angle and at the north pole they are almost parallel, having little or no effect at all. It is this uneven distribution of solar energy that is the principal moving force of wind and weather.

In the tropics, ground temperatures and Earth radiation are high and the air blanket warms most quickly. Warm air expands, rises, cools, condenses and, having shed its moisture, rises yet again, building into thunderstorms that pump energy high into the troposphere. Warmth and water vapour surge up until they bump against the weather ceiling at around fifteen kilometres. Here the updraft ends and the air turns and flows outwards in the only direction possible, toward the poles.

If our planet were a smooth featureless sphere and did not spin on its axis every twenty-four hours, the circulation of the atmosphere would be simple. The air around the equator would rise and move toward the north pole in the northern hemisphere and toward the south pole in the southern. And its place would be taken by cooler lower air flowing towards the equator from both directions. All surface winds would blow from the poles and all upper winds toward them.

But it is not that simple of course. Complications abound.

Even at the equator, land and ocean areas absorb and radiate heat at different rates through air of variant humidity. Deserts, grasslands and forests all have their own peculiar heat budgets and winds. However fast and straight their initial flow may be, they are slowed by friction and deflected in all directions by mountains, coasts and the lie of both land and sea. But the most devious influences are provided by the rotation of Earth and the tilt that produces our seasons.

This bias is a complex and disconcerting one, most easily illustrated by an orange and a ballpoint pen. Hold the orange with its stalk scar pointing up and place the tip of the pen on this north pole position. Now rotate the orange slowly from left to right, as Earth spins, and draw a perpendicular line directly from the pole down to the fruity equator. The pen will move in a straight line before your eyes, but the mark on the orange skin will veer away to its right, toward the southwest. While a similar line drawn directly upward from the south pole, will curve away to its left, toward the northwest.

This is the Coriolis Effect, named after a nineteenth-century French physicist who noticed that a moving body linked to a rotating body is deflected from its path by inertia. Which means that all movement in the northern hemisphere is deflected to the right, while in the southern hemisphere deflection is to the left.

This rule applies equally to all objects moving in any direction above any point of Earth's surface. Even to shells fired from cannon, for which appropriate ballistic corrections have to be made. And it applies without reservation to all bodies of air in motion, in the process of becoming winds.

Heat at the equator sets the engine in motion. Air rises and flows towards the poles. The further it travels, the more it is deflected, until by the time it reaches a latitude of about thirty degrees and cools and descends, it is blowing at right angles to its original direction. Which means that in both hemispheres in these middle latitudes, cool winds tend to blow directly out of the west, contributing to a river of wind that circles the globe in the latitude of the prevailing westerlies.

Back at the equator, the rising air has left a gap, the equatorial trough, an area of low pressure that has to be filled in order to restore equilibrium. And it is filled by cooler air flowing in along the surface from subtropical zones. At first these returning airs run directly towards their goal, but they too are soon deflected, toward the right in the north and to the left in the south, producing winds that are predominantly easterly rather than polar. These strong and consistent flows compensate for the westerly movement at higher latitudes and form the vital and reliable trade winds. Northeasterly in the northern hemisphere and southeasterly in the south.

This picture of general circulation is satisfyingly lawful and orderly, but unfortunately it is never that simple. For a start, there is more than just one equator around Earth's fat middle. In addition to the fixed equator of rotation, marked with such confidence on every map, there is a much more moody and realistic 'equator of the winds'.[351]

This subtle meridian moves. It drifts north and south with the seasons and bends sympathetically about the continents and seas. Wherever it lies, it has come to be known as the 'doldrums', from the Old English *dol*, meaning 'dull'.

In the sanitised language of the new science, this region is referred to as an 'intertropical convergence zone', but it remains the same infamous belt of low and uniform pressure whose fitful calms left sailing ships drifting at the mercy of stealthy currents. Its only saving grace is that the rising warm air is filled with water vapour which it dumps in heavy rains and short-lived squalls.

The doldrums follow a month or two behind the Sun and their width and precise position are highly variable. Seasoned skippers knew when and where the trade winds dipped to almost meet each other, and a few days of grasping at every stray cat's paw of wind would suffice to get a ship safely across the equator. Others were left four weeks on end with limp and fretful sails.

Another mobile meridian lies in the disorderly zone between the trades and the westerlies. This is where equatorial air sinks and warms, and skies are usually clear. Most of the world's major deserts – the Sahara, Arabian, Kalahari and Great Outback – lie along this belt of dry, heavy, settling air. At sea this fairweather zone is known as the 'horse latitudes'. A place where the ocean is often stagnant and left 'panting quietly, like a great heat-tormented beast'.[498] A demanding creature into whose maw the early explorers were often forced to dump their cargoes of dead and dying livestock.

It is from these ocean deserts that warm air masses sometimes flow towards the poles, bumping up against advancing colder fronts, producing a broad band of turbulence in each hemisphere and spawning an endless procession of high and low pressure systems. These disturbances are known as cyclones or anticyclones and they troop around the planet, carried along by the westerly

flow in mid latitudes, giving rise to wild and unpredictable conditions, to the extremes of both wind and temperature that we call
'temperate' weather.

This turbulent westerly river alters its course with the seasons,
pulling in polar air in the winter, freezing the United States,
Europe, Russia and China. And then relenting long enough to
allow warm air to dominate for the short, hot, humid summers of
the region.

In the southern hemisphere, there are no great land masses to
brake and modify the winds and in the 'roaring forties' these
regularly whip up some of the most ferocious weather conditions
on Earth. The momentum and energy of this southern circumpolar
whirl, where pressures are uniformly and consistently low, make
it almost a flywheel of the atmospheric engine, influencing weather
conditions everywhere.

It is joined in this stormy endeavour by four other powerful
winds. Four gigantic, super-Amazons of the heavens whose existence was not even suspected until forty years ago.

On November 24th, 1944, one hundred and eleven B-29
bombers were sent from Saipan to hit industrial sites near Tokyo in
the first high-altitude bombing mission of the Second World War.
As the planes approached Honshu at 10,000 metres and turned
into their run from west to east, they were suddenly swept forward
by a 250 kilometre per hour gale that carried all but sixteen of their
vast load of bombs harmlessly out to sea.

On other subsequent occasions, pilots reported that flying west
towards Japan, they encountered unexpected turbulence and such
powerful winds that they made no progress at all. 'Islands that they
should have passed long ago remained stationary below them as
though the whole scene had frozen,' and they were forced to turn
back before their fuel was exhausted.[537]

After the war, scientists from all over the world began to probe
these high and invisible rivers with instrumented balloons. They
discovered four principal torrents of thin air that howl endlessly
around the world at speeds of up to 500 kilometres per hour,
usually between ten and fifteen kilometres high – and called them
the 'jet streams'.

Jet streams occur where hot and cold air masses meet. They are
generated by the inherent shear of crisp, polar cold against sultry,
tropical heat. The main ones lie against both upper edges of the
westerly belt, at discontinuities in the troposphere along the polar
and subtropical margins. Like rivers, they are wide and relatively

shallow, moving slowly at the edges and at half the speed of sound down their tubelike cores.

Sometimes the two jets in each hemisphere touch and join into a sort of 'supercharged gulf stream of the upper sky'.[351] But always they spin and undulate with the Moon and the seasons, blowing completely around the world, forming the volatile, serpentine backbone of the greatest wind systems on Earth.

Below the jet streams and beyond the westerlies in both hemispheres, lie the comparative calms of arctic and antarctic. Winds here are most often mild and easterly, but in the north the situation is disturbed by the fact that the geographic pole rests on a stable and comparatively warm, ice-covered ocean. The 'wind pole', the coldest and most turbulent area at the pole of pressure, lies some distance away in dry Siberia.

This eccentricity, when added to the bias imposed by the relative preponderance of land in the northern hemisphere, produces a meteorological variety and caprice which is quite unknown in the south. And it is this whimsy, this waywardness, that has come to be reflected so richly by the dominant rhythm of the winds in life and human memory.

Wind and Weather

Living, as we do, at the bottom of an ocean of air, it is easy not to look up. To see only our immediate environment in two-dimensional terms. It is probably no accident that the first ones to question this view, to see something others missed, were a father and son who watched the weather go by from mountain peaks near the top of the world.

Vilhelm and Jacob Bjerknes were Norwegian meteorologists

who, during the First World War, organised a network of weather stations across their country. By releasing instrumented balloons and comparing observations, they became aware that the air was not homogenous, that it moved across the face of Earth in deep waves. They identified distinct chunks of air, independent volumes with definite properties. And, with the War still much on their minds, it seemed natural to them to describe the turbulent boundaries between these moving masses as 'fronts'.

The discovery that local air did not suddenly become cool or warm, moist or dry, but was completely replaced by a new and different volume, was vital. It gave rise to modern meteorology.

Circulation of the atmosphere as a whole, begins with inequities in the distribution of heat. Warm air masses appear, either as descending currents in the horse latitudes or by association with the heart of large land areas in summer. Cold masses accumulate around the poles or over the snow-covered wastes of northern Canada and Siberia. And as each mass of either kind moves, it changes.

Generally speaking, warm air moving toward the poles over cooler surfaces becomes bottom heavy and increasingly stable – unless it happens to be forced up by rising land, when it triggers off rain. Cold air always transforms rapidly as it moves away from the poles over warmer land or sea, becoming unstable and turbulent, bringing bright periods and squally showers. But the length of time any air mass retains its original characteristics, depends on its age and size and how far it travels before running headlong into a mass at a different temperature, forming one of those atmospheric seams or fronts.

Fronts are easy to recognise, once you know what you are looking for. They produce sudden changes of air, bringing a rapid rise or fall in temperature, a wind or humidity shift and a whole new aspect to the sky. Often the boundary is sharply drawn as a long straight line of cloud. Or it may be marked by a rumble of thunderstorms, a line squall, or just a roll of soft white cumulus.

Cold and warm fronts look quite different. A cold front hugs the ground. It edges forward beneath the warmer air in its path, prying it up and rolling it back like a toe tucked under carpet. Light clouds form and dance nervously in the early surge before the great wave itself strikes. Then cold winds rush in, squeezing moisture out of the air, and rain pours down. Half an hour later, it is all over. The front has passed and a new mass of air with its own kind of weather, swallows up the sky.

32

Warm fronts are far less precipitate. It is easier to see them coming. They send out envoys of high-flying wispy cirrus, often days in advance. The leading edge of light warm air rides up high over the cooler air in its path, like the curled tip of a snow ski more than a thousand kilometres long. Behind it comes a bank of descending cloud, dragging its heavier tail on the ground, bringing light rain or mist that can persist for days – usually until the next mass of air arrives.

And so the cycle goes. There is a rhythm to it, with warm and cold air often alternating as though they held each other in thrall. In the middle latitudes, where most people in the northern hemisphere live, they stream by in an endless procession, carried along by the circumpolar vortex of prevailing westerlies, at times even seeming to dominate and disrupt that mighty flow.

This is what the Bjerknes couple saw. They realised that they stood at a major atmospheric intersection, a sort of superfront between tropical and polar air, a place of permanent warm and cold front activity.

In winter, when there is a large difference in pressure and little exchange between the two great global systems of air, the upper westerlies flow along the dividing line in an undisturbed circumpolar ring. In the southern hemisphere, where west winds blow clean around the world with nothing to stand in their way, this polar front remains relatively simple. But in the north, where the westerly fetch is interrupted by the transverse chain of the Rocky Mountains, the vortex breaks and not just North America, but Europe and most of Asia are drenched in spray.

When the difference in pressure on either side of the polar front decreases, as it does in summer, the westerly jet slows and expands, throwing itself into wider undulations.

These are called Rossby Waves after another Scandinavian meteorologist, this time a Swede, who first described their motion. The undulations are known as ridges when they veer toward the pole carrying warm tropical air, and troughs when they bulge toward the equator, bringing cold polar fingers of air to play on warmer regions. The waves tend to build up downstream of the Rockies, and their number varies from two to six. When there are five or six, the troughs are just 5000 kilometres apart and the system begins to break down. The waves overtake each other, curling at the crests into storm. The ribbon of wind parts, meanders meet and kidney whorls of active air fly off on their own. Polar depressions roll anticlockwise down to the south, and warmer wheels of high pressure intrude northwards, turning in the opposite direction, spinning between the colder lobes like cogs in some great weather machine.

Left to themselves, the huge closed systems lose momentum. They weaken and die. But they are so big, bringing surface winds of such different and distant origins together, that they have a profound effect on local weather. Depressions may be 3000 kilometres across and the high pressure systems even larger, but in a global sense these meteorological monsters are nothing but transient eddies. Each plays its small part in the poleward transfer of heat that is the real and enduring business of general atmospheric circulation.

The existence and position of mountain barriers running west to east, has a particularly profound effect on wind and weather. The Himalayas, for instance, prevent even the most vigorous cold fronts from invading India and southeast Asia. Italy is largely protected by the Alps and Spain is fortified by the Pyrenees. Only north America has no east-west range to block the southerly flow of cold north winds, with the result that an occasional *norther* brings a tidal wave of polar air down to the Gulf of Mexico, dropping temperatures by as much as 15° Centigrade in a single hour. In the Texas Panhandle they say, 'the only thing that stands between us and the north pole, is a three strand fence with one strand missing'.

Where mountain barriers do exist and the moving air is stable, it may return to its original level on the far side, flowing down the lee slopes as strong warm winds of low humidity. Unstable air climbs and cools, leaving its moisture on the peaks or in lens-shaped clouds pendant to them. In both cases, the downwind result is a sudden hot dry flow of air with an appetite for moisture. These are the *föhn* winds, from the Gothic *fôn*, meaning 'fire', because they bring a real risk of conflagration.

In Alpine villages, where almost everything is made of wood, smoking is banned and even cooking fires are extinguished when the hot dry air rolls in. During summer months however, little can be done to prevent general desiccation. 'Leaves crumble to dust at the touch; corn, wheat and other cereals look as if they had been scorched by fire; apples are baked while still hanging on the trees.'[507]

The effects of the *föhn* in winter can be even more dramatic. It usually begins with a 'föhn wall', a line of high dark pancake clouds above the peaks that build and grow until they seem to devour themselves. The mountains begin to stir and rumble in their sleep as the air descends, increasing in temperature by 1° Centigrade for every hundred metres that it falls and compresses, breaking the power of the frost. The hot wind literally eats the snow, siphoning it straight off the slopes as if by magic, changing ice crystals directly into invisible water vapour.

In Canada and Montana to the east of the Rockies, they call their snow-eater by its Indian name, *chinook*, and usually find it welcome. In the darkest days of February, with temperatures well below zero and snow drifts deep on the plains, the *chinook* comes in like the cavalry, intent on lifting winter's siege. Or, in the words of a mountain poet, 'like a scented virgin come to seduce the gods of winter'.[408]

The *chinook* howls across the lee slopes, gathering dynamic warmth as it descends, wolfing down the snow, whipping the ice off ponds without even melting it, and licking up every' drop of standing water. On January 22nd 1943, Rapid City in the Black Hills of South Dakota, lived up to its name. A *chinook* lifted the temperature there from −20° Centigrade to +7° Centigrade in just two minutes. Before long, cattle that were 'moaning in distress from cold and hunger, their noses hung with bloody icicles', could lie down for the first time in weeks, knowing that now their bodies would not freeze to the ground.[56]

The family of falling winds include the *maloja*, which plummets into Switzerland; the *yama oroshi* of certain steep valleys in Japan; the *zonda* in the lee of the Argentinian Andes; the lusty black *reshabar* out of the high Caucasus; the *northwester* at Canterbury in New Zealand; the *autan* of the Corbières in France; and the malevolent *Santa Ana* of southern California.

In addition to shaping the flow of already moving air into lee waves that sailplane pilots ride like giant roller-coasters, mountains are able also to make their own indigenous winds.

During warm mornings, the air in deep valleys expands and, finding itself constricted by the walls, flows uphill along the valley floor, climbing the mountain as it goes. Additional heating further up the slope keeps the system going and cold air folding back along the ridges, piles up over the valleys, feeding the upward current. After the sun sets, mountain air cools rapidly, contracts and with the help of gravity, cascades back down onto the valley floor, often producing intense frost pockets.

Valley, upslope or anabatic winds; and gravity, downslope or katabatic winds; are almost metronomic in their regularity. In every Alpine valley they are known by local names, pet ones descriptive of the reassuring sound of the mountain's steady breathing. People there only start to worry when the breathing stops, when local winds fail and bigger storms are brewing.

On lakes ringed by the mountains of northern Italy, the alternation of winds each day 'seems as fixed as the law of the Medes and Persians. They act upon schedule time, and regulate both trade and pleasure.'[56] On Lago di Garda, the *ora* blows from the south between ten thirty and three, carrying sailboats to Riva at the lake's northern limit. And in the late afternoon, it obligingly makes way for the northerly *sover*, bringing the same boats back home on another following wind. On the lake of Como, similar facilities are provided by the *tivano* and the *breva*, while at Maggiore it is the *tramontana* and *inverna* that oblige.

But not all mountain winds are as kind. One of decidedly ill repute is the *mistral*, from the Latin *magistralis*, meaning 'masterly'.

This is a strong dry cold north wind funnelled through the valley of the Rhone between the Alps and the Cevennes, directed with malevolence at the French Riviera and the Gulf of Lyon. Howling like a jet stream, sometimes at 150 kilometres an hour, 'it has the ill-natured habit of scattering roof tiles about, knocking down chimneys, blowing small children into canals, tumbling walls onto the unsuspecting natives, upsetting wagonloads of hay and sometimes overturning freight cars'.[56]

One of the first to experience and describe the *mistral* was the Greek geographer Strabo, who condemned it as 'an impetuous and terrible wind which displaces rocks, hurls men from their chariots, breaks their limbs and strips them of their clothes and weapons'. It is still the scourge of soldiers and farmers and makes life miserable for harbour pilots in the port of Marseilles. The people of Provençal towns do what little they can by laying out their streets at right angles to the wind and building doors and windows only on the southeast side of their homes.

A close relative to the *mistral* is the formidable katabatic *bora*, child of Boreas and scourge of the Adriatic. This bleak wind whistles through the valleys of the Julian and Dinaric Alps, bursting violently out onto the coasts of Istria and Dalmatia, where it falls 'as if hard things were thrown to the ground', closing down the railroads and forcing ships from Trieste to flee for shelter further down the coast. It blows for forty days or more each year, bringing snow and ice in its train, whipping up short steep waves, atomising their crests into a spindrift mist that local fishermen call *fumarea* (fumarija) – 'the smoke'.[251]

The French novelist Stendhal, while he was consul to Trieste in 1831, wrote 'It blows a *bora* twice a week, and a high wind on the other five days. I call it a high wind when I hold on to my hat and a *bora* when I am in danger of breaking my arm.'[482]

Other near relatives are the cold *tehuantepecer* which sets in suddenly on the Pacific coast of Mexico and Panama; the *gregale* of Malta; the Balinese *klod*; the fitting *blaast* of Scotland; the deceptive *wisper* of the Rhine; and the furtive *williwaw* that leaps out of mountains that guard the Straits of Magellan.

It was the latter which ambushed circumnavigator Joshua Slocum in 1898 on a day that 'to all appearances promised fine weather and light winds. While I was wondering why no trees grew on the slope abreast of the anchorage, half minded to lay by the sail-making and land with my gun for some game . . . a *williwaw* came down with such force as to carry the *Spray*, with two anchors down, like a feather out of the cove and away into deep water. No wonder trees did not grow on the side of that hill! Great Boreas! A tree would need to be all roots to hold on against such a furious wind.'[449]

The greatest gust of any wind in the world, outside of a tornado, seems to be the 371 kilometre per hour recorded on April 12th, 1934 at the peak of Mount Washington in the northern Appalachians. This took place during a gale that averaged over 200 kilometres an hour for that day and was boosted to its record speed by the funnel effects of the mountains around the weather station. And it is no surprise to discover that the windiest place in the world is another mountain margin in East Adelie Land on the edge of Antarctica, where an exaggerated katabatic wind pours polar air down a steep slope to the sea, maintaining an average of sixty-four kilometres an hour every day and night throughout the year.

Apart from wind and weather produced and influenced by mountains, there are a number of local effects that depend on the different way in which air behaves over land and over water.

Along the shores of oceans and lakes, particularly during the summer months, winds follow a typical daily pattern. At dawn the air is still and the water mirror-calm. Then, as the sun rises and warms the land, air there lifts and cooler sea air flows in to replace it. This sea breeze at its best has a depth of about 1000 metres, moves at a speed of up to twenty kilometres an hour and can penetrate fifty kilometres inland. In places it is substantial enough to be deflected by the Coriolis Effect, turning an onshore wind into one that blows parallel to the coast.

By mid afternoon, this refreshing breeze with its tang of sea air, begins to decline. As the sun sets, it is calm again and only residual wavelets lap the shore. Then, after dark, the whole situation is reversed. The land cools more quickly than the water and air settles heavily upon it, edging out into lower pressure areas offshore.

This land breeze gradually picks up to seven or eight kilometres an hour, returning to the sea with a motley burden of dust and pollen, soot and soil. Hot evening air is sucked out of coastal cities and replaced by the smell of inland forests, of freshly ploughed fields and new-mown hay.

It is hardly surprising that these short-lived welcome winds should have been given some of the softest fondest names in the wind dictionary. They include the *imbat* that touches the hot coasts of Tunisia; the *datoo* which brings cool Atlantic air from the west across Gibraltar; the *vento de baixo* in Portugal; the *medina* of Cadiz; the *kapalilua* in Hawaii; the *kadja* that worships each day at the foot of the mountain they call 'the navel of the world' in Bali; the kindly *doctor* who postponed the white man's early grave in tropical West Africa; and the southwesterly *libeccio* that tempers Neapolitan summers efficiently enough to have been celebrated by John Milton in *Paradise Lost*.

There is a sea breeze called *ponente* that occasionally penetrates all the way from Tyrrhenia to Rome and 'comes trickling into the narrow streets like a refreshing drink', bringing everyone out to stroll 'save the bedridden and the people in jail'.[332] And in Valparaiso, the *virazon* sweeps away the heat of Chile's Atacama Desert, leaving the sky without a cloud, the atmosphere transparency itself. 'The Andes seem to draw near; the climate, always mild and soft, becomes now doubly sweet by the contrast. The evening invites abroad, and the population sally forth.'[325]

These are evanescent movements of air, but in some parts of the world land and sea winds are longer-lived, alternating on an annual rather than a diurnal basis. These are the *monsoons*, from the Arabic *mausim*, meaning a 'season'.

The mechanism which drives these is complex, but is in essence the same. When summer temperatures over large land masses build up day after day, the hot rising air leaves low pressure areas that extend for many millions of square kilometres. Cool moisture-laden air rushes in from the ocean to take its place and, because the pressure differences are so great, it keeps on doing so for months on end, overriding all small diurnal variations. And in winter, when the land cools off, the process is reversed and strong dry winds blow consistently from the diametrically opposite direction.

When conditions last this long, generally speaking anything over thirty days, it is seen to be something more than just weather. It is recognised as one of the major patterns of behaviour that are thought to be characteristic of Gaia.

It gets to be called a climate.

2. THE GEOGRAPHY OF WIND

In the early days, wind and weather were synonymous.

The two words became almost interchangeable in English, a relationship which survives still in names such as 'weathervane' for a device which does no more in itself than indicate wind direction.

When winds had a visible purpose and moved ships and mills or winnowed the grain, they were held in great esteem. People prayed or whistled for them or even, if it seemed expedient, bought one from an aged crone who sold the best ones cheap.

Today its social status has declined, but the importance of wind remains paramount. No other single factor has a greater influence on climate or our well-being.

Prevailing Winds

Weather is what we experience.

Climate is an abstraction, a fiction peculiar to the human mind, something it pleases us to deduce from the weather. It is the average condition at a particular place over a long period of time.

What we call climate, has a lot to do with closeness to the equator. The word itself derives from the Greek *klima*, meaning 'a zone of latitude', which is why we still sometimes speak of sunnier 'climes' when referring to places in or close to the tropics. And why most geographers recognise four principal climatic zones, dividing the world up along latitudinal lines, into tropical, subtropical, temperate and polar regions.

Any classification, however, is arbitrary and unnatural, and this one runs into particular difficulty by concentrating on temperature alone. Moisture is a better indicator, drawing harder sharper lines between habitats and, combined with temperature, provides a measure which effectively defines many of the natural vegetation types. But the most satisfactory system of climatic classification is one which is based on atmospheric circulation. On wind, which is ultimately responsible for both local temperature and humidity.

At any one time, 97 per cent of all water on Earth is in the oceans and most of the remainder is locked up in ice sheets and glaciers on land. A mere 0.0001 per cent is held in the atmosphere, sufficient in itself for only ten days' rain. But there is a constant influx of moisture into the air as a result of evaporation from both land and sea. This moisture is released when air expands or cools sufficiently to allow condensation to take place, producing a mean annual global rainfall of eighty-six centimetres. Local evaporation is, however, seldom the major source of local precipitation. For example, only 6 per cent of all rain that falls in Arizona originates in that state. The rest is carried there by wind, mainly from the sea. And the most dramatic example of this airborne transport is provided by the monsoon.

In their classic form, monsoon winds blow across the Arabian Sea for about six months from the southwest, and then reverse themselves and blow instead from the northeast for the following six. This impressive alternation was hinted at by Aristotle, who described it in his *Meteorologica* in the fourth century BC. But it was Edmund Halley, he of the comet, who proposed the first scientific explanation for the monsoon in a famous memoir on the trade winds presented to the Royal Society in 1686. Halley correctly identified pressure differences over land and sea as the driving force behind the winds, but with the information at his disposal was unable to complete the picture. More recent work shows that the Asian monsoon is a special case deriving from a complex global pattern of airflow.[513]

One of the vital properties of water is its high capacity for storing solar heat. And one of the most important consequences of moving air is the way in which it stirs the sea, creating turbulence and eddies which mix the water and distribute heat throughout the great ocean mass. Even the water from the bottom of the deepest trenches returns to the surface at regular intervals. As a result, the temperature of a water surface varies less than that of the land. The oceans act as a heat sink or an enormous flywheel, holding potential energy and releasing it at leisure later.

In autumn, as the continents cool, heat loss from the land is rapid. Round about October, Asia begins to breathe out. Cold high-pressure air builds up over the interior and slides toward the equator and equilibrium. As it moves, it is deflected to the right by Earth's rotation, producing the cool dry northeasterly trade wind. This blows steadily from November to March, pouring down from the Himalayas, bringing air that jaded Indians describe as 'clean and fine' or 'pure and most delicious'.[19]

This cold air mass, moving from the northeast along the surface, is balanced by warm air from the ocean that moves back northward at a height of about 12,000 metres. It is deflected at height to the east, creating the intense winter jet stream over Asia and Japan that was discovered by the United States Air Force during the latter years of the Second World War, and which still spawns the low pressure winter storms that lash North America.

By spring, the potential energy that powered the winter wind has dissipated. The northern Indian Ocean has finally cooled to match the temperature of the land. The system is in equilibrium and April is a month of calm on land and sea. But the Sun is moving north again and by May, continental temperatures begin once again to rise. A new pressure gradient is created, this time with the ocean lagging behind, and air too begins to move back north. In its passage it is deflected by Earth's rotation, this time to the east, and the true wet southwest monsoon is born.

The birth can be spectacular. It is usually described as 'bursting' on the land. The first Indian prime minister, Jawaharlal Nehru, told of the arrival of the rains in Bombay. 'They came with pomp and circumstance and overwhelmed the city with their lavish gift. There was a ferocity in this sudden meeting of the rain-laden clouds with land. The dry land was lashed by the pouring torrents and converted into a temporary sea. Bombay was not static then; it became elemental, dynamic, changing.'[44]

Once over India, the warm moist airstream is checked by the

Himalayas and forced upwards, cooling and producing exceptionally heavy rain. The inundated land attracts and holds more heat, producing even greater convection, exaggerating the differential between land and sea and perpetuating the southwesterly monsoon flow. The creation of rising currents over India and China seems also to have an effect on the upper airflow, contributing to a complex easterly jet stream that lasts from July to September. This is the highest of all the great air currents, reaching a peak around eighteen kilometres over Pakistan, and falling and warming like a gigantic dry *föhn* wind over the summer deserts of Saudi Arabia and northeast Africa. Parts of it even go on to cross the equator, merging with the winter westerlies of the southern hemisphere.

It becomes clear that the monsoon is much more than a glorified sea breeze, bringing water to half the people of Earth. Its pendulum swing over the Arabian Sea may be influenced by local heat and pressure gradients, but it is also an important wheel in the global atmospheric machine, with a considerable influence on climates a long way from southeast Asia.

Australia has its own mini-monsoon which blows from the northwest, out of Indonesia, early in the southern summer. Parts of the outback near Port Hedland know 160 days in a row when temperatures are above 38° Centigrade. These conditions produce widespread atmospheric upwellings, corresponding disturbances in the upper airflow and an inrush of cooler, moister masses from over the Timor Sea and the southern Indian Ocean. Then the deserts get some of their rare falls of rain.

And something similar happens in the eastern Mediterranean in July and August when air is sucked off the Balkans down south toward the hot Sahara. These are the *etesian* winds, from the Greek *etos*, meaning a 'season', which despite a clear sky, whip the blue Aegean into a stormy froth and keep old women in their island villages knitting thick winter sweaters for astonished and grateful summer tourists. Along the Turkish coast, they call these winds *melteme*, the 'bad tempered ones'.

Even Europe has a sort of 'monsoon'. It is less obvious in Brussels than it is in Bombay, but nevertheless exists. There is a continuous series of official records of weather in London collected every day since 1670, and further indirect evidence from a variety of private weather diaries going back as far as 1340. Analysis of these shows a tendency for a certain type of weather to recur with fascinating regularity around the same date each year. The most interesting of these seasonal singularities is one which reveals a sharp increase in westerly winds about the middle of June. An invasion of moist maritime air at this time brings steady rain and 'summer monsoon' weather to most of central Europe in too many years for it to be purely coincidental.[269]

Gaia spins on her toes in front of the erratic radiant heater of the Sun. And yet despite its flux and her perennial wobble, conditions at the surface of Earth remain remarkably stable. And they do so largely because our planet is water cooled. Five hundred million million tons of moisture are sucked in and out of the air and blown about the world in a way that keeps the heat budget beautifully balanced.

Winds quench their thirst by evaporation, picking up 84 per cent of their moisture from the oceans and dropping a good part of this directly back into the sea or on to the nearest bits of land. The rest is moved a considerable distance, mostly towards the equator in low latitudes and towards the poles in middle latitudes. The classification of climate on the basis of air movement, clearly reflects these facts.

In the tropics, climate and weather are almost synonymous. There is little variation in either. Local differences depend on the degrees to which moist equatorial calms are disturbed by wind. Where there is little or no disturbance, just a gentle drift of moist air punctuated all year round by thunderstorms, the weather is more or less constantly wet and rain forest thrives. This is the Equatorial climate type, characteristic of the central Amazon and Congo basins and the larger islands of Australasia.

Where the doldrums are disturbed in winter by dry easterly trade winds, rainforest is replaced by grassland or savannah. Or, where the usual summer rains are augmented by monsoon conditions, broadleafed semi-deciduous forest may persist. These are the Tropical climates, found in Venezuela, the *mato grosso* of Brazil, the plains of central and east Africa and the Indian subcontinent.

The subtropics are similarly fractured by wind.

Where warm air descends in the 'horse latitudes', permanent high pressures produce a hot Desert climate with results best seen in the Atacama, Sahara or Kalahari. Where westerly winds swing towards the equator long enough to touch the desert margins with winter rain, they produce a Mediterranean climate, known also from California, Chile, the Cape and the southwestern tip of Australia. And where winds drop altogether, allowing convection heat and thunderstorms or moisture to be sucked in from the sea, the climate becomes Subtropical, as it does in Florida, south Brazil and coastal China.

In the middle latitudes, roughly between forty and sixty degrees from the equator, are the westerly winds and the so-called Temperate climate. It rains all year round in these regions, but weather is so variable that it is difficult to describe in precise terms. Hence Mark Twain's forecast for a typical New England day: 'Probable nor'east to sou'west winds, varying to the southard and westard and eastard, and points between; high and low barometer, sweeping round from place to place; probable areas of rain, snow, hail and drought, succeeded or preceded by earthquakes with thunder and lightning.'[493] In other words, if you don't like the weather, just wait a few minutes.

In fact, the prevailing westerly winds do at least make some generalisation possible. Broadly speaking, oceanic islands, all small land masses, and the western coasts of the continents, can be grouped together. All are exposed to the full force of the usually wet winds, have heavy rain, high humidity and relatively uniform temperature. While the interior and eastern parts of the great land masses tend to be drier, with a greater range of temperatures, both diurnal and seasonal.

Beyond the limits of the westerlies in the northern hemisphere, there is an extensive subpolar zone. Here, despite the fact that there seem to be just two seasons, winter and July, there is enough warmth and moisture to produce the stunted growth characteristic of tundra. In the south, there is not enough land to have any kind of climate at all until the true polar zone of Antarctica.

The Polar climate in both hemispheres is one of sporadic winds and low precipitation.

There are of course transitional areas and places whose special topography puts them beyond the reach of such a broad classification. And no allowance has been made for elevation, which can have the same effect on local climate as an increase in latitude. But on the whole, the system works. It puts emphasis in the right place, concentrating on the effects of wind on the weather. In every zone, it is the quality of the air and the direction and duration of its movement that shapes a climate and everything in it.

"After all,' said Gertrude Stein, 'anybody is as their land and air is. Anybody is as the sky is low or high, the air heavy or clear, and anybody is as there is wind or no wind there. It is that which makes them and the arts they make and the work they do and the way they eat and the way they drink and the way they learn, and everything.'[459]

Winds of Occasion

Air is normally invisible.

So too are the 'viewless winds' that move it 'with restless violence round about the pendant world'.[433] But they leave tracks.

When air expands and cools, water vapour condenses into minute droplets which are light enough to be kept airborne, suspended by the swirl of other moving molecules. Fifty billion of these tiny tears would scarcely fill a teacup, but they supply the pigment with which wind fills in its footprints, painting great diagrams of atmospheric flow across the sky. They are the substance of clouds.

'Clouds were once the thrones of gods; on them the angels knelt and saints took their rest.'[13] It was a pillar of cloud which led the Jews out of Egypt and another that covered the congregation in the tent of the tabernacle when their god revealed himself to his people. This so impressed them that 1300 years later the apostle Paul, in his first letter to the licentious Corinthians, reminded them that 'all our fathers were under the cloud', meaning that they were in a state of grace rather than under suspicion.

Later the concept of clouds became less sublime and they changed to symbols of transgression. And it was not until the beginning of the nineteenth century that anyone really looked at clouds as natural phenomena in their own right.

The first to do so was an amateur meteorologist called Luke Howard who, 'for the benefit of Agriculture and Navigation', applied the Linnean system of classification to atmospheric phenomena.[216] Howard invented the major cloud genera, drawing on Latin for *Cumulus*, meaning 'a heap or pile', for the common fairweather forms; *Stratus*, meaning 'a spread or layer', for the ones stretched out like blankets against the light; and *Cirrus*, meaning 'a curl or tuft', for wispy filaments of the upper sky. And he qualified these with the terms *nimbus*, for the ones bearing rain; and *alto*, for the higher varieties. Within this small vocabulary, he was able to characterise all the panoply of cloud effects and create an excellent natural taxonomy, based on both structure and be-haviour, that now recognises ten genera and twenty-six species of cloud.[534]

Cirrus is the largest genus, and forms the élite amongst clouds, cool and lofty, nothing but silver lining. These are the 'eyelashes of the sun', visible from 300 kilometres away, holding the warm colours of the day long after the lower clouds have all gone grey. Their substance is diffuse, consisting largely of snowflakes that form around ten kilometres high and grow as they fall, cascading down through the upper levels of the troposphere, whipped by the wind into long streaks of icy spume.

The five recognised species include *Cirrus fibratus* – which is threadbare and loosely woven; *Cirrus uncinus* – in which the fila-ments are distinctly hooked; *Cirrus spissatus*, which is a little thicker, like a white carpet with frayed edges; *Cirrus floccus*, looking like loose tufts of cotton; and the high and fortified turrets of *Cirrus castellanus*.

Closely related to these wing-feathers of the wind, are two slightly more earthy genera. *Cirrocumulus*, composed of regular wave patterns of ice grains laid down on the leading edge of a warm front, and diversifying into tufty *Cirrocumulus floccus*; little lens shapes of *Cirrocumulus lenticularis*; battlements of *Cirrocumulus castellanus*, and the wonderful rippled *Cirrocumulus stratiformes* that produces a wall-to-wall 'mackerel sky'. And *Cirrostratus*, the milk cloud, a high silky veil of crystals in just two forms – hairy *Cirrostratus fibratus* and the slightly spooky *Cirrostratus nebulosus* that throws a halo round the Sun.

Below five kilometres are the wet blankets of the heavens. The highest of these is *Altostratus*, thin enough to just let the Sun shine through, as if seen through ground glass. It is known only from the type genus, but beneath it are several species of *Stratocumulus*; most notably *Stratocumulus stratiformes* in a continuous canopy, like the ceiling of a tent with irregular folds of light and dark running right across its fabric. When these thicken further, they may become *Stratocumulus castellanus*; but if they split and frac-ture, the result is more likely to be *Stratocumulus lenticularis*.

Beneath them all lies the shapeless homespun substance of pure *Stratus*, the most stable of clouds. At its best, a dour blue-grey drizzle, touching the ground as *Stratus nebulosus* or Scotch mist – the 'Tory fog of the lower levels'.[351] At its worst, it becomes the sombre genus *Nimbostratus*, whose only species has a habit of producing continuous rain or snow.

The *Stratus* clouds are all stolid sorts, undisturbed by wind or rising currents, formed when air lies at rest over a cooler layer near the ground. Exactly the opposite of conditions which give rise to the more dynamic *Cumulus* clouds, the radicals of the family.

Cumulus are vertical clouds, puffs of steam generated by hot rising air, the caps on winds that blow straight up from ploughed ground or car parks. They usually appear in the middle of a sunny morning, starting as stray wisps of *Cumulus fractus* and then turning to better-defined *Cumulus humilis*, that spring up like fluffy cabbages in a field of summer sky, all at the same altitude. Sometimes they gather along the shore in the way of sheep grazing together at a fence line, marking out the margin of the land. Always they roll and grow and change, bursting with evolutionary vigour.

These common and relatively humble species of *Cumulus* mutate, given half a chance and enough moisture, into more ambitious *Cumulus mediocris* and, as they grow vertically, into distinctly bumptious *Cumulus congestus*. And if such upwardly mobile clouds spread far enough and climb high enough, they give rise to a new genus – *Altocumulus*, that includes some rare and showy species. Such as the laminated *Altocumulus stratiformes*, which makes wonderful sunsets; the heavy pile carpet effect of *Altocumulus floccus*; and, when the shape of the land beneath is just right, the intriguing *Altocumulus lenticularis* which are flattened into almond or flying-saucer shapes and keep getting their pictures in the papers.

All these more volatile species are produced by rising heat. They start with random and competitive eddies, but go on to grow and coalesce, finally becoming incorporated into energetic systems complex enough almost to be called metabolic. And the most lively and organic of all such sky creatures are the ones that grow rampant superstructures and move to the top of their profession in the genus *Cumulonimbus*.

Thunderheads occur in two species, bald or hairy. *Cumulonimbus calvus* has a high smooth pate, but the more majestic *Cumulonimbus capillatus* is crowned with an elaborate coiffure that can, at its most impressive, be drawn out into a large sharp peak or anvil-shape that is the mark of a mature thunderstorm.

A thunderstorm is a living thing, an organism with a definite and easily recognisable form. It stands on a foot of cool hard rain, with a heel of drizzle and a toe of rolling squall, feeling its way slowly forward step by step. Drafts of warm air rise all about its body, flaring out at the head into an anvil of ice and hail. The anatomy is characteristic and well-defined. Morphologically, it is divided into what can almost be called functional body parts. Behaviourally it is unique in the non-living world, defying the tendency of most inorganic systems to slide into inertia. Thunderstorms go on doing something, moving, exchanging material and information with their environment. They even reproduce themselves by giving rise to daughter storm cells near the edges as the mother clouds move past maturity into senility and decay.

In its prime, a thunderstorm has a voracious appetite for fresh warm air. This is sucked into its heart where it rushes up to the head at speeds of over 100 kilometres an hour, rising to the level of the icy anvil ten kilometres up in less than five minutes. And then pouring down again through the fabric of the storm, creating immense shear forces and electrical fields.

In 1938, during a gliding competition over the Rhön mountains in central Germany, five contestants flew into such a cloud in search of lift. They were sucked directly into the violent centre and as their sailplanes broke up around them, all five parachuted free. 'The consequences were dreadful. Instead of falling gently downwards, the parachutes were filled to bursting point by the wind and carried upwards. Higher and higher they soared into increasingly colder layers of cloud. However hard they tried to steer with their arms and legs, they could not escape the howling force of the gale. Huge raindrops soaked their bodies in a few seconds. Hailstorms lashed their faces . . . Only one man, severely injured, escaped with his life.'[289] It is difficult to imagine the ordeal of the others. At a height of over ten kilometres, they must have been frozen, tossed about like living icicles, stabbed at by lightning, until the storm grew tired of its game and discarded their bodies.

Even large aircraft fare little better. During the Second World War, a squadron of eight fully laden bombers heading north from Australia flew into a line of *Cumulonimbus*. Only two came out the other side.[537]

Sometimes, usually when there are a number of thunderstorms lurking along a squall line close to the polar front, one or more of them will develop dark drooping pouches that identify it as a particularly dangerous species – *Cumulonimbus mammatus*. The thunderheads on these storms often grow turrets that rise above the anvil, breaking through the high jet stream to puncture the stratosphere, producing inversions and creating extraordinary pressure differences. Strong up and down drafts develop and, possibly as a result of massive electrical discharges which accompany these, pockets of air begin to turn and a dark funnel dips down from the base of such a cloud, spinning like a twisted cornucopia, producing the most violent and predatory of all windstorms – a *tornado*.[500]

The name comes from the Latin *tornare*, 'to turn', and there are vivid descriptions of early turning winds in Pliny, Seneca and Lucretius. The Romans were also aware of the steady electrical discharge that accompanies such disturbances and called the most luminous ones *prester*, which may account for the name given to a medieval king who is said to have ruled over a fabulous country somewhere in Asia. Though he was believed to be immortal, he could be seen by his subjects only three times each year, nevertheless, 'Prester John endures for ever, with his music in the mountains and his magic in the sky.'[365]

The earliest recorded report of a tornado is probably one from about 600 BC, described by the prophet Ezekiel as a 'whirlwind come out of the north, a great cloud, and a fire unfolding itself, and a brightness was about it, and out of the midst thereof as the colour of amber'.[21] Sand-filled twisters with lightning playing about them are still seen in Israel and Australia and from as far afield as Moscow and Fiji. They are familiar to residents around the Black Sea as sounding *trompa*, and to those on the Baltic as lusty *skypompe*. In France they call them *tourbillon* and in China *piao*. But the major breeding ground of the most aggressive tornadoes seems to be on the Great Plains of North America.

In 'Tornado Alley', which runs through Kansas, Oklahoma and Missouri, over 700 a year twist out of the base of thunderclouds that form most afternoons in spring and early summer, as hot humid air from the Gulf of Mexico clashes with cold dry polar masses in the north. The majority of these hasty little winds are small and short-lived. An average tornado lasts for less than ten minutes and travels no more than a few kilometres at a speed seldom greater than fifty kilometres an hour. It takes about fifteen seconds to pass by on a narrow path usually less than 150 metres wide, and it has a tendency to leap along, sometimes with its foot barely touching the ground. But despite its size, it leaves devastating tracks.

The air pressure inside a tornado is abnormally low, dropping as much as 150 millibars, which is equivalent to an instant increase in altitude of 1500 metres. As though New York were transferred in a flash to the height of Denver, or London swept up into the thin air well over the summit of Ben Nevis. This partial vacuum produces enormous pressure differences. On the inside of a one-metre-square pane of glass for instance, it may be more than a ton higher than the outside. Small wonder then that tyres burst, corks pop out of empty bottles and houses sometimes explode.

A pilot flying over Waco in Texas during a tornado in 1953, saw 'plate glass windows on both sides of Main Street burst outward in progressive waves, opposing brick walls meet each other as they crashed on to the lanes of slow-moving cars, roofs fell into the wall-less interiors of the stores. A theatre and a six-story furniture mart burst at the seams like slow bombs, both immediately collapsing into twisted heaps of wreckage.'[351]

The same thing happens to creatures that get caught in the funnel and blown up like balloons. There are numbers of reports of chickens not only grossly inflated in this way, but plucked completely clean. It is possible that the birds are denuded as air sealed in the quills of the feathers, actually explodes when pressure falls suddenly around them.

Although a tornado itself moves relatively slowly and it is easy to outrun one in a car, the wind in the pendant stem reaches frightening speeds. These have never been measured directly for the simple reason that anemometers caught in the path of a twister are invariably destroyed. But radar readings taken from a distance indicate speeds of at least 300 kilometres an hour.[452]

In June 1953, three transmission towers were destroyed by a tornado that hit Worcester in Massachusetts. Engineering calculations, derived from previous testing to destruction on the steel in these structures, suggest that the wind velocity would have needed to be greater than 550 kilometres per hour.[42] Aerial photographs of a track left in 1955 by another tornado that passed at a measured speed through the North Platte Valley in Nebraska, show ground marks left by something carried in the swirl that must have been moving at 780 kilometres per hour.[496] And it is suggested by other meteorologists working on the dynamics of what they call 'super' or 'inconceivable' tornadoes, that winds in the critical circle of these might at times approach the speed of sound at 1220 kilometres an hour.[1]

These upper estimates may be too high, but there is little doubt about the force generated by the combination of high speed and low pressure. Sand and gravel are blown about with such force that they enter human bodies like bullets. Wheat straws are routinely embedded like blowgun darts deep into the bark of trees. A 100-kilogram steel beam over two metres long, fastened at each end with heavy bolts, was torn free and hurled 200 metres to be driven through the trunk of a cottonwood tree six metres above the ground.[530] And in one well-documented case, a pine plank was shot right through a solid iron girder supporting the Eads Bridge in St. Louis.[344]

There are stories of an egg found with a bean blown into the yolk through a neat hole bored in the otherwise uncracked shell, of undamaged candles embedded deep in the plaster walls of rooms, and even of a flower forced into the fabric of a plank of wood.[139]

There is no question of the lifting power of these giant vacuum cleaners. A house in Oklahoma was picked up, turned ninety degrees and set neatly back down again sideways to the road.[246] Another in Kansas was lifted so gently that the owner, 'unaware of what was going on, but having a hunch that all was not going well, opened his front door, stepped out and dropped twenty-five feet through empty space to the ground'.[275] Others, less lucky, are carried high into the air with their homes and never seen again.

The spire of a church in Kansas, complete with weathercock, was deposited twenty-five kilometres away from its base. Wells and streams are frequently sucked dry. Five coaches of a train travelling through Minnesota in 1931, were lifted off the track and left twenty-five metres away in a ditch.[275] And an engine on the Union Pacific Railroad was picked up, turned around in midair and set down on a parallel track, facing in the opposite direction.[225]

Tornadoes kill hundreds every year. 'Human beings and animals are often beheaded, torn to pieces, impaled by flying timber, maimed and mutilated, carried aloft, stripped of their clothing and spewed out of the funnel, their naked bodies made hideous with a jet black incrustation of mud. Mud fills the eyes, ears, nostrils of survivors and victims while wounds are not only impacted with mud, but filled with sticks, pebbles, leaves, nails and splinters.'[56]

These wayward winds seem also to have a sense of the ridiculous. A house can be ripped apart while the kitchen cupboard, filled with crockery, is carried off and set down somewhere else without a single broken dish. A mattress was sucked through the window of a farm house in Kansas without even waking the child still sleeping on it. A lighted kerosene lamp was moved 300 metres and deposited right side up, still burning, with just a slightly smoky chimney.[351]

There are records of horses being carried three kilometres and set down unharmed, and of others left considerably surprised astride a barn.[506] There is a tale of one man picked up by a tornado who, 'while travelling skyward, stretched out his hand which came in contact with the tail or mane of a horse (his indefiniteness may be pardoned in the circumstances). He grasped it firmly but during his aerial excursion he became separated from the horse, and landed grasping a handful of horse hair in one hand and his hat in the other.'[275]

Milkmaids have been left with nothing but the bucket as a wind carried their cow away; and on one memorable occasion a whole herd of cattle were seen drifting off together, 'looking like gigantic birds in the sky'. An incident later described by another meteorologist as 'the herd shot around the world'.[215]

When a tornado crosses water, it becomes a waterspout and continues to lift and throw things about, but perhaps because water is seldom as warm as the land or inclined to offer as much resistance, it seems less ferocious. A tornado that formed at Norfolk in Virginia in 1935, went through several transformations. Shortly after starting to demolish the town, 'it crossed a creek, sucking up water until the bottom was exposed and then gouged a channel in the mud. As a waterspout, it lifted small boats on to the shore, ripped off part of a heavy pier and – a tornado again – destroyed several buildings. Crossing Hampton Roads it turned into a waterspout again, then into a tornado which flung rolling stock off the tracks in a railroad yard. As a waterspout again, it sucked up another creek and as a tornado damaged some aeroplane hangars on shore. When last seen, it was heading up Chesapeake Bay as a waterspout.'[353]

True waterspouts, which actually form over water, can go through the same metamorphosis on approaching land. But most seem to keep to the ocean or to larger lakes, where they acquire their own distinctive aquatic character. This differs from tornadoes in that the foot is more narrow, averaging about twenty metres across, but is surrounded at the surface of the sea by a constant cascade of water that is three times as wide. And this is not salt water falling back from the updraft, but pure fresh water produced by the cooling and condensation of inflowing air. Higher up the spout, air and water seem to rise around the outside of the whorl, and flow down along the axis in the interior.[410]

The White Star liner *Pittsburgh* was hit by a huge waterspout in mid Atlantic in 1923. So many tons of water were dumped on her so suddenly that her bridge was wrecked, the crow's-nest and her officers' quarters flooded and she had to heave-to in a calm sea while the damage was repaired.[275]

Sailors have never liked waterspouts. In the sixteenth century, a man would be sent forward with a black-handled knife to try and cut the throat of a threatening 'waterdragon'. Later mariners believed it was necessary, and effective, to break or let air into the spout with a well-aimed shot from a cannon.[231] But the truth is that they represent a relatively minor hazard with nothing like the energy or invective of a tornado. Thousands are seen each year, most of them grazing gently in the hot moist calms of the doldrums and very few ships or sailors ever fall foul of them.

The best documented of all spouts was a perfect specimen that formed in front of a crowd of thousands off Martha's Vineyard in Massachusetts in 1896. 'A whirling funnel began to bulge down from the cloud, while at the same moment, and before the funnel was more than a projecting knob, the water on the surface of the sound beneath began to boil furiously and to rise up in a whirling mound.'[36] By the time the two met, the waterspout was 1000 metres high, its column 250 metres wide at the head, forty metres at the centre and seventy-five metres broad at the foot. The cascade surrounding it was 200 metres in diameter, over 100 metres high and a substantial part of it fell on several yachts becalmed nearby. It was all fresh water.

The fact that the two ends of the spout appeared independently and later met, shows that there was a vortex connecting them from the start, which only became visible when pressure within it had fallen far enough for condensation to occur. Tornadoes make these connections more quickly due to the dust and debris which give them substance almost from the start. When neither solid matter nor water vapour is available, the result is a 'bulls-eye' or 'white squall' that is totally invisible and strikes, like clear air turbulence on the air routes, out of nowhere.

The British barquentine *Bel Stuart* fell foul of such a fiend off Nova Scotia during the last century.

All hands being on deck after supper, noticed a peculiar change in sea and sky, and were discussing it, when, without a moment's notice, the sea forward seemed to swell up and swept the bark across her bows, carrying away her foretop-gallant mast, jib, jib-boom, foretop-mast stays, and her maintop-gallant mast, with all their accompanying sails. In a moment, as it seemed, the bark, with all her sail set, in a fair wind, with a moderate sea, was left a comparative wreck.[132]

It is just such a small vortex of heat over water, unattended somewhere in the doldrums, which gives birth to the mightiest of all wind storms.

There is no difference between hurricanes, cyclones and typhoons. All are disturbances that begin as low-pressure areas at hot spots in the tropical oceans in late summer and grow into great spirals of hungry air. In the north Atlantic they are called *hurricanes*, in memory of the Mayan storm god *Hunraken*; in the Indian Ocean they have been known as *cyclones*, from the Greek *kuklos* meaning 'circular', ever since the President of the Marine Courts at Calcutta first used this term in the middle of the nineteenth century; in the China Sea they call them *typhoon*, from *ty fung* or 'great wind'; in the Philippines it is *baguoi*; in Japan *reppu*; and *asifa-t* in the Persian Gulf.

All start with hot humid air rising from the ocean, spiralling upwards, cooling and condensing. Clouds develop, rain falls, energy is released which warms the air and reinforces the updraft. A low pressure area develops and more moisture-laden air rushes in at an angle determined by the rotation of Earth. The spiral spins faster, creating a centrifugal effect and a partial vacuum which also draws air from the upper atmosphere down into the core. This air is warmed by compression as it descends and adds further to the heat pool in the growing heart of the depression. Soon the system has a mass and a momentum of its own. It graduates from a depression to a tropical storm and begins to move, like a rapidly spinning top, slowly off in a definite direction. At first this is westward with the Sun, but as the storm spins, clockwise in the southern hemisphere and anticlockwise in the north, it slides away from the equator and is gradually deflected in a long parabolic curve back east.

Tropical storms have winds which blow at least sixty kilometres per hour. At this point, they get christened from an alphabetic roster that ever since the Second World War has consisted entirely of girls' names, but the catalogue has recently been liberated to allow David and Frederic to share the blame for outrageous behaviour with Betsy and Camille. When wind speeds reach 120 kilometres per hour, the system, regardless of its sex, is upgraded to hurricane, cyclone or typhoon status.

A full-fledged hurricane is a vast self-sustaining heat engine 100 times larger than a thunderstorm and 1000 times more powerful than a tornado. An ordinary summer afternoon thunderstorm has the energy equivalent of thirteen Nagasaki-type atomic bombs. Most hurricanes have at least 25,000 times that potential for destruction. One storm in 1928 dropped 2500 million tons of water on Puerto Rico in just two hours, but this was in fact only a fraction of its total capacity. An average hurricane precipitates about 20,000 million tons of water a day, which represents the energy equivalent of half a million atom bombs.[400]

The core of this awesome being is its 'eye'. The condensation of billions of tears in hot air rising around it, provides the energy that drives the engine. The average eye is about twenty kilometres in diameter, surrounded by a coliseum of clouds rising 10,000 metres to a ceiling of blue sky. The air in it is unnaturally calm and frequently filled with swirling birds trapped in an aerial cage that can carry them along for thousands of kilometres before it subsides. Low pressures in the eye strain the eardrums, bring the taste of blood to the mouth and suck the surface of the sea up into a hill three or four metres above its normal level. Relieved from restraint by the surrounding winds, the waters rush in from all directions, clashing and throwing up great plumes of spray. They 'pile up on every side in rough, pyramidal masses, mountain high . . . boiling and tumbling as though they were being stirred in some mighty cauldron'.[476]

The average life of an Atlantic hurricane is about nine days, during which time it may travel 5000 kilometres or more. One of the first signs of its approach, are warning waves which travel ahead of the storm at speeds of over fifty kilometres an hour. These heavy swells have an ominously slow beat, pounding on distant shores at three or four a minute, compared to the average wave rate of seven to ten in a sleeping sea.

The next symptoms are fingers of high and icy cirrus cloud all trailing off in one direction, as though gathered in there by a giant hand somewhere below the horizon. Then the barometer falls, smooth sheets of altostratus obscure the Sun and the rains begin. A bar cloud, a heavy black wall of cumulonimbus, closes in like a curtain of doom announced by bursts of wind. The intervals between the squalls shorten until wind and rain are almost continuous and the hurricane itself arrives, allowing no respite until the false calm of the eye some 150 or 200 kilometres further on.

Wind speeds are never as high as those in a tornado, but they are sustained for hours on end at averages of up to 150 kilometres an hour. At sea this is enough to generate waves over twenty metres high. The biggest ever actually recorded was one twenty point four metres from trough to crest which took fifteen seconds to pass a weather ship lying close to the track of a hurricane in 1961. But calculations suggest that waves of twenty-five metres are probably not uncommon, and one of thirty-four metres, said to have been encountered by U.S.S. *Ramapo* in the north Pacific during a typhoon in 1933, is not impossible.[109]

When hurricane winds blow, wave tops of all sizes are swept away in sheets of spray so dense that a sailor cannot tell where ocean ends and atmosphere begins. On land, crops are laid waste, trees uprooted and all but the strongest buildings destroyed. Winds tend to be most severe in the 'danger quadrant', that part of a storm lying on the right front edge of its track in the northern hemisphere. The highest wind ever officially recorded during a hurricane was a gust of 295 kilometres an hour from the infamous 'Long Island Express' that swept the eastern seaboard of the United States in 1938. But supporters of Camille, who savaged the Gulf of Mexico in 1969, claim a gust of 347 kilometres per hour.[275]

Hurricane Betsy of 1965 is classified as the greatest natural disaster ever to be inflicted on the United States, producing well over a thousand million dollars worth of damage. But her ravages pale into insignificance compared with an anonymous cyclone of only moderate strength that hit the coast of Bangladesh in 1970, killing an estimated 300,000 people. The difference lies in the effects, not of direct wind, but of the storm surge those winds can produce.[145]

As any circular tropical disturbance moves across the land, the hill of water in its oceanic eye crashes on to the shore like a tidal wave, inundating anything that lies within four or six metres of normal sea level. The Bay of Bengal, where the Ganges enters the sea, is a natural cul-de-sac for such storms and their effects at the head of the bay are horribly exaggerated. The ocean first recedes for twenty kilometres, leaving the coastal shallows completely exposed. Then it comes roaring back in a solid wall of water fifteen metres high that carries everything in its path. A cyclone at Calcutta in 1737 destroyed 20,000 boats and drowned 200,000 people in this way. And the 1970 disaster merely amplified this loss by its effect on a higher coastal population. If nothing can be done to build suitable defences or change the present pattern of land use in the area – and neither measure seems likely – it is almost inevitable that the next serious cyclone there will take a million lives.

An ironic addition to the winds that cause disasters, are the natural disasters that produce their own winds. When an earthquake struck Tokyo and Yokohama in 1923, starting fires which destroyed both cities and killed 160,000 people, the horror was compounded by *tatsumaki* or 'dragon whirls'. These were fire-induced whirlwinds that coiled down out of the immense plume of smoke and left a further 40,000 dead.[150]

Something similar happened on the Indonesian island of Sumbawa in 1815. Immense pressure built up inside the volcano Tambora and the top 1250 metres blew off and left a crater ten kilometres in diameter. Fifteen thousand people died in the eruption and the subsequent tidal wave. More still perished as a result of tornadoes that formed in, and dropped out of, the great cloud that hung over the island for weeks.

This eruption threw more debris into the atmosphere than any other ever known. The cloud of dust turned day into night for hundreds of kilometres around. Madura, 450 kilometres away, was plunged into darkness for three continuous days and, once the high jet stream began to distribute the debris more democratically around the equator, sunsets all over the world were extravagantly coloured for the rest of the year.

Nothing happens in isolation in our atmosphere. Little goes on anywhere in Gaia's being that is not very soon known everywhere. Air carries messages and the nervous system of the winds circulates the news, constantly rearranging Earth's crazy-paving pattern of heat and cold and making changes in the weather.

Winds of Burden

On moonless nights at sea or out in the country far from city lights, it is sometimes possible to see a pale ghostly beam in the sky.

About two hours after sunset, a very faint triangle of light appears in the west. This grows as the night progresses until it forms an enormous diffuse pyramid, almost as bright as the Milky Way. At midnight, like Cinderella, it vanishes, only to reappear in the east a few hours before sunrise. Because it seems to manifest only along the track of the Sun, it has become known as 'zodiacal light'.

Early theories suggested that it must be caused by reflection from a thin cloud of particles surrounding the Sun itself, but *Pioneer 11* carried a light meter that continued to record the ghostly glow well beyond the orbit of Saturn 1500 million kilometres away. It is becoming clear that space is far from empty, that the gaps between the planets are punctuated with particles which probably have nothing at all to do with our Sun.

Even the spaces between suns are decorated with debris, with dust and gas arranged into dramatic nebulae that rise like great clouds of matter along the spiral arms of the galaxies. And we, in our little solar system, swing round through this cosmic smokescreen picking up bits and pieces in our net of gravity.

The astronauts tell of minute grains that ping 'like birdshot pellets' as they strike the shell of a spacecraft. It goes on all the time. Every day Earth collides with more than 100 million tiny meteors, and something like 100 tons of extraterrestrial material comes pouring into our atmosphere. Most of these intruders burn themselves out as 'shooting stars' in friction with our air, but a large part of the crust of our planet is nevertheless made of cosmic silt, about a ton each day settling slowly on the surface. Enough, it is said, to double Earth's weight by the year AD 3000 million million.

The rest bounces back into space, but enough remains in suspension in the atmosphere to produce profound effects. An Australian meteorologist, comparing data from seven different control points around the world, points out that heavy rains fall with surprising regularity thirty days after Earth passes through particularly dense meteor showers like the Geminids, Ursids and Quadrantids.[47]

This is not as crazy as it sounds. Condensation takes place only with great difficulty in clean air. Moisture needs a suitable surface on which it can form. This can be Earth itself, as in the case of dew or frost, but in free air condensation almost always takes place around nuclei such as salt or smoke or dust. Without some form of pollution, fairweather clouds become unlikely and rain clouds totally impossible. A fall of meteoritic dust would provide ample opportunity for condensation and the thirty-day delay is very close to the time it takes for such particles to drift 100 kilometres down into the weather zone.

There is of course little danger of prolonged drought in the periods between showers of meteors. The air near Earth will never become too pure for condensation. The cleanest natural air, under a downdraft from high altitude, has 300 to 400 particles in each cubic centimetre. That is 200,000 in every lungful. Even in mid Pacific, far from any land mass, the count is over 1000 per cubic centimetre. Above forest this rises to 50,000 and in cities to 150,000 increasing into the millions at times of smog. And all this space dirt, this soil in suspension, is cultivated and circulated by the wind until every square kilometre of Earth surface seems to hold at least one speck of dust from every other square kilometre.[275]

Some hold a great deal more.

Roughly twenty million square kilometres of Earth surface are covered in a soft pale yellow substance composed largely of irregular quartz particles between 0.01 and 0.05 millimetres in diameter. Most of this mysterious soil lies in the northern hemisphere – in the Mississippi Basin, in bands across Europe that include the Rhine and the valley of the Danube, on the steppes of Asia and around the River Huang of Mongolia and China.

It is called *loess*, from the German *lose*, meaning 'friable', and it is light and crumbles easily between the fingers, but the irregular shape of the particles locks them together like pieces in a jigsaw puzzle, holding vertical or even overhanging cliffs in place. In China, deposits of *huang-tu*, the 'yellow earth', are immense – covering the land to a depth of almost 500 metres and staining the rivers. In the northern provinces of Honnan and Kansu, more than ten million people live in underground towns that have been carved completely out of this substrate. Houses, walls, beds and furniture, factories and schools are cut directly into the loess, providing clean, vermin-free dwellings, cool in summer and warm in winter, leaving the land above them free for cultivation.[450]

The origin of this benison is still in dispute, though most geologists agree that the winds have played a major role in depositing the fine yellow silts where they now lie. The largest loess fields are found near major rivers, but they show no suggestion of sedimentary bedding. The materials seem to be largely of glacial origin, ground small by the mills of ice and left along the retreating front in barren lands with little rain or vegetation to consolidate or keep them there. And it is from these outwash areas, mainly in what is now the Gobi Desert, that the winds seem to have made their selection. The finest dust was blown away to girdle Earth, but the particles of intermediate size were carried eastward into the great river valleys, settling slowly from the air and locking into place.

During the winter monsoon, a henna-coloured haze, rising as high as 3000 metres, still rolls out of Mongolia, reddening eyes in Peking and disrupting navigation as effectively as fog on the aptly named Yellow Sea.

The fresh angular uneroded quality of the loess grains suggests that they were not blown very far or fast. It is even possible that they were never blown at all, but simply fell, right out of a meteor shower. If this is so, then their only connection with any Ice Age could be that both were caused by the passage of the solar system through an area of space especially rich in cosmic debris.

There is normally no need to look beyond the bounds of our planet for the source of airborne material. Every ploughed field and sandlot is a stockpile waiting to be raided by the winds, which carry off 800 million tons of dust a year, but they seem to reserve their special efforts for the margins of the desert.

In the northern spring, when a series of low pressure areas moves east along the Mediterranean, people along the coast feel the hot breath of the Sahara. Air tumbles down from the interior, warmed still further as it falls, and strikes the barren plains almost vertically, putting millions of tons of dust into the air. This is the *sirocco* of Morocco, the *chili* of Tunisia and the *ghibli* of Libya. In Egypt they call it *khamsin*, believing that it lasts for fifty days, and at its hottest and dustiest everyone refers to it simply as *simoom*, the 'poisoner'.

Herodotus, the 'Father of History', reports that an army despatched by the king of Persia to attack Ethiopia in 525 BC was engulfed by the *simoom* and never seen again. And that another small nation, somewhat aptly called the Psylli, was so enraged by this evil wind that they declared war on it and marched out in full battle array, only to be rapidly and completely interred.

Dust storms seem to occur in three basic shapes – the most dramatic of which is the 'wall'. This is preceded by air that is abnormally calm, hot and heavy. Then the horizon is blocked by a yellowish mass which gradually increases in size as it approaches, until it becomes an opaque wall over 2000 metres high. 'Suddenly the wind rises; your skin is cut by fine particles of sand; they get into your eyes, ears, nose and mouth, and you are forced to find shelter.'[100]

A mining engineer in old Abyssinia fell foul of such a wall and found the desert billowing like an angry sea around him. 'All that we could see about us was in motion. Breakers rolled onward like wind-blown water. The sand was whipped so fast from their crests that within a few moments the troughs under them became filled, and new waves rose up where they had been. The process was continuous, and the effect was that of a slow-rolling sea.'[56]

Similar sights are produced by the 'black rollers' of the western United States, the 'brickfielder' that assaults the southeast coast of Australia, and the *haboob* in the Sudan. But one of the most unpleasant is the *karaburan*, the 'black blizzard' of the Gobi Desert. This first appears on the horizon as a blue-black line that broadens and shoots out arms, plunging the world into cold and dark, howling and flinging sand and gravel. Until it departs, almost as suddenly, leaving the desert swept clean and travellers 'queerly dazed, as after a long illness'.[200]

The second form of dust storm is the 'column', which is a solid whirling tornado-like structure caused by conflict between two rival currents of air. These are very common in all hot areas, peaking around noon, and move, often in long files, swaying almost gracefully as they advance. Their structure is complex, consisting usually of a downward spiral of air inside a sheath of several separate upward flowing vortices. 'One could see them,' recalled Loren Eiseley of his childhood in Nebraska, 'hesitantly stalking across the alkali flats on a hot day, debating perhaps in their tall, rotating columns, whether to ascend and assume more formidable shapes. They were the trickster part of an otherwise pedestrian landscape.'[122]

In India they are known as 'dust devils', in France as *trombes giratoires*, in California as 'sand augers', and in the Sahara as *waltzing jinns*. The Swiss traveller John Burckhardt described the Abyssinian variety as 'prodigious pillars of sand . . . stalking in with majestic slowness . . . their tops reaching to the very clouds'.[56] Most are short-lived, but there seems to be some substance to the old formula which puts their longevity at one hour for each 300 metres in height.

And the final form is the dust 'sheet', an amorphous all-enveloping cloud of fine sand that covers everything. 'Nature suddenly seems to be on fire; the sky is covered with an immense copper-tinted veil. Suddenly a storm begins and drowns the scene in thicker and thicker mist. Gusts follow gusts. A fine dust penetrates the tiniest crevices, mixes with food and drink and gets into the eyes, nose and parched throat of the unfortunate traveller.'[100]

These sheets are always the product of large, longer-lived weather conditions. In Iran there is one that blows constantly from June to September and is known as *bad-i-sad-o-bist-roz*, the 'wind of one hundred and twenty days'. Lord Curzon described this gale, which is dry and hot and laden with dust and salt, as 'the most vile and abominable in the universe'.[56] It rolls the desert out like a carpet over the land, burying farms and villages in their entirety, just as it once obscured great Palmyra and Ur of the Chaldees.

While fine sand and dust blow long distances, it is more common for the coarse grains to bounce along close to the surface, or to creep and flow almost like moving water. These grains, with a diameter greater than 0.06 of a millimetre, come to rest when the wind drops or grows turbulent as it swirls around an obstacle in its path. Once started, this local accumulation itself acts as an obstruction and attracts more and more sand, until it becomes a dune.[15]

The principal requirement for dune formation is a supply of sand, so they tend to be most abundant along the coasts of lakes and seas with onshore winds. A large number also form on or near the floodplain of rivers which supply the necessary silt. Desert dunes are fed largely by quartz particles produced by the wind itself as it disintegrates exposed sandstone.

The classic dune is crescent-shaped, with its horns facing downwind, and deep concave slopes. These are called *barchans* and are characteristic of desert areas with limited supplies of sand, and winds which blow constantly in one direction. They move slowly in that direction, often across bare surfaces or *serirs* carpeted with pebbles too large for the wind to carry. The biggest natural dunes of this type are thirty metres high and 300 metres long, but sometimes barchans merge with each other, becoming even longer and less symmetric, building 'hairpin' formations.

Where prevailing winds have access to larger supplies of sand, more than can be handled by a barchan 'fleet', they build *transverse* dunes, which are wavelike ridges or corrugations at right angles to the wind. And where the winds are variable, these dunes become *longitudinal*, parallel to the major wind direction, stretching in desert regions for as much as 100 kilometres. These dunes sometimes start when barchans become anchored by a patch of vegetation and grow long tapering points known as *seifs*. Very large longitudinal dunes with flat tops, produced perhaps by the convergence of several seifs, are known as *whalebacks*.

All these are continental dunes. Coastal ones can have such forms, but also include *foredunes* that lie on or parallel to the shore.

The buildup of coral or beach sand in dunes is one of the most common ways in which land can reclaim itself from the sea. The existence of whole areas of low-lying country, such as Florida, the Baltic coast and shores along the Gulf of Guinea and eastern Malagasy, are proof of the constructive power of prevailing winds. But sometimes they work against human interests. The ancient ports of Acre, Tyre and Laodicea were all silted up and moved inland. Often the wind has human help. On the coast of New Zealand, the introduction of sheep which grazed on dune grasses, freed the sands to march inland and invade farming communities, who were forced to fight back with programs of replanting and stabilisation.

The problem of the interior, in desert lands, is a different one. In the Sahara, for instance, dunes propelled by the northeast trades are gradually filling up Lake Chad and encroaching on agricultural land on the margin of the Sahel. Reclamation of these lands rests traditionally in the hands of local experts called *riaha* or 'men of the wind', but they fight a losing battle. 'Sand,' in the words of meteorologist Sir Napier Shaw, 'is a sort of snow that never melts.' Its formation 'goes on at a steady rate and its gradual increase over cultivated areas is inevitable because there is no means effectively available for its removal by any natural process. Sand may be converted into cultivable soil by the admixture of other suitable ingredients as in Egypt by the periodic mixture of the mud of the Nile; but if the increase of sand is greater than the corresponding supply of the other ingredients which human activity can provide, progressive deterioration of the soil, encroachment upon the rivers and their ultimate submergence, are inevitable.'[439]

All types of dune become 'fossilised' when climates change, covering them with other soil or vegetation, and it is possible for archaeologists to learn something of past climate from their shape and structure. Wind direction can be detected by the dip of sand beds, with the necessary reservation that the effective, the dune-forming, wind may not have been the prevailing and more gentle one. Crossbedding implies winds of variable direction. Sporadic gales show up well as laminates of coarser grains. The existence of a mobile barchan dune is a clear indication of a dry climate and past lack of vegetation. Asymmetric hairpin dunes are usually a sign of a slightly more humid environment with some vegetation, and often occur as a result of human degradation of a habitat.[441]

The breakdown of a landscape is, however, more often revealed by the removal than by the addition of surface soil and detail.

Wind with sand in suspension is highly abrasive, frosting glass, blasting the paintwork off automobiles, sawing down telegraph poles and sculpting rock into extraordinary shapes. Variations in texture and composition of different strata produce differential weathering and result in the knobs, statues, pillars and arches which decorate many deserts. Nature's airbrush on its own is responsible for the architectural spires which turn canyons in Utah into oriental cities, for the rocky mushrooms which sprout from the floor of Death Valley, and the perfect spheres that dot the Kharga Oasis in Egypt like shrubs in an ornamental garden.

Many of these works are autographed by the wind and recognisable as 'ventifacts' by smooth grooves and flutings which show the direction of prevailing blasts. In central Asia, these U-shaped furrows or *yardangs* are cut into the desert floor itself and may be six metres deep. Sometimes just one side of a cobblestone is bevelled or it may be turned and polished along two planes into a pyramidal form. On occasion, stone may resist alteration altogether by moving.

In a dry lake bed called the Racetrack Playa in California, are a tribe of rocks, varying from pebbles to half-ton boulders, that travel in different directions, sliding rather than rolling, leaving conspicuous tracks across the ground. Thirty of these rocks were kept under observation for seven years between 1968 and 1974, and during this period twenty-eight of them moved. The greatest distance travelled was 262 metres, 201 of which were in one session during March 1969, by a stone weighing 250 grams. Another weighing twenty-five kilograms travelled a total of 219 metres. Some of the paths were erratic, on occasion even describing a loop, but the net direction of travel was northeasterly, in accordance with the prevailing wind. It seems that a combination of high wind, a little rain on the fine clay floor of the old lake bed, and the flat bottoms of the rocks themselves, together contrive the conditions necessary for their periodic migrations.[437]

In other places, wind-induced movement is so drastic that entire landscapes disappear altogether.

The Great Plains of the United States have scanty rainfall and a sparse prairie vegetation. It is a marginal area for farming, but in the early 1930s numbers of eager settlers broke up the surface with their ploughs. Deprived of the tough grass that normally protects it, the fallow soil was defenceless. Two or three dry years later added to the problem and between 1933 and 1937, great clouds of silt and dust took to the air. People stood in the doorways of their homes and watched their farms blow by, leaving much of Oklahoma, Kansas and Nebraska in a 'dustbowl', at places fifty metres deep.

The Libyan desert is another such 'blowout or windhollow', in this case 300 metres deep. Both put millions of tons of dust into the air, scattering the short wavelengths of light and colouring sunsets for thousands of kilometres around.

The *harmattan* or northwest trade which blows across the Sahara, is always filled with reddish dust from a gigantic plume which rises to 5000 metres between November and March, even making navigation difficult along parts of the coast of West Africa. Ships as far offshore as the Cape Verde islands find themselves enveloped in a 'mist of fog of dust as fine as flour, filling the eyes, the lungs, the pores of the skin, the nose and the throat; getting into the locks of rifles, the works of watches and cameras, defiling water, food and everything else; rendering life a burden and a curse.'[56] This may well have been what early mariners referred to as the 'Sea of Darkness' and it could be what lay behind 'the darkness which may be felt' that fell upon plague-ridden Egypt.

Under the right weather conditions, usually when polar front depressions move over the edge of the Sahara and air is pulled up into the zone of high altitude winds, intercontinental dust storms develop. Dust deposited on Barbados in the Caribbean and in Florida has been shown to be of Saharan origin, but the most impressive fallout takes place on occasion over Europe.

Dry red fogs from the Sahara are carried polewards and descend as far north as Cornwall and Devon, producing showers of mud so red it is sometimes taken for blood. Homer and Virgil tell of such falls, and Gregory of Tours records that in AD 582 a 'shower of blood' so terrified Parisians that they 'rent their garments'. A rain of mud blocked gutters in Provence in 1846, and in 1859 over 30,000 square kilometres of Germany were covered in bright pink snow. 'Blood rains' were widely reported in Portugal and Spain in 1901 and an estimated two million tons of fine red mud poured down all across Europe in April 1926.[100] More recent deluges include the embarrassment of tidy Arosa in Switzerland in 1936,[166] and the 'great fall' of March 29th, 1947 over Luxembourg.[465]

The quantities of dust involved on these occasions was impressive, but small compared to the amount hurled into the air, and falling slowly back to Earth, from volcanic eruptions.

Vulcan, blacksmith to the gods, keeps his forge going constantly, deep underground, building up enough heat and pressure to create molten rock. Most of the time this is contained by Earth's cool crust, but on roughly 500 separate occasions since history began, a conduit has formed to the surface, releasing gas and lava, throwing debris and dust high into the atmosphere.

The most explosive eruptions are those in which a plug forms at the surface, holding back the magma until the pressures become catastrophic. These are known as Plinian eruptions, after Pliny the Younger, who described such an outburst by Vesuvius in AD 78, which left Naples under a pall so thick it seemed 'the last eternal night of story had settled on the world'.

On August 27th, 1883, a small island in the Sunda Strait between Java and Sumatra blew itself apart with the loudest sound on Earth. The blast was heard across 4500 kilometres on Rodriguez Island, and even barometers in London registered the concussion, and continued to do so up to nine days later, when the echoes of the report were making their seventh circuit of Earth. A sea wave almost 100 metres high swept outward in all directions, snapping lighthouses like matchsticks, drowning 36,000 people, leaving a Dutch warship three kilometres up a Sumatran valley and rocking other vessels at their anchorages nearly 8000 kilometres away on the coast of South Africa.[151]

This was Krakatoa. It was not the first time it had erupted, but on this occasion the explosion pushed a gigantic column of smoke and dust right through the ceiling of the troposphere up to a height of fifty kilometres. The heaviest material fell within the first 500 kilometres, carpeting the land and the surface of the sea, and either blanking out the Sun or turning it blue and green.

We saw a green sun, and such a green as we have never, either before or since, seen in the heavens. We saw smears or patches of something like verdigris green in the sky, and they changed to equally extreme blood red, or to coarse brickdust reds, and in an instant passed to the colour of tarnished copper or shining brass.[475]

Ten days later and more than 10,000 kilometres away in Hawaii, a clergyman reported that the sky near the noon Sun was glaring white instead of blue and that this bright halo was surrounded in turn by rings of pink, red, orange-rose and brown. This effect, now known as 'Bishop's Ring', continued to be seen for the next two years and was caused by the same high cloud of fine dust that produced brilliant sunsets right around the globe.

Nearly the entire western half of the horizon has changed to a fiery crimson: as time goes on, the northern and the southern areas lose their glory, and the greys of night contract, from the northern end first, most rapidly; the east is of normal grey. The south now closes in and a glow comes up from the west like that of a white-hot steel, reddening somewhat as it mounts to the zenith.[475]

In Paris and London right through the following winter, this glare was bright enough to allow people abroad in the streets at night to read their newspapers.

Dust of any size in the troposphere, the weather zone, is washed out in a matter of weeks, but debris in the stratosphere persists for far longer. After Krakatoa, it was possible to estimate the height of stratospheric veils only by observation of their effect on Sun and Moon light. In recent years, most notably after the eruption of Gunung Agung on Bali in 1963, direct sampling by high-flying aircraft has shown that there is a distinct concentration of dust particles between twenty and twenty-five kilometres high in a zone now known as the 'aerosol layer'.[346]

Particles of different sizes remain in the stratosphere for different periods. Calculations show that dust grains of 0.005 millimetres in diameter will fall into the troposphere within ten weeks, while those of 0.001 millimetres may remain in the stratosphere for almost two years. But these figures assume a constant settling rate in still air and, given the existence of movement in the upper atmosphere, there is no reason why the smaller particles of dust from a volcanic eruption as explosive as Krakatoa's, should not remain in orbit for ten or even twenty years.

The initial concentration occurs directly over the eruption, but this reservoir is very soon disrupted by wind. Dust from Krakatoa took just two weeks to circle the Earth at the equator and then began more slowly to drift towards the poles. The flotsam from the Balinese eruption in 1963 made its first circuit in four weeks. In both cases, it was carried westward by the equatorial jet stream and within four months of arriving in the stratosphere formed a uniform veil covering the whole wind zone.[198]

It has not always been possible to track the shifting sands of the upper atmosphere, but the eventual deposit of volcanic dust on Earth's surface, most vividly on polar ice sheets, provides a dramatic 'fossil' record of the upper winds at the time of the eruption. Analysis of these shows that volcanoes in middle and upper latitudes spread their pall mainly eastwards in the westerly stream, and that the effects of the cloud are limited to those latitudes and points closer to the poles. But eruptions near the equator produce worldwide veils. In both cases, the dust persists longest over the poles.[115]

The first person to wonder about the effects of these curtains on our climate seems to have been that inventor of both the lightning rod and the American Republic, the ubiquitous Benjamin Franklin. 'During several of the summer months of the year 1783, when the effects of the Sun's rays to heat the Earth in these northern regions should have been greatest, there existed a constant fog over all Europe and great part of North America.' The rays of the Sun were, he noticed, 'rendered so faint in passing through it that, when collected in the focus of a burning glass, they would scarcely kindle brown paper.' He attributed this dry fog to 'the vast quantity of smoke, long continuing to issue during the summer from Hekla, in Iceland, and from that other volcano which arose out of the sea near the island, spread by various winds over the northern part of the world'. And because the summer was cool, 'the surface was early frozen. Hence, the first snows remained on it unmelted. Hence, perhaps the winter of 1783–84 was more severe than any that happened for many years.' He concluded that it seems 'worthy the inquiry whether other hard winters, recorded in history, were preceded by similar permanent and widely extended fogs'.[442]

Such an inquiry has now been made by the veteran climatologist Hubert Lamb.[270]

Any veil of dust in the atmosphere is going to screen Earth from solar radiation and affect our heat balance. After the Balinese eruption in 1963, the temperature of the stratosphere above Ascension Island rose to 3° Centigrade above any value recorded before, and surface temperatures for the following year over the whole world below the veil were half a degree lower than expected. This is a small enough fluctuation, but over a wide area it can have considerable effects. There is only a 6° Centigrade difference between the average temperature of Interglacial and Ice Ages.

Analysis of volcanic activity since the year 1500, shows that there have been similar effects following all eruptions involving a high output of dust. After the great Tambora eruption of 1815, the annual mean temperature dropped by about 1° Centigrade and 1816 became famous in Europe and much of North America as 'the year without a summer' or, more picturesquely, as 'Eighteen Hundred and Froze to Death'.[463] This was the year that the westerly flow of winds became blocked and polar air penetrated unusually far south, producing famines and food riots in Ireland and Wales.

Similar effects followed eruptions in 1693, and in 1783 in Japan, 1811 in the Azores, 1831 on Graham Island, 1835 in Nicaragua, 1875 in Iceland, 1888 in Japan and 1902 in the Caribbean. The first of these involved several volcanoes that put up an extensive dust veil and was followed by seven cold summers. The arctic ice pack came abnormally far south, completely surrounding Iceland in 1695, while in Scotland the impoverished countryside was virtually depopulated.

Lamb has calculated a dust veil index involving the density, extent and duration of a volcanic cloud. Krakatoa in 1883 is used as the baseline, with an index of 1000. Compared to it, the combined effect of Mediterranean, Philippine and Ecuadorian eruptions in 1831 score 1750; Coseguina in Nicaragua and a group of related Chilean eruptions during 1835 give a combined index of 3000; and Tambora in 1815 rates a massive 4000. The highest scorer in recent years was the Balinese volcano of 1963 with about 800, but that has now been superseded, not by the well-publicised eruption of Mount St. Helens in Washington in 1980 – which occurred in a high latitude and produced a sidelong blast – but by the eruption of a remote volcano in Mexico in 1982. It is too soon yet to have a rating on it, but it looks like being the most significant dust-producer since Krakatoa.

El Chichon erupted on April 4th, 1982 with a tight vertical blast that put sixteen million tons of debris forty-two kilometres up into the stratosphere. It followed close on the launch of a satellite specifically designed to monitor ozone and water vapour in the upper atmosphere. The orbit of this satellite passes directly over the volcano, making El Chichon the best observed eruption ever. The dust and vapour cloud drifted westward into the Pacific in the equatorial jet, with its heaviest layer at about twenty-six kilometres. Within a month, the veil had not only circled the globe, but spread to thirty degrees either side of the equator, with thinner fringes of the aerosol spreading as far as the poles.[182]

By the summer of 1982, El Chichon's cloud was already producing stratospheric warming and preventing the usual solar heat from reaching the surface. The southeast trade winds, which normally blow for thousands of kilometres across the equatorial Pacific from May to September, failed altogether, accentuating an increase in air pressure over Australia and Indonesia, producing drought, devastating bush fires and famine. And by Christmas, El Niño, 'the child', had arrived.

El Niño is a massive surge of warm water that, once every decade or so, builds up in the eastern Pacific. It sweeps down the coast of South America bringing high temperatures and heavy rain, pushing back the cold rich Humboldt current. As a result, the anchovy fail to make an appearance, sea birds starve and fishermen face disaster. 'The child' caused greater havoc in early 1983 than at any time this century, prompting the declaration of a state of emergency in Ecuador.

The unusually warm sea temperatures of the eastern Pacific in the spring of 1983 also had the effect of deepening an area of semi-permanent low pressure in the Gulf of Alaska, bringing rain to California. In the first week of March, the entire coast was lashed by unusual storms and flooded by record tides.

It is difficult not to connect the eruption and its veil of wind-borne dust and sulphuric acid vapour with the changes in climate that followed. Clearly, not all unusual weather can be blamed on volcanoes; but it seems significant that the period from 1430 to 1850, which was cold enough to have been called the Little Ice Age, coincided with a wave of volcanic activity in which great eruptions were abnormally frequent.[403]

There seems to be a rhythm to volcanic eruptions which coincides with a 179 or 180 year cycle in sunspot activity. It is possible that both are produced by tidal stresses set up in our solar system as a whole and that these give Gaia a little drunken wobble once in such a while.[181]

The number of eruptions in the last 180 years, perhaps twelve with a dust veil index greater than 1000, is higher than for any similar period of time in the preceding 10,000 years; and seems, from what we can tell of frequencies during the last two million years by looking at 'fossil' ash deposits, to be similar to that which prevailed during the devastating Quaternary Ice Ages.

There are other factors in the climatic equation which seem to be pushing us toward a generally warmer, rather than a cooler, epoch. But there are always short-term fluctuations in the longer cycles and it would be unwise to disregard the Vulcan warnings or to underestimate the vital role played by global winds, by the breath of Gaia, in shaping our destiny.

PART TWO
Wind and Time

People have always known that weather was changeable, but since the invention of the barometer and thermometer in the seventeenth century, there has been a growing tendency to see the climate, the sum of weather, as constant.

At a conference in 1935, the forerunner of the World Meteorological Organisation recommended that climate be regarded as the average of weather over a period of thirty years, and suggested that the years between 1901 and 1930 be regarded as the 'normal climate'. But the choice of this period and the word 'normal' now seem unfortunate.[272]

We are beginning to realise that the limited parameters of temperature, pressure and humidity are not enough to tell the whole story. They accumulate over periods of thirty years into reassuring, apparently steady, blocks of statistics that seem to confirm the constancy of climate and the power and propriety of the scientific method. But they also successfully mask other shadowy and perhaps more fundamental changes – the sort of things that have always been apparent to people everywhere, who persist in seeing not just weather, but climate too, as an unreliable, shifting, fluctuating thing.

Everyday conversation, as recorded by diarists like Samuel Pepys in the seventeenth century, has always been sprinkled with references to 'the greatest storm in living memory' or 'so deep a snow that the oldest man living could not remember the like'. People still say things like that and are rightly unimpressed when meteorologists tell them that statistics show the weather in question was not really that unusual.

Since 1935, a few climatologists with broader vision – notably Hubert Lamb who founded the Climatic Research Unit at the University of East Anglia in England – have realised that the first thirty years of this century were highly abnormal. And that the years 1931 to 1960, which were promptly selected as the 'new normal period', were more unusual still.

Lamb and others like him have returned to parameters that make more general environmental sense. To grain harvests, fish catches and great vintage years. To births and deaths and pilgrimages, and times of feast and famine. To floods and fires, the construction of new windmills, the price of hay, and the growth of mistletoe. These are things that everyone understands. They have always been the cues, sometimes quite subtle ones, to which ordinary people respond, and which are now proving to be more sensitive and realistic indicators of what is actually happening to the weather. The evidence is that climate never was normal or constant, that it and we are subject to continual change.

And with this growth of new awareness, comes a realisation that we might have done better had we paid more attention to an instrument older than the barometer or thermometer. To a simple device that was once the only mechanical indicator of weather change – the windvane.

The simple origins of the vane are apparent in the derivation of the word from the Old English *fane*, meaning 'flag'. The first one was probably a rudimentary boundary marker, a piece of wool maybe, tied to a convenient tree. Or perhaps a feather pennant, like the one Tahitians still attach to the rigging on their canoes to detect changes in wind direction. But it was not long before things became more sophisticated.

By the first century BC, the Macedonian astronomer Andronikos had built a waterclock inside a tower on the edge of the market-place in Athens. But this was no ordinary building. It was octagon-al, and each side faced towards the source of one of the eight major winds, and was decorated with a sculpted frieze showing that wind and its attributes in human form. On the top of this 'Tower of the Winds', which can still be seen in the old Agora on the slopes of the Acropolis, stood a huge bronze figure of the fish-tailed sea god Triton, who swung around to point his staff at the name and face of the wind responsible. And by the first century AD, all the smarter villas in Rome had followed suit, building their own private windvanes, some of them connected directly to a pointer in a windrose on the ceiling of the room below.[464]

The weathercock seems to have made its appearance in the ninth century, when Pope Nicholas I decreed that it was to be used on abbeys and churches to remind the faithful of St. Peter's dilemma. During medieval times it was replaced on castles and civic build-ings with metal banners bearing coats of arms, and from the fifteenth century with scenes from rustic life. Today there seems to be a growing vogue for swordfish, whales, prancing horses or just arrows, coupled with an unfortunate tendency to forget that the whole thing is more than a roof decoration. We need to be reminded that an awareness of the change in wind direction, with time, and without any other information, is in itself an astonish-ingly powerful tool.

In 1857, a Dutch professor of physics realised that wind blew along isobars, lines of equal pressure in a weather system, and that the area of low pressure was always to the wind's left. Buys Ballot's Law says: 'Stand with your back to the wind and the pressure will be lower on your left hand than your right.' In the southern hemisphere, the reverse is true, but everywhere the rule of the road in the air is dictated by the movement of the air itself which bends isobars into the familiar dartboard rings on a weather map. Pilots in the northern hemisphere always fly to the right, with the wind, when going around a thunderstorm, while those in the south, if they value their lives, go left.

What this means for the rest of us, is that it is possible for anyone standing on the ground, just by watching changes in wind direction, to locate and track any storm, front or low pressure area. Because winds travel anticlockwise around a depression in the northern hemisphere, a wind from the south is the first warning of a storm moving in from the west. And if this south wind veers (that is, changes in a clockwise direction) to come from the west, it means that the storm centre is passing north of you and your local weather will soon be warm and bright. If the south wind backs (changes in an anticlockwise direction) to the east, you are in the wet sector of the cyclone and can expect cold and rain or snow.

This simple system can be refined by using natural weather-vanes, the clouds at different altitudes, to detect wind direction, and therefore the position of pressure areas, in the middle and upper air. And because low pressure air at altitude is always cold, and high pressure relatively warm, this makes it possible to find out where cold and warm fronts lie, and which way they are likely to move. And once you know the wind direction at two different levels, you can predict the tendency of that warm or cold air to descend on you.

Assuming still that you are in the northern hemisphere, if the upper winds veer (change direction clockwise with respect to the lower winds), it is going to get warmer. If they back, it will suddenly turn cold. Applying Buys Ballot's rule, if you stand with your back to the lower wind, and high clouds appear from the left, the weather will deteriorate. If the high clouds advance from the right, the chances are that it will improve.

A skilled meteorologist standing in his own garden, with nothing but a windvane and a little time, can turn these, and a host of other wind cues, into a detailed weather map and forecast. But the beauty of these simple signs, is that they are available to anyone,

and have become formalised in time into ritual sayings and fragments of weather lore, which are both accurate and relevant.

In England they say:

> A veering wind, fair weather,
> A backing wind, foul weather.

And in Scotland, with a slightly different accent:

> West wind, north about,
> Ne'er hangs lang out.

These are small demonstrations of local knowledge and short-term
prediction, but they are based on long experience of the consequences of change. No individual acting on, and benefiting from,
such traditional wisdom, need ever know how it works. It is enough
that it does, that it has survival value, for it to become part of a
culture, changing the lives of those involved.

There are other signs, to be sure.

Virgil noted that:

> A bee was never caught in a shower.

Sailors swear by:

> Rainbow to windward, foul fall the day;
> Rainbow to leeward, damp runs away.

And no ploughman worth his pay ever harnessed a team before
consulting a flower called the weatherglass *Anagallis arvensis*:

> Now, look! The weatherglass is spread –
> Against a rainy hour.[237]

But wind, perhaps because it moves and breathes and seems to
have a life of its own, rising and falling, growing and dying with the
passage of time, has always played a very special part in our history
and awareness of the world:

Nature, with equal mind,
Sees all her sons at play;
Sees man control the wind,
The wind sweep man away.[9]

3. THE HISTORY OF WIND

Not long ago, human social evolution was thought to have passed through three main stages.

The first was described as 'savagery', in which we lived in small bands, hunting and gathering a variety of foods. The second was 'barbarism', which involved growing some of our own. And finally, and gloriously, but only for the favoured few, there was 'civilisation'.[72]

This exalted state has never been very clearly defined, but it seems to be generally agreed that it involves living together in large numbers and eating food produced by others. And this dependent condition appears to have befallen us independently on only one occasion, in a process that began around 12,000 years ago in the rich river valleys of the Tigris and Euphrates. From there, the urban habit spread to the ends of an ancient fertile crescent on the banks of the Nile and the Indus, to the Hwang Ho in China, and somehow also to the forests and highlands of Central and South America.

Those who like tidy explanations have been quick to point out that all of these early civilisations lie along an isotherm, a line joining points at which the mean annual temperature is 20° Centigrade. Wherever this 'ideal' temperature coincided with reasonable humidity and arable land, they suggest, civilisation was inevitable. So we have the Phoenicians installed in comfort in Carthage, the Egyptians in Memphis, the Assyrians in Nineveh, the Babylonians in Babylon, the Persians in Persepolis, the Chinese in Ningpo, the Aztecs in Teotihuacan, the Maya in Tikal, and the Inca in Cuzco.[320]

The growth of civilisation elsewhere, in Greece, Rome and colder northern lands, becoming possible only with techniques of climate control that maintained high levels of creative energy by keeping temperatures in people's homes close to the magic 20° Centigrade.

The geographer Ellsworth Huntington believed that civilisation was triggered not so much by equable 'ideal' climates, as by ones that provided stimulus by being changeable. 'The civilization of the world varies almost precisely as we should expect if human energy were one of the essential conditions, and if energy were in large measure dependent upon climate.'[227]

He suggested that the centres of civilisation had moved not because of growing human ability to control climate, but because the climates themselves had changed, becoming more uniform. He went to considerable lengths in a series of books to justify his belief in a climatic 'pulse', that was in a process of constant change in tune with Earth's physiology.

The notion that a challenging climate leads to greater achievement, is one that receives wide support.

An analysis of the content of folk tales told by fifty-two preliterate tribes, shows that stories involving the successful accomplishment of labours or difficult tasks, are more common amongst people who live in temperate areas. And such tales reach a peak of popularity in places whose mean monthly temperatures vary by at least 11° Centigrade – but by no more than 20° Centigrade.[297]

Of fifty-three historical rulers whose titles include designation of 'the Great', forty-nine held power at times when the climate of their countries was passing through a transitional cold phase.[517]

Social stability tends to increase in colder climates, perhaps, because 'anybody who doesn't spend a lot of time in planning for future contingencies – selecting leaders who will organise things so that snow will always be plowed from the streets and food will always be readily available – is in danger of being missing from the population come spring.'[57]

There is, in warmer climates, less incentive to curl up with a good book. People in Florida and Louisiana buy and read fewer books than others of equivalent income in Oregon or Maine. And people everywhere show a preference for light fiction in the summer.

For a climate to be invigorating and lead to a high level of human arousal, temperature and pressure and humidity are not in themselves important. What is vital is that they change, often and at irregular intervals. Which, in principle, means that there needs to be frequent passage of a number of weather fronts, alternating highs and lows, bringing sharp fluctuations in local conditions and in human physiology.

Huntington defines such changes as 'storminess', which he attributes to low barometric pressure accompanied by inblowing winds that give rise to changes in temperature and humidity, producing 'all sorts of stimulating variations', but not necessarily rain. He points out that Colorado, which is one of the stormiest parts of the United States, has only a third as much rain as Georgia. He condemns California as 'too uniformly stimulating', and extols the English climate as one which has its share of invigorating surprises, but 'never reaches such extremes as to induce the nervous tension which prevails so largely in parts of the northeastern United States'.[227]

There is a large and growing body of evidence to show that the climate of Mesopotamia is now considerably less stormy than it was during that critical formative period 12,000 years ago. The path of the westerly stream of low pressure centres that now passes over northern Europe, at that time flowed along a line thirty degrees of latitude further south, passing directly over Memphis and Babylon. But it is difficult to prove that the slow northward drift of this storm track led to the progressive decline and fall of the river valley civilisations and their successors over the next 10,000 years.

Arguments rightly rage over whether empires like the Roman one fell because of moral corruption, economic weakness, inevitable senescence, or the effects of climate. There are even those who suggest that it never fell at all, but simply expanded its culture over the whole world, concealing it under a number of new social and political disguises.

It is however clear that cultures, like individuals, are sensitive to weather. And that wind, as the dominant factor in much weather, has put some fascinating twists and turns into the stormy path of human history.

Winds of Passage

We are Pleistocene people, born sometime during the last 600,000 years in an Ice Epoch.

An analysis of sediments from deep ocean beds shows that winds in glacial times were stronger than they are now, carrying large particles of red desert sand and ash much farther out to sea. A steeper temperature gradient between poles and equator, resulted initially in higher winds, cloudier conditions, and heavier rainfall, but as each glaciation progressed, leading to cooler seas and locking up more water in the ice sheets, the air became less humid and winds drier and more gritty.[374]

In short, each of the long Ice Ages in the most recent Epoch produced climates as stormy and challenging as any new young species with a bulging brain could hope for.

By the time the most recent icing began, about 125,000 years ago, there were communities who were adroit enough to be digging haematite out of deep mine shafts in southern Africa, and human enough to want to decorate themselves with the bright red pigments they found there. *Homo sapiens* had already arrived.[43]

That influential glaciation started slowly, but set in with a vengeance between 70,000 and 60,000 years ago, throwing ice sheets across most of Europe, Asia and North America in less than a thousand years. This is thought to have been initiated by some alteration in the tilt of Earth's axis, but whatever the cause, the result was a vicious and extensive circumpolar flow in the northern hemisphere, with prolonged and extraordinary cyclonic storms, bringing bitter northerly winds.[181]

Further south, conditions were changeable, at first even a little warmer than before, with prevailing southerly winds flowing in to fuel the cyclones and feed the growing ice caps that mushroomed to heights of over 2000 metres above Hudson Bay.

In the southern hemisphere, antarctic ice packs extended close to South America and New Zealand, and the westerly flow cooled and moistened Australia and South Africa far more than they do today. The movement of water from the oceans to the land produced, about 50,000 years ago, a drop in sea level of more than eighty metres, opening land bridges from Asia over the island stepping stones to Australia, and across the Bering Sea to North America.[274]

The disappearance of Neanderthals from Europe remains a mystery, but the breezy, sapient people who took their place, seem to have thrived on the challenge.

Bison and other wild cattle, several kinds of elephant, rhinoceros, and herds of horses were abundant on the edge of the advancing ice, and the groups of hunters with their new tools took a mammoth toll, even pushing some of the larger creatures into extinction. They celebrated their success and their awareness in vivid murals on the walls of caves and shelters, sometimes including self-portraits that show them standing, a little stiffly like guests at a wedding picture, dressed ice-age fashion in skins decorated with ivory beads that trace the form of a windproof pullover with a high round neck.

The mountains of ice began to melt less than 20,000 years ago, returning their water to the oceans and allowing the land to rebound. The lowlands of Finland and northern Sweden, which lay under one of the heaviest loads, are still rising at a rate of about a centimetre every year.

Change was at first rather slow, but then there was a sudden and dramatic warming which, in less than a thousand years, formed the North Sea, isolating Ireland and Britain from the European mainland, flooding the coastal plain of the eastern Mediterranean, and severing the land bridges to America and Australia.

Some of this inundation was sudden and cataclysmic enough to have drowned early settlements and left a number of people with persistent oral traditions that tell of a 'Great Flood'. North America lagged behind the rest of the world because of the size and depth of its ice cap, which finally broke up only when the sea, probably in less than a century, rose up and took back Hudson Bay.

Throughout this warming period, westerly winds were strong and steady, carrying storms of free moisture into areas like the Sahara that are now trade wind deserts. The Asian monsoon was stronger too, hurling summer rains up against the high Himalayas. Lakes grew and permanent rivers flowed on the plains, nudging the early farmers and pastoralists into their first experiments with cultivation and domestication.

Fertility was added to an already invigorating sequence of warm summers and cold post-glacial winters, building up the vital equation that was to lead eventually to the first river-valley civilisations. There was a flowering of both crops and cultures, an evolution of weapons into tools, transition from skin and stone to weaving and pottery, but one factor was still missing.

Starting about 8000 years ago, the storm belt moved away back north, leaving Mesopotamia to the mercy of the horse latitudes.

The desert reasserted itself, pastures dried up, herding and farming became difficult, finally impossible, anywhere except along the banks of the great rivers. People were thrown together into larger groups requiring greater organisation, priests and bureaucrats emerged to guide and govern, refugee herdsmen found themselves compelled to work for others for a living – and it was this factor, the existence of dependent and indentured labour, that made cities and civilisation possible.

It may have been the challenge of wind which produced technologies that went into city-building, but it was a failure of that same wind which created the slaves who did the actual work.

While all this was going on in the one fertile crescent of the near east, wind was moving people elsewhere in other important ways.

Settlements in Europe and western Asia were small and scattered. Raw materials had a restricted distribution and there was an extensive network involved in the trade of grindstones, flint and amber. People – known to us by their products as the owners of linear, cord-impressed and beaker pottery, or as the builders of megaliths – travelled widely, and did so most effectively by water.

Riverbanks have always been powerful influences, diverting travellers from their paths and leading them off at a tangent. But running water is more than a passive guide. It offers an irresistible free ride, either back into familiarity or on into the unknown. Rivercraft were the first of all vehicles, shrinking the Stone Age. With them, areas that could be explored only by generations of effort, became accessible to casual drift. The rivers were our first highways, but sooner or later all of them led to the sea, to surf and sea breezes, and before very long, to sail.

In Thor Heyerdahl's words: 'Man hoisted sail before he saddled a horse. He poled and paddled along rivers and navigated the open seas before he travelled on wheels along a road.'[207] Even when the oceans were at their lowest ebb, there was a water gap between Gibraltar and Morocco of ten kilometres, and another between Borneo and the Celebes eighty kilometres wide. Both were crossed before the end of the most recent glaciation.

Rock engravings at Tassili in the Algerian Sahara, show hippo being hunted from reed boats when the desert was still a great lake, some 6000 years ago. The bones of deep sea fish in middens on the Scottish and Swedish coasts, suggest that craft of some kind, possibly made of bark or skin, were being used there during the same period. Fifteen alderwood logboats, each 4000 years old, have been found in a peat bog in Denmark. The earliest known sail is a simple trapezoidal one hung from the mast of a reed boat shown in a predynastic rock engraving, perhaps 5000 years old, in Nubia.[302]

The period of post-glacial warmth during which the storm sequence was restored to its northern track, was a quiet one. Seas were calmer and forests grew much closer to the open Atlantic coasts of Scotland and the Hebrides than at any time since. Conditions were ideal for trade and communication by boat between the continent and Corsica, Sardinia, the Balearic islands, and even the outer Orkneys.[272]

The megalithic movement, and perhaps even some of the stones themselves, spread through Ireland, Cornwall, Brittany and the Mediterranean, along with a southward trade in tin and a westward movement of amber and bronze. The fact that skies were quiet and clear a lot of the time, may have been responsible for the construction of a large number of stone circles that seem to be lunar and stellar observatories, and for the general acquisition of considerable navigational skills. There is no evidence that anyone travelled far enough west to reach Iceland or North America during the late Stone or early Bronze ages, but at the other end of our Eurasian cradle, some impatient early mariners were already flexing their curiosity and unfurling their sails.

The first human migration into the Pacific area involved an Australoid people and seems to have taken place largely overland. A cremation burial at Lake Mungo in New South Wales shows that settlement in Australia was well advanced at least 32,000 years ago. And there are deposits of waisted-blade stone tools 26,000 years old at Kosipe, 2000 metres up in the central highlands of Papua. The aboriginal inhabitants of those areas and of nearby island Melanesia are direct descendants of these early immigrants.

The rest of Micronesia and Polynesia seems to have been settled by more Mongoloid people who began to move out along the island chain of southeast Asia about 5000 years ago. By the time they reached the limits of the continental shelves of the Philippines, Borneo and New Guinea, these 'Austronesians' found themselves on the edge of an ocean that covers a third of Earth's surface. But they were well-equipped to meet its challenge, because they had already made one of the world's great inventions.

To a simple canoe, either dug out of a log or built up of sawn planks, they added a balancing outrigger. It sounds simple, as obvious to hindsight as the lever or the wheel, but it gave them the ability to operate their craft under a wide variety of conditions.

Add to this a deep and asymmetric hull to resist leeward drift, and a triangular lateen sail that could be swung from one end of the boat to the other when changing direction, and you have one of the most efficient small sailcraft ever devised.[30]

There are thousands of islands in the Pacific, but the total area of what Captain James Cook later called 'these detached parts of the earth' is less than 200,000 square kilometres in an ocean that covers 165 million. Most of this land is concentrated in lumps around New Britain, New Caledonia, the Solomons, Hawaii and Fiji, leaving the rest scattered in coral specks that even on a large scale map look like printer's errors. And yet, within 2000 years, those early sailors had visited almost every one.

The chances of them doing so by accident, are remote. 'We notice the rivers,' says Thor Heyerdahl, 'no matter how slowly and smoothly they may flow through the land. But we do not see the ocean currents and are therefore apt to forget the greatest and mightiest of all streams; they have banks of water and flow invisibly through the sea. The largest river with its source in Peru is not the Amazon, flowing eastward through Brazil, but the Humboldt Current flowing westward through the Pacific.'[207] This is the stream that carried the balsa raft *Kon-Tiki* to its crash landing on the Tuamotan atoll of Raroia in 1947, supporting Heyerdahl's theory of an American origin for at least some of the Polynesians.

But his is becoming a very lonely hypothesis.

The Humboldt current feeds the south equatorial flow, which passes north of Polynesia at up to forty kilometres a day. The only eastward drift in the central Pacific is a narrow counter current some six degrees north of the equator that touches nothing but a small group of islands on the Christmas ridge. A computer analysis of over 100,000 simulated drift voyages, using extensive wind and current data for the whole Pacific, shows that most of the ocean's islands are inaccessible to accidental drift. It is possible, as Heyerdahl showed, to drift into Polynesia from the east, but only to a limited area. And then only if the raft in question has some sort of steering device and starts, as *Kon-Tiki* did, by being towed some distance offshore.[282]

All the major voyages that we know were made by the early settlers seem, from archaeological and anthropological evidence, to have begun from the other direction. They were very much uphill, against the currents and the prevailing winds, and were the result of deliberate navigation under sail.

The distinction between accidental drift on a current, and contrived movement at a different angle under sail, is an important one. It represents a vital advance in an organism's response to wind, which is nicely illustrated by the earliest sailors of all.

Five hundred million years before Ferdinand Magellan entered the Pacific in 1520, another Portuguese sailor circumnavigated the world. It still does. There is just one species, found in all oceans and known since the fifteenth century as the 'Portuguese Man-of-War'.

It has other names – *Physalia physalis*, 'blue bottle' and *agua viva* – but the fact that it carries a crest that resembles the fore-and-aft lateen sails of the tough little caravels sent out by Henry the Navigator to extend the limits of the known world, gives it a place of honour in both nautical and biological circles.

It is in fact a jelly fish. A creature without head, tail, limbs, mouth, gills or body cavity, that looks and behaves like an individual, but is actually a colony of larval and adult animals that cling together and have, between them, contrived to develop a transparent, sky-blue, air filled float that acts precisely like a sail and carries this enterprising community wherever the wind blows.

The float is a comparatively simple structure, a membrane surrounding a bladder of air produced by special gas glands. It is attached, however, to a tangled darker blue tissue mass that is anything but simple. Part of it is the original polyp, now surrounded by a crowd of daughter buds, some of which are protective and sensory, some of which take in and digest food, some consist entirely of a trailing tentacle which may be up to fifty metres long, and a few are little sexual adults. These groups of specialists form the sense, digestive, feeding and sex organs of the creature's body, but it is very difficult to decide just where individuality lies.[486]

Lacking the ability to produce true organs, these relatives of the corals have escaped from the limitations of their simple body structure by converting whole individuals into organs and incorporating them into societies which function like higher organisms. They have the sort of coordination and unity of purpose we normally associate with individuals, though their behaviour is more like that of a well-integrated orchestra. There is no central nervous system, but electrical potentials are conducted through the whole community in concerts of rhythmic and harmonious activity.

The actions of the members of the colony are controlled by, and subordinate to, 'colonial will'. But when the time comes to multiply, the 'sex organs' – those colonists specialising in reproduction – simply leave, as though they had reservations all along about communal living, and had never become fully integrated.[307]

Physalia is, in essence, a superorganism. The product of a line of evolution that took polyps, like the sea anemones which are fixed to the bottom, cut off the 'flower' disc with its flowing tentacles and let this float free, upside down, just beneath the surface.

All swimming coelenterates have some sort of buoyancy device, but *Physalia* has broken through the tension of the water surface and begun to lead a double life. Colonies drag their stinging tentacles through the rich plankton layers just beneath the ocean currents, and they lift another part of their structure up into the air to take direct advantage of the wind that moves the water.

There is one distantly related species *Velella velella* – with a fixed fin for a sail – but the man-of-war is the more accomplished sailor, with an astonishing ability to change and trim sail by adjustments in muscular tone which erect or collapse the float and alter its sailing posture.

When there is little or no wind, *Physalia*'s sails are deflated or lie flat on the water, and if the sun is hot, the colony contracts and starts a rhythmic rolling motion. This may be triggered very simply by the differences in osmotic pressure produced by evaporation on one side only, but the result is that the whole float is kept wet and protected from desiccation.[533]

When the wind blows, the man-of-war sets sail. The float is pumped up to take full advantage of the breeze and the colony works as a well-trained crew, trimming the sail by fitting its curvature precisely to the wind. And the most wonderful thing of all, is that *Physalia* does not just go wherever the wind blows, but sets its own course.

The tentacles stream out behind like a drogue or sea-anchor, and the rest of the colonial appendages are arranged in a clump or bulge on one side of the sail, which means that the colony, the hull of the vessel, is asymmetric. It floats with its long axis, and therefore its sail, at an angle of forty-five degrees to the wind. With the result that it travels downwind, but at an angle of about forty-five degrees to the wind. In nautical terms, it sails on a broad reach.

Physalia travel at about six metres a minute in a wind blowing at four knots – 120 metres a minute. Which means that even in a light breeze, they can cover almost ten kilometres a day. But one of the hazards facing a sailing colony is the likelihood of getting the long trailing tentacles fatally entangled in floating weed. Windrows of weed, most often *Sargassum*, which also has air-filled floats, tend to form lines parallel to the wind. But these lines are distorted by currents which are deflected by Coriolis force – to the right in the northern hemisphere and to the left in the south. With the result that objects floating passively downwind frequently become entangled in the weed, while those that travel at an angle stand a better chance of getting through.[531]

Observations of sargasso weed at sea shows that, in the northern hemisphere, bottles and lightbulbs accumulate on the lefthand upwind side of each line. Looking downwind, flotsam comes from the left rather than the right of such weed patches. So anything that has a tendency to sail to the left, rather than directly downwind, is less likely to become trapped.

And, as luck and provident evolution will have it, recent studies of *Physalia* show that the majority of colonies in the northern hemisphere bulge to the right of their floats, which means that they sail at an angle to the left, between the windrows of weed. In the southern hemisphere, the opposite situation exists and the majority of colonies are right-sailers.[308]

The existence of left and right-sailers in both populations is useful, it guarantees wide dispersal by sending parts of each brood off in different directions, but the bias in each hemisphere has other interesting effects.

In the north, the winds and waters move in a predominantly clockwise direction around both Atlantic and Pacific oceans. With the result that the left-sailing men-of-war move outward away from ocean centres, while a smaller number gravitate in toward the centre. This produces two more or less distinct populations, with the majority of colonies living and feeding in richer coastal waters.

In the southern hemisphere, the oceans turn anticlockwise and a predominance of right-sailers produces a similar effect.[532] The genetics in either case seems to be in favour of sending *Physalia* on an outward tack, and the fact that *Velella* – the smaller, less mobile relative, has come quite independently to have a mechanism with a similar bias, suggests that this pattern is one with significant survival value.[116]

The first Polynesians had a similar bias.

Captain Cook, who was one of the last to see the big voyaging canoes in action, reported that they were much faster than the *Endeavour*. He timed one cruising at seven knots and thought that they could 'with ease sail forty leagues a day or more'.[23] That is 200 kilometres, or an average of about eight kilometres an hour. Most authorities agree that the boats should have been able to cover between 150 and 250 kilometres a day 'on any point of sailing where they could lay a direct course to their objective without having to tack'.[302]

The last clause is the crucial one. Unlike *Physalia*, which is limited to travel at an angle of forty-five degrees (135 degrees off the wind), the canoes could sail at right angles across the wind and even make direct progress upwind at angles of up to 130 degrees (fifty degrees off the wind). With a certain amount of discomfort, they could also tack, zig-zag across the wind and travel toward a completely windward destination. But the easiest, swiftest and most comfortable sailing attitude for them was to reach – to sail at right angles to the wind.

The first voyagers seem to have set sail in a spirit of high curiosity. 'Fired by the lust for adventure and the desire to see new lands, canoe after canoe set out and ranged the seas. Fear of storms and shipwreck left them undeterred. The reference in an ancient song to the loss of a man at sea as a "sweet burial" expresses very well the attitude.'[135]

With no definite goal in mind, it is unlikely that they would have gone through the painful business of trying to make progress upwind. So it is reasonable to assume that the natural bias of an Austronesian canoe can be taken to be largely downwind, with a distinct preference for the easy balance enjoyed while reaching – sailing at an angle of ninety degrees to the wind.

The prevailing winds in the areas where most voyages seem to have begun, are southeasterly. Which puts the initial choice of course anywhere between southwest and northeast.

Assuming that the jumping-off point lay anywhere along the Moluccan–New Guinea–Melanesian island arc, which runs from the northwest to the southeast, the natural choice for anyone coming from Asia or Australia with eyes on the open Pacific, would have been to let the canoes carry them swiftly and easily in a northeasterly direction.

For those few who may have set out from a point in the northern hemisphere, the same situation with respect to the prevailing northeasterly trade winds, will have been resolved by heading out roughly toward the southeast.

Both courses put such voyagers right in amongst the Solomon and Caroline Islands, which is precisely where recent excavations have been turning up archaeological evidence of the Lapita culture. This is thought to be the oldest in Oceania, to have started in

Melanesia around 2500 years ago, and spread northeast into Micronesia and later eastward into Polynesia proper.[30]

This pattern of migration assumes no nautical aptitude other than the ability to bend a sail to the wind, but the early Polynesians almost certainly had other and far more impressive wind-skills.

David Lewis, while attached to the Australian National University, sailed 20,000 kilometres through the Pacific, many of them without instruments, under the instruction of island navigators.[283] He discovered that the veteran landfinders followed star paths, oriented by the Sun, made back sights, allowed for leeway, and used dead reckoning in much the same way as modern navigators do, but only when they knew where they were, and where they were going. When exploring a new area, or when displaced from an old one by a gale, they turned to other systems.

One of these Lewis calls 'expanded target' navigating. Palm trees on a typical atoll only become visible from a canoe at a distance of about fifteen kilometres, but birds based on that island are often encountered feeding out at sea up to seventy-five kilometres from shore, and as dusk falls they begin to return home. Frigate birds *Fregata* abandon their leisurely patrols, soar even higher and set off in a definite direction, probably homing by sight. Boobies *Sula* stop circling and diving and fly low and arrow-straight for the horizon. Dark noddy *Anous* and white-bellied sooty terns *Sterna* weave in and out or swoop over the wave crests, all fluttering in the same direction. And at daybreak the whole movement takes place in reverse.

Birds give an island extra breadth, but clouds expand it vertically, building up over land in unmistakable ways. Cumulus tend to dip in 'V' shapes down towards an island, or to be locked to it while other clouds pass by on either side. Reflection from the shallow lagoon in an atoll gives a greenish cast to the underside of any low cloud, while a solid landmass tends to darken it. Even when there is no cloud at all, islands 'loom' over the horizon, showing up as pale, shimmering columns of reflection in the air, which are visible to experienced navigators even on moonlit nights.

But the system most favoured by the Polynesians is steering by swells on the surface of the sea.

Swells are simply waves that have gone beyond the winds that produced them. The alignment of water patterns is slow to grow and once moving, possesses an inertia that resists later and more transient changes in wind direction. To remain perceptible after travelling long distances, they must have their origin in strong or persistent weather systems, such as the trade winds or the wester-lies, which lie in a fixed and known direction from the point of navigation. Having come a great distance, their length from crest to crest is also long and they move by with a slow and stately undulation which is quite distinct from the short, steep, breaking waves produced by local weather conditions.

On Puluwat atoll in Micronesia, they recognise three swell patterns. The 'North Wave' is a long, majestic ground swell that actually comes in from the northeast, where it seems to be generated by the trade winds of that hemisphere. The 'Big Wave' is shorter, steeper and more energetic, and arises close by as the product of intermittent easterly winds. And the weaker, less regular 'South Wave', which is an echo of the distant southeasterly trades.

Thomas Gladwin, an American anthropologist who worked on Puluwat, explains,

> Because the Big Wave has a more pronounced character and passes with greater frequency, it is preferred to the North Wave. Either is more useful than the unreliable and often weak South Wave. However, the choice of a wave depends not only on which one is running strongly, but also on the course to be held. In general it is easiest to steer by waves which are either at right angles or parallel to the travel of the canoe . . . Puluwatans steer by the feel of waves under the canoe, not visibly.[160]

Amongst the Santa Cruz islands of Melanesia, there are also three dominant swells, varying in relative prominence. These are identified as *hoahualoa*, the 'Long Swell' from the southeast trades; *hoahuadelahu*, which seems to be produced by the north-west monsoon; and *hoahuadelatai*, the 'Sea Swell' from the more distant northeast trades.

Sometimes all three may be flattened and distorted by a local gale, but it is rare that a skilled observer cannot find one or another of these perennial patterns. On one occasion, after a navigator who was guiding Lewis was thrown off course in a storm that came at the end of a solid month of westerly gales, he was still able to detect an underlying easterly swell.

Lewis says,

> the swell pattern is almost always a complex one, with several systems that differ in height, length, shape, and speed moving

across each other from different directions at the same time. It follows that every island navigator must select those swells that he considers most significant and reliable, and though there are patterns that are generally recognised throughout each navigational area, there can also be a personal element in this selectivity.[283]

Holding course by recognition of such swells seems to be a matter more of feel than sight. Young trainees are taken out to sea and told to lie in the water on their backs, relaxed and floating so they get to know the 'feel' of the waves. Some navigators still prefer to lie down inside the canoe's cabin and call their directions out to the helmsman, but most of them by day or night, choose to stand, carefully balanced with legs slightly apart, waiting until the pattern they want becomes prominent. It seemed to Lewis that they were plumbing the swell, feeling its effect in subtle shifts from the vertical, detected largely by the pendulum swing of their own testicles.

When any of the deep water swells reaches the shallows or passes an island, it is distorted and forms new patterns which can also provide useful information.

Two basic things happen to a swell as it meets land. Part of it bounces back directly off the reef or island, forming reflected waves of far shorter length. Part of it is refracted, dragging where it touches the shore, until the broken edges of a long ocean swell are actually in line with the coast and wrap around an island, meeting up in its shadow to produce cross seas and a brisk local chop. The rest of the swell passes by unaltered. But all the deep water patterns experience similar distortions and the resulting confusion of refraction, reflection and secondary waves of a variety of lengths, forms a complex interference pattern around any island or archipelago that a computer would find hard to analyse. The navigators take it in their stride.

The art of locating islands by 'land waves' seems to have been refined almost beyond belief amongst the Marshall Islands of eastern Micronesia.[528]

Four main swell patterns are recognised there, coming roughly from the cardinal points of the compass and being reflected in each quadrant as backwash waves that are perceptible thirty kilometres or more away from an island. These secondary waves react with incoming primary swells to form complex nodes, roots, knots and holes in the sea – that have been plotted over the course of generations and recorded in stick charts of palm ribs bound together with coconut fibre, with cowrie shells attached to the ribs to represent the position of the islands.

There is one of these constructions on display in the Museum of Mankind in London that is so complex, so elegant, so clearly intelligent, that it deserves to be ranked with other famous intellectual yardsticks such as the Rosetta Stone. These *rebbilib* are products of long local experience, but the Marshallese also have more abstract charts they call *mattang*, which show the general principles of wave interference and can be used as keys to the understanding of new conditions anywhere.

At night, a Polynesian navigator can feel the rhythm of complex water patterns through his boat, but on some dark nights he does not need to try. The eddies stir up planktonic crustaceans that mark each disturbance with a cool phosphorescent comment, setting up a luminous display that highlights the position of the nearest land as surely as the glowing pattern on a radar screen. The islanders call it 'underwater lightning'.

Wind consciousness of a Polynesian order is rare, if not unique, but what the island navigators seem to have done is to formalise a pattern of knowledge that depends ultimately on an unconscious sensitivity we all share. It may have to be taught afresh to each new generation, and brought to awareness in this way, but we come to it equipped with a natural feeling, a bias for wind signs as useful signals.

In a study of human orientation, Robin Baker of Manchester University found that even inexperienced city students were able to pick up directional clues from the wind when blindfolded.[17] The first Austronesian voyagers cannot have been any less sensitive.

The ability of marine turtles to return, despite contrary currents, across thousands of kilometres of ocean to nest on the same remote islands each time, may have nothing to do with celestial navigation or, as has been suggested, with their need to track down an ancestral breeding ground now separated from the coast by continental drift. They could be responding, as it seems we can, to the relative stability of wind-generated waves in surface waters.

Young turtles may have to train themselves, as we do, in the niceties of response, undergoing a kind of spatial learning. But they and we are natural wind freaks, bursting with the urge to travel, just waiting for the right kind of fix.

Winds of Trade

While Polynesia was being settled from the west, things were not going well back at the fount of civilisation.

A small salty lagoon on the west coast of the Crimean peninsula has, in the layers of mud deposited on its bed, kept faithful record of the rainfall each year since 2300 BC. These 'varves' show that Lake Saki, from about 2100 BC, was in an area of considerable drought that lasted for centuries.[55]

An equally precise and even longer climatic record is being kept in the White Mountains of California by some of the world's oldest living creatures, the bristlecone pines *Pinus aristata*. This goes back to 3431 BC and shows a marked decline in summer temperatures beginning in about 2500 BC.[271]

It was in this same period that the Australian deserts began to dry out, while in Japan, New Guinea and the Andes, the height of the timberline was reduced. In Europe, the northern limits of forest growth were cut back by up to 400 kilometres, and in North America the Barren Grounds were abandoned by both the caribou herds and their hunters. Everywhere the story was the same.

This worldwide downturn in climate marks the end of warmth that flowered following the last glaciation, and the beginning of a cooler time. The cyclone belts moved closer to the poles, bringing wetter times to northern Europe, with a resurgence of bog growth, but areas nearer the equator were left high and dry. The Nile normally floods each year in August and September as a result of summer monsoon rains in East Africa, but starting in the Second Dynasty, around 2800 BC, there were dramatic changes in its level.

There was a great famine, the first on record, between 2180 and 2130 BC, whose effects are vividly portrayed in an inscription from a tomb near Luxor which reads – 'All Upper Egypt was dying of hunger, to such a degree that everyone has come to eating his children.' And another from Lower Egypt, dated 2000 BC, saying –

'The river of Egypt is empty, men cross the water on foot . . . The south wind drives away the north and the sky has only one . . . Foes are in the east, Asiatics are come down into Egypt.'[271]

Something similar clearly happened on the steppes of Asia, because starting at about this time there began a series of nomadic explosions that hurled mounted people in wave after wave down on all the margins of the continent. The English historian Arnold Toynbee describes what took place as 'a dynamic interaction between the Desert and the Sown'.[487]

The nomads were pastoralists, very much climate-dependent, a fierce hard people with a warrior tradition, and the effect of their cavalry on farmers and city people was devastating. They overthrew the Assyrian Empire, and came back to harry the Persian one that replaced it. Sarmatians, Scythians, Aramaeans and Cimmerians swept down through the Carpathians and the Caucasus into Europe and the Balkans, and through the Pamirs into India. These incursions were all brief and bloody, depending on the leadership of individuals whose charisma could hold a coalition of independent tribes together long enough to make a successful raid and get back to the steppes with their booty. They built no empires of their own, but everywhere they went, they broke up the old order, and left the even more powerful legacy of an Indo–European language that survives still in Celtic, Germanic, Italic, Slavic, and Baltic tongues.

These pressures from the interior, and the destruction of overland trade routes during the second millennium, led to a corresponding increase in seafaring.

By 1800 BC, the Minoans controlled a maritime empire based on their palaces in Crete, with a flourishing trade in oil, grain, wine, copper, timber and tin between Levantine, Egyptian and Libyan ports. This was serviced by the world's first commercial civilisation, a coastal mix of merchants, traders and sailors who carried both goods and ideas around the Mediterranean.

Their dominance lasted until 1450 BC, when a huge eruption of the volcano Thira on the island of Santorin, overwhelmed the Aegean and put up a dust pall five times as dense as that of Krakatoa, changing the climate of the entire northern hemisphere.

Global temperatures were lowered for several years, forests died and there were devastating fires across Canada that turned pinelands into tundra. In Europe, this was the time of *Ragnarök*, the twilight of the northern gods, and the dreadful *Fimbulvinter* in which snow was driven from all quarters by a biting wind, and three winters followed hard upon each other's heels, with no sign of a summer between them. When the wind finally dropped and the air cleared, the seas were left open for the Phoenicians, who turned from sporadic piracy to build a great trading empire based on sail.[272]

Trade involves giving and receiving, it is a two way process. And trading under sail implies winds blowing in both directions, or at least ships manoeuvrable enough to make it seem as though they do. While the ceremonial long boats of other nations grew into cumbersome galleys and military triremes, the Phoenicians concentrated on rounded, stubby, practical little merchant sailboats.

By 1200 BC, the coast of Syria from Cilicia to the Egyptian border, was studded with Phoenician cities connected by regular routes to Cyprus, the Greek mainland, Sicily and Spain. The traders founded a colony at Cadiz in 1100 BC and by 841 BC had become rich enough to set up a new capital city at Carthage and extend their routes more ambitiously out through the dreaded 'Pillars of Hercules' at Gibraltar to outposts in West Africa and the Azores. And by 595 BC, under the sponsorship of the pharaoh Nekho, it is said they even sent out an expedition which completely circumnavigated Africa.

The Greek historian, Herodotus, says that this took almost three years. 'When autumn came, they went ashore, wherever they might be, and having sown a tract of land with corn, waited until the grain was fit to cut. Having reaped it, they again set sail; and thus it came to pass that two whole years went by, and it was not until the third year that they doubled the Pillars of Hercules, and made good their voyage home.'

They were not, however, the first to ply the Indian Ocean.

Thor Heyerdahl, intent as ever on demonstrating seaborne contact between early civilisations, has pointed out that the hieroglyphic sign for 'ship' in the earliest known Sumerian script is identical to the ideogram for 'marine' in ancient Egypt. Both use the outline of a sickle-shaped reed boat with high bow and stern, a design he suggests was meaningless in terms of river navigation, but essential for dealing with breaking surf and open sea.

Eratosthenes, librarian at Alexandria before this priceless store of papyrus scripts was burned, described the island now known as Sri Lanka as being twenty days' sailing from the mouth of the Ganges, implying that ancient Egyptian traders and mariners had travelled at least that far and returned to tell the tale. Perhaps in the process stimulating some sort of reciprocal trading venture that would account for the large number of early Indus seals with their characteristic pictographs that have turned up in excavations of ancient Sumeria. There seem to have been agents from India settled more or less permanently in the Sumerian port cities.

'No part of coastal Asia,' says Heyerdahl, 'can be considered out of reach, even in the second millennium BC . . . Indeed, the seasonal reversal of wind and current directions in the monsoon area would favour all sorts of communications in the Indian Ocean.'[207]

He is absolutely right about the winds. The so-called trade winds are fine if you happen to be going one way, setting out to explore new routes, but for two-way traffic there is nothing quite like the Indian monsoon. Nowhere else in the world is there such a reliable, predictable, reversible pattern of natural propulsion.

To be fair, the monsoon is a modified trade wind, able to draw on the regular trades of both hemispheres by virtue of the southward extension of India, that makes this ocean the only one that has relatively little water area north of the equator. The regular northeast trade of the northern hemisphere carries cool air off the continent in winter, blowing across the Arabian Sea and the Bay of Bengal, straight down the line of the coast of East Africa, curving to the east at a point well beyond the equator. Between November and March, fleets of outbound dhows still sail from the Gulf, Pakistan, and India to Mogadishu, Lamu, Mombasa and Dar es Salaam, flicking out at the tail of the wind to the Comores and Malagasy, carrying cargoes of spices, carpets and salt fish for sale in African bazaars. In April they rest and sleep and patch their sails, waiting for spring in the north and a time when Asia warms enough to draw the southeast trades once again out of the lower Indian Ocean.

By early May, India is dominated by a high pressure system and the southern trades, curving north of the equator to come sweeping in from the southwest, arrive in Burma. By late May, the monsoon front of warm moist air covers Sri Lanka, and in June all of India and Pakistan enjoys its rain. In May, June and July the dhows, loaded now with cargoes of charcoal, grain and mangrove poles, pick up the incoming monsoon at meeting points all along the African coast, joining the flow on time-honoured schedules that were set at least 3000 years ago.

Knowledge of this regular reversal is continuous for the people of the Indian Ocean, but seems to have been lost to Europe with the decline of the river valley traders. When Alexander the Great constructed a fleet to sail down the Indus in 326 BC, his admiral Nearchus coasted laboriously back through the gulfs to the Euphrates when he could, just by waiting a short while, have taken direct advantage of the northeast monsoon.

Greek traders from Arabia Felix (now the Yemen), hugged the coasts all the way to the delta of the Indus, instead of sailing comfortably with the dhows, directly across the Arabian Sea. One unfortunate soul, blown by accident all the way down to Hippuros (in Sri Lanka), and not realising that the wind was just about to change and bring him back, abandoned his boat and returned overland instead.

The Arab seamen had no intention of giving away their wind lore, which remained entirely in their hands until Hippalos, once a pilot to Alexander, hit upon at least part of the secret by discovering the timing of the southwest monsoon. Pliny, writing two centuries later, calls the wind itself Hippalos. The discovery of a trustworthy deepwater wind of passage immediately let loose a flood of traffic in June and July bound from Africa, Aden and the Red Sea to the Malabar coast. And it cannot have been long afterwards that the ship captains, waiting in eastern ports for a favourable return wind, discovered that it came all at once, and kept on coming, from November on.

By the time that Claudius came to the imperial throne in AD 41, trade with India was booming. A round trip from Rome, including the crossing of Egypt by camel, could be made in a year. 'India,' said Pliny, 'had been brought near by gain.'

Lions, tigers, rhinos and elephants appeared in the Roman circuses, parrots were kept as pets, ivory and turtleshell were carved into ornaments, food was liberally peppered, and everyone who could afford to do so, wore silk and pearls. In about AD 60, a magnificent merchant's guide – a sort of early Roman Michelin – with details of all the harbours, markets, tides and winds of India, was published as the *Periplus of the Erythraean Sea*.[226]

Bigger and better ships were financed and built to operate on this profitable route, including one mentioned with awe in the *Periplus*, which had *seven* sails and 'lashed the sea into foam'. And these increases on the Indian Ocean led to equivalent advances in the Mediterranean. In the second century AD, a ship fifty-five metres long and fifteen metres wide, 'with a crew like an army, passengers of both sexes, and corn sufficient to supply all Attica for one year, brought a large annual profit to its Egyptian–Greek owner'.[508]

Perhaps even more important, the trade brought with it a greater awareness of distant places, of the remote ends of Africa and Asia. The *Periplus* even discusses the Zanzibar Channel, Mount Kilimanjaro, the Malay Archipelago and Tsin, which we now know as China.

Europe after Rome, maintained only desultory contact with Asia, being somewhat preoccupied with barbarian invasions, the growth of Islam and the crusades. But when the Middle Ages were over and the Renaissance began, the separate civilisations began to look each other's way again. It is no accident that the first globe was designed, Christopher Columbus reached the West Indies, Amerigo Vespucci charted part of the South American coast, and the first sea route to India was explored, all in the same decade – the last of the fifteenth century.

It was Vasco da Gama in 1498, more than two millennia after the first Phoenician circumnavigation of Africa, who reopened the route and effectively started the colonial era. And almost everything that happened from that moment until the invention of the steamship by John Fitch in 1790, was touched in some way by the tyranny of wind or the actions of those who rode it.

The 'space capsule of the Renaissance' was the three-masted ship. This skeleton-built, non-edge-joined, vessel was not merely the most important single development in the history of sailing, it was one of the most influential of all developments in human history.[301]

Ships with three masts had sailed before, skeleton frames were invented by prehistoric skinboat builders, but the addition of these two features to a planking technique (sometimes called carvel building) which allowed more weight and greater resistance to stress, created a broad-beamed, deep-hulled ship that could go anywhere, and did.

It took courage and a wind-sensitive skill. Alan Villiers, perhaps the last of the great square-rig skippers, says,

I often think that Columbus' great achievement was not in stumbling more or less fortuitously upon the fringing tropic islands of the great American continents, but in blazing the sailing way across the North Atlantic. His trade winds run, from the Canaries to the West Indies, became the classic way to make westing in that ocean, and his passage back to Europe in the Gulf stream drift beyond the trades – north of them – was the best way to sail eastwards too.

In the beginning the only rules were – get away from land and stay away, using the west winds to travel east and the trade winds for changes in latitude or, where possible, for going west. Everywhere else, it was a question of slogging it out and hoping for the best.

The initial goal for everyone was the spice islands of the East Indies, which meant getting round the Cape of Good Hope. But as soon as any southbound ship crossed the Atlantic equator, it ran headlong into the southeast trades. So right from the start, the technique was to let the trades carry the ship west to the coast of South America and then follow the Brazil current southward, 'rolling down to Rio', to pick up the westerlies. Without this natural bias, which persisted well into the days of the great steel barques of the early twentieth century, it is unlikely that Brazil would now be as developed, or Angola as neglected, as they are, despite the fact that both were settled at the same time by the same colonial power.

Once in the westerlies and round the Cape, the southeasterly trades gave easy access to anywhere in the Indian Ocean. And the existence, and an awareness of the existence, of the monsoon, eliminated problems north of the equator. Exploration in the Pacific, without the wind and wave sensitivity of the Polynesians, was mostly a matter of finding small specks of land in that vast area. A problem which Captain Cook, certainly the most successful of the later pathfinders, simplified by carrying Tupaia, a dispossessed high chief and navigator-priest of Raiatea. When he went with Cook to Batavia in Java, a distance of over 10,000 kilometres from his home, despite the ship's circuitous route, 'he was never at a loss to point to Tahiti, at whatever place he came.'[23]

Trade winds north and south of the equator carried the first traders easily from Sydney to Seattle and San Francisco, but getting out of the Pacific was another matter. The only logical way was in the westerly flow, but 'Cape Horn is too far south for comfort at any time and in winter is hell.'

Running the easting down called for constant knowledgeable alertness, because half the run from Australia towards the Horn is inside the ice-line and all of it lies along the track of a noisy procession of gales that march constantly across the bottom of the world. No captain could ever allow a wind to get the jump on him and most lived and slept, when they could, with their ears constantly tuned to any sounds of change.

Their ships could speak to them in ways no power-driven vessels ever did or could. No throbbing sea-punching propeller thundered behind them, no noisy shaft deadened the sound of wind and sea, no pulsing engines, no clang of fireman's shovel on steel stokehold floor, no whining old pump disturbed the natural sounds that were the square-riggers orchestra – a harmony of wind, ship and sea. The ship herself called any good master when she needed him, and he was up on deck in a bound – on deck and in charge.[499]

They made some amazing passages.

During their brief reign, the famous tea clippers guaranteed ninety days from Batavia to the channel ports via Cape Town, and some made it in eighty round the Horn. A good run was thirty fearsome days from Australia to the tip of South America, another twenty-five relatively easy ones to the line, and a further twenty-five or thirty days, depending on the doldrums, back home. The record for the outbound run from the English Channel to Melbourne, was sixty days in the beautiful great grain-carrier *Thermopylae*.

To begin with, during the sixteenth and seventeenth centuries, the basic three-master underwent only small modifications in sail and steering, and changed very little through the hectic days of Portuguese, Spanish, Dutch, British and French struggles for dominance in world trade. All the nations were very conscious of the fact that the creation of wealth through trade lay at the root of political and military power. Going into the seventeenth century, Holland alone had 10,000 merchant vessels at sea. But after the heady early days of the spice trade, when the biggest profits went to the fastest and best-defended East Indiamen, merchant ships settled into normal competition in which the advantage went to those carrying the largest cargoes and employing the fewest men.[300]

These factors produced the last major advance in sail design, the introduction of schooners with fore-and-aft rigging, which meant a small crew could handle the same sail area without going aloft. And the need for large capacity, moulded some of these workhorses into monstrous four, five, six and even seven-masted giants that ran the coal, iron and timber trade routes of the late nineteenth century.

After 1850, and the opening of the Suez Canal, the cargo steamer began to edge its way into long-distance trade. But sailing vessels were able to offer highly competitive rates and kept most of the bulk business – jute and rice from India, nitrate fertiliser from Chile, and wheat from California – until late in the 1890s, delaying the demise of sail until the coming of the first efficient high-pressure marine engines.[173]

In the climate of economic uncertainty following the First World War, there was a brief resurrection in the fortunes of sail. A few of the world's oldest sailing ships were pressed back into service and earned fortunes. They faded once more when industry resumed its stride, but there are still those who believe that we will live to enjoy again the days of trade by sail. Even if it consists only of computer-controlled solid sails that can be unfurled on the mastheads of giant oil tankers to make them more fuel-efficient themselves.

Others hope for a more classic revival.

The moderate sized, four-masted, fore-and-aft schooner – brought up to date perhaps with the use of modern materials and equipped with an auxiliary engine – probably represents the cleanest, most efficient mover of goods that has ever been devised. After four centuries of refinement, the thoroughbred lines of such vessels became so tempered in their prime by wind and wave that they seemed to almost sing.

It would be fine to hear that music played again, though it will be hard to find the right musicians. You cannot just break great traditions and expect to mend them later on at will.

Winds of War

The renowned Scottish anthropologist James Frazer put an old problem very simply:

> To the mind of primitive man, natural phenomena assume the character of formidable and dangerous spirits whose anger it is

his wish to avoid, and whose favour it is his interest to conciliate. To attain these desirable ends he resorts to the same means of conciliation which he employs towards human beings on whose goodwill he happens to be dependent; he proffers requests to them, and he makes them presents; in other words, he prays and sacrifices to them; in short, he worships them.[147]

We now recognise other more complex factors involved in the growth of human awareness of the natural world, but basically he was right. We still pray for divine intervention in the weather – and the favour in greatest demand now, as then, is good weather for us, and bad for them.

When the Sumerians settled in Mesopotamia in about 3200 BC, they brought their gods with them. There were three. *Anu*, who was powerful but somewhat remote, and his two active arms – *Enki*, 'Lord of the Waters', and brave *Enlil*, 'Lord of all the Winds'. *Enlil* was installed on the terraces of the Ziggurat, a great temple in the religious centre of Nippur, and it is he who was credited with responsibility for the destruction by storm and subsequent capture of the city of Ur.

By the time that Sargon, 'Ruler of the four quarters of the World', conquered all Mesopotamia in 2350 BC and established the Akkadian empire, *Enlil* the wind god was the leader of the celestial trio and too powerful to be abolished. So he was simply adopted into the Semitic religion and known instead as *Bel*, 'Master of the Land'.

Bel survived the Canaanite invasion in about 2000 BC, but when Hammurabi moved the centre of authority to Babylon (literally 'God's Gate') in 1728 BC, a 'new' *Bel* was appointed and called *Marduk*. Elsewhere, on the coast of Syria amongst the Semitic Ugarit, *Bel* or *Baal* was still supreme and known as 'Rider on the Clouds'. And he stayed up there triumphant until about 1250 BC, when the Israelites advanced into Palestine worshipping a new storm god, *Yahweh* the Lord of Sinai, who had appeared to Moses in thunder and lightning.

This was a more personal kind of god, an active warrior who made a point of destroying his enemies by fire or smote them with scorching desert winds. He took over from *Baal* and was built up by the literary prophets into the 'One God' based on the temple in Jerusalem. But despite a change in style and address, there is much in *Jehovah* of the old storm god, who could summon up winds at will. Which he soon did, in the guise of a wild northeast *gregale*, smashing the fleet that Jehoshaphat, son of David, sent unwisely against Tarshish.[240]

Meanwhile, somewhere north of India, a different but rather similar pantheon emerged. The oldest literary documents in any Indo–European language are the Sanskrit hymns of the *Rig Veda* which identify two major deities – *Varuna*, the earth mother and *Dyaus*, the sky father. Beneath them are a complex of minor deities, but when a group of Vedic tribes travelled west and settled in the shadow of Mount Olympus, the only one they seem to have brought with them was the sky god, whose name by now was softened to *Zeus*, the 'bright or shining one', creator of the wind and weather.

While the various Greek tribes about Olympus were welding themselves into some sort of Hellenic cohesion, the Medes and Persians were building a rival empire across the Aegean in Iran, 'the land of the Aryans'. They began to subjugate the outlying Greek cities in Asia Minor and under Darius took India, Egypt, Libya and Crete, and were only narrowly defeated by the Greeks themselves in Marathon in 490 BC. Ten years later, a new and larger force under Xerxes sacked Athens and prepared to deliver the coup de grâce to a heavily outnumbered Greek fleet, that lay waiting for the decisive battle in the narrow straits between Attica and the island of Salamis.

Command of the allied Greek forces in 480 BC was in the hands of a wily political strategist called Themistocles, who had been elected three years earlier, outmanoeuvred his opponents in government, and managed to carry out a fleet-building program that produced 180 new triremes. These, and an equal number of older boats, were all he had to set against a Persian fleet of 1400 ships. But he knew he could also call on the services of Zeus, god of the weather, and some canny local knowledge.

The Greek captains were unhappy about being trapped at Salamis and wanted to fall back to the larger southern gulf of Argos, but Themistocles made them wait, and he made sure that Xerxes got to hear of this dissension in the Greek ranks. During the night before the battle, some of the Greeks made a feint to the west and Xerxes detached a quarter of his fleet to seal that end of the strait. Themistocles kept all his fleet hidden behind a small island in the narrowest part of the strait itself, and prayed.

The day of the battle dawned fine and the Persians out in the gulf waited for the Greeks to make the first move. Nothing happened until mid morning, when Themistocles sent a handful of his ships west to test the Persian rearguard and make a lot of noise in the process. Xerxes ordered the bulk of his fleet directly into the eastern end of the straits in hot pursuit, and fell directly into the trap.

As soon as the Persian ships were strung out in a vulnerable line, the 310 Greek triremes burst out of their cover and attacked at ramming speed. And with them, as Themistocles knew it would, came the *Etesian* wind – a blustery, powerful, gusty wind that howls across the Aegean in autumn, whipping down from the hills of Attica and getting into its stride around noon.

The Persian ships were taller, more top-heavy, tired from a night of manoeuvre, and totally unprepared for the combined onslaught of both Greeks and weather. In the narrow straits there was no room for movement, oars became locked and ship upon ship piled up in a lethal tangle. The Persians were sunk or routed and what was left of Xerxes' fleet withdrew, unpursued to Asia Minor.

Salamis was the turning point in the war, and the following year the campaign ended. The despotic Persians gave up their expansionist aims, the political and intellectual independence of Greece was assured, and the unfolding of Greek civilisation began. Without that timely wind, we may all have been deprived of the sculpture of Praxiteles and Polycleitus, the drama of Sophocles and Euripides, the writing of Xenophon and Thucydides, the philosophy of Socrates, Plato and Aristotle, and the perfection of democracy.

Dyaus Pitar, the 'shining one', went on to become *Jupiter* to the Romans, 'radiant father of the heavens', god of rain, wind, storm and thunder. And on more than one occasion joined forces with Mars and the legions to defend the Empire.

When Marcus Aurelius, the last of the adopted emperors, came to the throne in AD 161, the only threat to Rome came from the Parthians whom he soon defeated in Syria, and a group of unruly German tribes. When the Quadi and the Marcomanni insisted on crossing the Danube in AD 174, the legions were sent against them, but found themselves hard pressed, outnumbered and short of supplies. The empire was on the brink of ignominious defeat when a sudden and severe thunder storm struck the enemy lines and the frightened Germans turned and ran. From that moment until the partition of the Roman Empire, the troops of Jupiter came to be known everywhere as the 'Thundering Legion'.

Following the division of the Roman world into eastern and western empires, there was perennial tension between the rival emperors, culminating at the end of the fourth century AD in the appointment of Eugenius in the west. His introduction of heathen cults led to war with Theodosius in the east, who had just made Christianity the state religion. At the decisive battle between them at Aquileia on the Frigidus in AD 394, a 'miraculous wind' is said to have routed the ungodly Eugenians by turning their own weapons back against them.

The eventual decline of Rome coincided with a drought that lasted until about AD 800, setting off *Völkerwanderungen*, a time of great migrations. Most of these wanderers were displaced nomads and barbarians from central Asia, moving west, bringing dissension and occasional plague. Ostrogoths and Huns, Avars and Bulgars, Khazars and Magyars, who all left their marks on eastern and central Europe. But it was not until the Middle Ages, and a warmer wetter climate, that a Mongol people were able to come out of the steppes in strength rather than confusion, under the dynamic leadership of Genghis Khan.

Temujin, 'the Blacksmith', was made supreme ruler of his people in 1206 and announced his campaign for 'the conquest of the world'. And he very nearly did it. By the time Genghis Khan died in 1227, the Mongol Empire extended from Korea to the Caspian Sea. His sons and grandsons carried the campaign on to Germany and Poland, and sent two armies of half a million warriors each to take China. When Khubilai Khan came to power in 1260, there was little left to do on the mainland except consolidate his hold on China, so he turned his gaze east to Japan.

The Mongols had never had to deal with an ocean before, but undeterred, Khubilai issued orders for the construction of a thousand ships. This took five years and in 1274 the fleet set sail with 40,000 troops.

The Japanese had no ships at that time, so the invaders were able to make an unopposed landing at Hokozaki (Hakata) Bay on the southern island of Kyushu. They fought a short way inland without too much trouble, but when dark fell made the mistake of deserting their beach-head in favour of spending the night afloat out of reach of ambush. During the night, a violent gale blew in from the north, the only quarter to which the bay was exposed, and the fleet was forced to struggle out into open water. Just 300 ships and 20,000 men made it back across the sea to Korea.

The Khan was furious and ordered another thousand ships. Six years passed, in which the Japanese had time to fortify their coasts. In 1281, the second armada was ready. In addition to his Korean fleet of 1000 ships with 50,000 men, Khubilai now had a further 3500 captured Chinese junks, carrying 100,000 troops and all their horses and supplies. The combined fleet gathered off Iki island in early June, an accomplishment that in itself ranks as one of the greatest achievements in nautical history, and headed for Kyushu.

Part of the fleet made once again for Hokozaki and anchored there, lashed together by chains and planks for mutual support, while they began a bombardment of the shore fortress with heavy catapults. The rest made a successful landing further south at Imari and started to fight their way round to take Hokozaki from the rear. It was a sound tactic, which would have succeeded if the Japanese had not spent the six years since the last attack preparing for just such an eventuality. The Mongols were forced to pay heavily for every metre gained, and after six bloody weeks, decided to change tactics and reunite the fleet. Then, on August 15th, the typhoon struck.

It hit both anchorages directly, parting cables, snapping masts, grinding the ships up against each other and the rocks.

Mountainous seas swept the wreckage ashore in tangles so thick that the next day it was possible for a man to walk hundreds of metres out to sea on shattered hulls, planks and spars. Most of the invaders drowned, and the few who did manage to struggle exhausted through the surf, were cut to pieces by defenders on the storm-swept beaches. More than 4000 ships and 130,000 men were lost to a force that had rolled irresistibly across Europe and Asia crushing everything else before it, only to be stopped dead now by 170 kilometres of sea and a wayward wind.

Fishermen of Takashima, which lies between the two harbours in which the Mongol fleet was destroyed, are used to finding ceramic dishes and barnacle-encrusted sword hilts in their nets. A sonar survey in 1980 located seventy-two wreck clusters on the bottom, and an archaeological expedition during the last few summers has begun to recover the vast hoard that includes anchors, iron ballast, vases, gunpowder jars, and catapult balls bearing traces of the blazing oil and cloth in which they were wrapped and fired by the attackers in 1281.[415]

The Khan died in 1294 before he could mount a third attempt, and Japan was not invaded again until nearly seven centuries later. The years between are bridged by *shimpu*, 'divine winds' that the Japanese believed were sent to rescue them from the fury of the Mongols in the thirteenth century, and could do so again against another invader in the twentieth. In 1944 however, they chose not to leave everything entirely to the gods or meteorological chance, but to form a 'Divine Wind Special Attack Force' that became popularly known as *kamikaze*, the 'spirit winds'.

Three hundred years after the destruction of the Mongol fleet, another 'invincible' armada set sail, this time from Lisbon. Philip II of Spain saw his mission in life to be the reunion of all Christianity under the Catholic faith and Spanish leadership. He made a start by marrying Mary Tudor, Queen of England, and when she died at an early age, paid court to her successor Elizabeth. But the new young queen scorned him and his religion, and encouraged her naval heroes to join other privateers in plundering on the Spanish Main. Philip manoeuvred, through France and Mary Queen of Scots, to overthrow Elizabeth and secure a Catholic successor to the English throne, but when Mary was executed in 1587, he decided that invasion was the only answer.

Like the Khan, Philip had no docks and no permanent navy, but he had recently conquered Portugal and, drawing mainly on her resources, was able to gather together a considerable fleet of 130 ships, with 2431 guns and 27,365 men. This assortment of men-of-war and converted merchant vessels left the Tagus on May 19th, 1588, but ran into such adverse winds that it was two months before they cleared the Bay of Biscay and headed north into the English Channel.

The summer of 1588 was a bad one, like many others that century. The southern spread of Arctic ice and polar water made much of the North Atlantic 5° Centigrade cooler than it is today, creating a thermal gradient that spawned a succession of severe cyclonic wind storms which, in the words of a modern climatologist, 'correspond to jet streams at the limit of, or beyond, the maximum speeds expected from modern experience'.[272]

By early July, the English fleet under Sir Francis Drake and Admiral Lord Howard, was waiting at Plymouth. They knew the Armada had left Corunna in northern Spain, and they made several forays and patrols into the Channel, but were beaten back again and again by the weather. Then finally, on July 19th, Spanish sail was sighted off the Lizard, beacons burned all the way along the coast and word was carried to Drake relaxing at a game of bowls on the grassy Hoe. 'We have time,' he is reputed to have said, 'to finish the game and beat the Spaniards too.'

All that night the English fleet warped and winched itself out of Plymouth harbour, and before daybreak managed to work its way to windward of the Armada. The wind was blowing steadily from the southwest and for the next three days, the story was the same.

The heavy Armada had to move with the wind, working steadily up the Channel, and the smaller English fleet darted in on their perimeter like sparrows mobbing a hawk, pecking at the Spanish ships with their light shot and staying out of range of the heavy guns. 'It seemed to us that we were anchored, and the English had wings to fly as and where they wished,' one of the Spanish commanders later wrote to his king. But the English fleet could take no real advantage. Although half of the Armada were not warships, they made a solid body like a moving town, that was extremely hard to break up.

After a week of such skirmishes, watched with awe by huge crowds all along the cliffs of Devon and Dorset, the attacks began to take a cumulative toll. No ship of the Armada was sunk, but most were disabled, had men wounded and water casks holed. And day by day they were being shepherded further into the narrow bottleneck at Dover by a fleet of little ships less than one third of their size.

On July 27th, France was in sight and the Armada decided to anchor off friendly Calais. The English response was a simple one, to send fireships – empty hulks, ablaze with pitch and powder, sailing down into the anchored fleet in the night. It worked.

Within half an hour there were a hundred and twenty disordered ships fleeing without direction in the night, afraid of collisions and shallow water, afraid of fire, afraid of the English, afraid of the devil . . . they wallowed on, not daring to raise sail because of the shoals and unable to anchor because they had left their anchors and main cables in the Flemish ooze.[189]

The English fleet fell among the stragglers at dawn and for two days pitched battles raged, with at least three Spanish men-of-war going down. Then the wind suddenly backed to the south and the ▪Armada took advantage of it to break off and sail away up through the North Sea and around Scotland bound for home. But though the English held back, bad weather followed and many of the Spanish vessels never made it to Scotland, others were wrecked off the Irish coast, and a total of fifty ships and 9000 men never saw Spain again.

Both sides afterwards made much of the wind's role in the battle. The English because it pleased them to think that God was after all a Protestant, and the Spanish because they preferred to acknowledge an unkind fate rather than their own incompetence. There is no question that the English fleet triumphed tactically and proved themselves better and more versatile seamen. But the fact remains that if the Armada had not arrived already weakened by sixty days of storm, and if the wind had at any time during their headlong flight up the Channel, turned to the east and given them a chance to bring their heavy guns to bear, the outcome may have been very different.

After almost 3000 years of naval engagements between ships propelled by oars, these battles were the first ever fought between machines driven entirely by the wind. They inspired new tactics that led inevitably to the expansion of a minor sea power into the greatest navy and the largest colonial empire of all.

But the winds of war did not blow their way exclusively.

If sudden contrary winds had not blown three ships with their sensitive cargoes away from New York harbour and north into the arms of a community already fired by Thomas Paine's pamphlet and heavily involved in a boycott of British goods, there would have been no Boston Tea Party in 1773 and no declaration of martial law.

The war which followed was probably inevitable anyway.

In January 1777, with George Washington up to his knees in mud along the Delaware and unable to bring up his artillery to contain the advancing troops of Charles Cornwallis, 'the wind veered to the northwest; the weather suddenly became cold; and the by-road, lately difficult for artillery, was soon frozen hard.' Washington, as a result, inflicted two swift defeats on the British at Princeton and Trenton, which turned the tide of the War of Independence and led directly to the surrender of the same General Cornwallis at the siege of Yorktown a few years later.

It is probably no accident that the names of Ares the Greek, and Mars the Roman, gods of war, derive from the same Indo-European root *mar*, 'to carry away or destroy', which produced the *maruts* – genii of the air who hustle the clouds and ride on the wings of the storm.

Or that the fortunes of all warriors in every war ever fought, have been closely linked with their commander's ability to predict, control or take advantage of, the vagaries of wind – which, said Charles Dickens, 'blows great guns indeed'.

4. THE FUTURE OF WIND

'Of all natural phenomena,' said James Frazer, 'there are, perhaps, none which civilized man feels himself more powerless to influence than the wind.'[146]

And yet we keep trying.

Starting in the third millennium BC, with the first great recorded drought, there was a movement into temperate forest areas. The migrants were pastoral and agricultural people and their initial response to the woods was to get rid of them, to create an artificial steppe or savannah with tilled land and planted crops. And they succeeded in removing enough of the natural cover to increase wind speed and change at least the local climate.

It has even been suggested that the *mistral* is 'the child of man, the result of his devastations. Under the reign of Augustus, the forests which protected Cevennes were felled in mass. A vast country, before covered with impenetrable woods was suddenly denuded, swept bare, stripped, and soon after, a scourge hitherto unknown struck terror over the land from Avignon to the Bouches du Rhone, thence to Marseilles, and then extended its ravages over the whole maritime frontier.'[322]

It is unlikely that the blustery northwest *mistral*, or any other wind, came into being totally as a result of deforestation, but the loss of tree cover undoubtedly makes a difference.

The Roman author Petronius, who wrote in the time of Nero, refers to winds on the Italian coast, but makes no mention of the *sirocco* which now makes life there quite difficult at times. Horace, on the other hand, a century earlier, goes to some length to extol the virtues of the North African coast as a land of fertility, luxury and wealth, where it was possible to ride from Tripoli to the Atlantic in the deep shade of trees.

That coast is now denuded, a place of hard brown hills. The forests where Rome collected many of the animals that performed in the Colosseum, have gone. And the *sirocco* is a fact of life.

About thirty-four per cent of the present equatorial zone is covered in tropical rain forest, but it will not be for very much longer. The Amazon Basin alone, an area the size of the face of the Moon, is being cleared at the rate of 3000 square kilometres every week. At that rate, it will be gone by the year 2000. And if it does go, we can expect local rainfall to drop by ten per cent immediately, perhaps more over a long period, and for there to be far-reaching consequences for the climate of Earth as a whole.

This change is likely to occur as the loss of forest results in a reduced absorption of solar energy, followed by reduced evaporation in a tropical zone which has become cooler, which would lead in turn to cooling in the middle and upper levels of the troposphere, producing a lower temperature gradient between the equator and the poles – and ending in a generally weaker pattern of air circulation.

The whole globe would cool, and there would be a particularly severe decrease in rainfall in all areas more than forty degrees from the equator, which are now fuelled by strong and generally reliable circumpolar westerlies. This scenario accurately reflects what climatologists believe took place following the denudation of the area that is now the Sahara Desert.[388]

There have been deliberate attempts to reverse this sort of process, to curb wind by planting trees and windbreaks. The first large scale efforts seem to have been those made in England and Scotland following the Culbin disaster, in which a violent northwesterly wind deposited over thirty metres of sand on farms and farmland near the Moray Firth in 1694.

These early planting programs were followed in the eighteenth century by others in Denmark, Norway and on the plains of Russia, and by more systematic shelterbelt experiments in recent years. We have come a little way, scientifically at least, since James Russell Lowell suggested that 'the only argument available with an east wind, is to put on your overcoat.'

But it is safe to say that we have only just begun to explore the true extent of our ability, by accident or design, to exert any real influence over the wind.

Old Winds

Early attempts to control the wind were straightforward, some-times earthy, often combative.

In the Canadian Arctic, when the wind blew for weeks on end and the Inuit began to despair of ever going out hunting again, they made a long whip out of seaweed and struck out in the direction of the gale, crying '*Taba*! it is enough.' Or, if this failed, they might build a special fire and select an old man, whose task it was to coax the demon of the wind in to the fireside to warm itself. All the men then urinated in a jar and threw this into the flames, on the assumption that no self-respecting demon would want to hang around after such an insult.[40]

At Point Barrow in Alaska, it was the women who drove the offending wind spirit toward the fire with clubs, while the men waited there with a bowl of water, to make the demon manifest in steam, and a large flat stone, to crush it to death.[352] In Greenland, one woman on her own was sufficient, provided she had been recently in labour. All she had to do was to go out into the storm, fill her lungs with air and come back in with the wind held captive.[119]

In South Africa, a Xhosa priest-diviner would lead a large number of his people up a convenient hill, and fill his mouth with a special potion which he spat right into the eye of the wind, while directing a mass demonstration of communal rage. In parts of southern India, the priest confronted a storm entirely on his own, armed with just a sword, a club and a flaming torch.[38]

The weapons selected for use against the wind, varied a great deal. In Borneo, the Kayan people drew their swords every time they heard thunder. In old Sumatra, it is said that the Batta rushed from their houses to meet a threatening wind, the rajah at their head, hacking away at the invisible foe with swords and lances. 'An old woman was observed to be specially active in defence of her home, slashing the air right and left with a long sabre.'[203]

Cornwallis Harris in the Ethiopian highlands, reported that 'no whirlwind ever sweeps across the path without being pursued by a dozen savages with drawn creeses, who stab into the centre of the dusty column in order to drive away the evil spirit that is believed to be riding on the blast.'[146] In New Guinea, the Kai people were content to tie a spear to the roof of a house in order to pierce the wind's belly; or, if absolutely necessary, threaten it with a club, saying 'If you enter my home, I'll beat your feet flat.'[359]

The Dayak adopted a more deterrent attitude, beating gongs during a storm, not so much to frighten away the wind spirit, as to apprise it of their whereabouts so that it would not inadvertently knock their houses down.[379] In the highlands of Scotland, the pragmatic approach was simply to toss a shoe, usually the left one, in the wind's general direction.[67] While the proud Cherokee faced an oncoming storm with one outstretched hand and just pushed it off in a new direction.[343]

These direct approaches had to be refined after the agricultural revolution, when it became necessary to distinguish between winds, and to petition actively for those vital to activities such as winnowing.

In old Persia they indicated what was necessary by rubbing *bad engiz*, a type of saffron, into a fine powder and tossing it into the air in golden clouds.[397] The Berbers believed that movement was attractive to a wind, and stuck a flag into a pile of waiting grain, or suspended a beetle from the branch of a tree so that it would wave its legs, beckoning to a passing breeze. And to be certain that no impediment stood in a wind's way, they would take care even to brush out every knot in their hair.[277] The Igliwa in Morocco still build a pile of stones to the west of the winnowing floor, or tie a ribbon to the end of a tall waving reed. And if things get really desperate, they gather in a group and whistle, or see who can tell the most outrageous lie about an already notorious liar.[516]

Sailors are at the direct mercy of the wind, and have always had to be especially careful, exercising a nice judgment in asking for just enough wind for the next voyage.

In parts of New Guinea, they struck particular stones with a stick, very conscious of the fact that to do so too hard could bring a typhoon.[342] At Murray Island in the Torres Strait, the correct procedure was to point coconut leaves at a particularly sensitive rock on the shore.[188] On Vancouver Island, there used to be a number of large wind stones, and all a seafarer had to do was move the right one, just enough.[41] In the Hebrides, there was a blue stone on the altar of Fladda's chapel that was always moist – a little extra water poured over it was regarded as worthwhile wind insurance.[89] On Gigha Island in Argyllshire, it was stones that covered water in the lucky well of Beathag. A prudent mariner took good care to remove the cairn, scoop up water in a wooden bowl and throw it three times in the direction of the desired wind before covering the well once again.[447]

Some people seem to have more aptitude for communion with the wind than others.

In New Guinea, when the wind blew hard through the territory of the Bogadjim on the north coast, they looked across to an island just visible on the horizon and said, 'Those Bibilis are at it again, puffing out their cheeks.'[191] On Kei Island in Indonesia, the women ashore kept a constant vigil and, if necessary, danced in a ring, swaying and chanting, to secure the safety of their men at sea.[383] In Australian Gippsland, there used to be a wizard known as 'Great Westwind' who would bind a wind to his will by tying a string around his head; and amongst the Kwakiutl of Canada, it was said that any twin could control winds with a simple movement of the hands.[146] But a Haida of the Queen Charlotte Islands had to shoot a raven, sweep it over the sea surface four times in the desired direction and then set the unfortunate bird up in a sitting position on the shore, facing into the wind with its beak propped open.[99]

The island of Sena off ancient Gaul was said to have a priestess – *la baragouin*, who could produce winds to order. Her successors later came to ply a similar trade on the Isle of Man. Sir Walter Scott told of meeting a wind witch at Kirkwall in the Orkneys in the early nineteenth century,

> We clomb, by steep and dirty lanes, an eminence rising above the town, and commanding a fine view. An old hag lives in a wretched cabin on this height, and subsists by selling winds. Each captain of a merchantman, between jest and earnest, gives the old woman sixpence, and she boils her kettle to procure a favourable gale.[290]

There were other old women in the Shetlands who sold their winds in the shape of knotted handkerchiefs.[404] And right through into this century, at Stonehaven harbour in northeast Scotland, an old lady sold wind charms consisting of a red thread tied with three knots – the first to be untied for light wind, the second for strong wind, and the third never to be disturbed on any account.[312]

But when it came to real control, there was nothing to touch the Lapps and Finns.

They also used knots in a magical cord, but it was sometimes necessary to shop around, because each wind vendor had command only over the particular wind that happened to be blowing at their moment of birth.[165]

Izaak Walton, bewailing the lack of the precise breeze needed to blow his bait into a fish's mouth, once said, 'I wish I were in Lapland, to buy a good wind of one of the honest witches, that sell so many winds there and so cheap.'[504]

And, more recently, Richard Dana revived the mystique –

Sailors sometimes see a strange sail heave in sight astern and, though they are beating into the wind, overhaul them hand over hand. On she comes with a cloud of canvas, all her studding sails out, right in the teeth of the wind, forging her way through the foaming billows, dashing back the spray in sheets from her cutwater, every sail swollen to bursting, every rope strained to cracking. Then they know that she hails from Finland.[90]

In some communities it was not the maker of winds, but the keeper of calms, the one who held winds captive rather than turning them loose, who was seen to be the most valuable citizen. On the Banks Islands of Melanesia, the favoured technique was to fill a giant clam shell with earth, stand a special oblong wind stone upright in the soil, cover it with powdered red ochre, and then surround the whole affair with a stockade, a fence of sticks with creeper twined about it. Only then was that wind considered to be safely bound.[84]

The Baganda in East Africa used a special *nantaba* fetish, which consisted of a gourd that had been held over the stump of a freshly-cut wind tree to capture its spirit. This was then stitched into a goatskin decorated with cowrie shells and kept in a temple guarded by one of the king's wives. When other winds blew, drums were beaten to confuse the captive and prevent it from joining its fellows. As long as it remained where it was, the people had access to a powerful spirit capable of ensuring fertility.[406]

Direct worship of a particular wind at its own altar, and the existence of wind spirits and wind gods, led inevitably to an association of wind with prayer.

Where worship involved the use of repetitive chants, the wind participated by animating flags and banners printed with the prayers, so that each eddy of air would carry an impress of its own orison. In Tibet, where prayer became mechanised in the form of sacred writing rotated in wheels, the air was persuaded to take over the whole ceremony by the invention of a bladed propellor that caught the wind and turned the wheels in endless litany.[519]

These word wheels were also the first windmills.

Milling was always done between two stones. There are Egyptian statuettes from the Fifth Dynasty in 2500 BC that show women kneeling astride stone *querns* in the shapes of saddles, grinding their grain with smaller hand-held stones. Things are still done this way in much of Africa, but by Greek times an ass or mule had been harnessed to a shaft that turned the heavy upper stone. And by 85 BC, there was at least one water-powered mill in Thessalonika.

Before the advent of Christianity, natural forces such as wind and water were viewed with some awe and there was a strong resistance to harnessing gods in the service of humans. Both early Greeks and Romans knew and used complex gearing devices and had hydraulic machinery for sacred functions such as opening temple doors and making music. Hero of Alexandria, in the third century BC, built an organ which was powered by a small wind rotor with four sails.[138] But both civilizations were slow to apply these ideas to the design of mills and pumps. There were political considerations. The work of grinding grain and hauling water was done manually, by slaves, and a familiar fear was expressed that the state would be endangered if this great force were to be released from their labours.[396]

But by the fourth century, under the Christian emperor Constantine, milling by water was an accepted practice, and the old aqueduct of Trajan, bringing water to Rome from Lake Sabbatia, was set aside for this use. From there, the practice spread rapidly throughout the Empire, with the construction of mills for Roman bread in Gaul and Britain. But where water was less plentiful, another solution had to be found.

No one knows who first conceived the simple and extraordinary notion of harnessing the wind to grind corn, but it seems to have taken place in Persia sometime around the sixth or seventh century. There are reports of windmills pumping water in the province of Segistan in 134 BC, but it was probably not until some centuries later that square towers, now called *panemones* – literally 'things of all winds' – were built on the windy plains of eastern Iran and

Afghanistan with slots that opened on to a bladed wheel whose shaft led down to a grindstone.[238] The design seems to represent a happy fusion of eastern and western ideas. The milling mechanism was copied from the Roman waterwheel, but the wind turbine is pure Tibetan prayer wheel.[28]

From Persia, the principle travelled east to China, where it was discovered that if the sails were 'feathered', set at an angle on the shaft, it was not necessary to have an enclosed wheel with slots channelling the air directly on to each blade. And as the invention moved west again, this free wheel seems to have been gradually tilted more and more by the use of gears until it eventually stood upright in the manner of a classic 'Dutch' windmill.

The Crusaders were apparently responsible for bringing the idea back to Europe, because before 1096 there were no such mills in the north, but by the end of the following century, they were common enough to be the subject of a new papal tax.[70]

This ready acceptance of the new wind engines is characteristic of the Middle Ages, whose glory is usually seen to rest only in its cathedrals and scholars, but which was equally notable for having built the first complex civilisation which relied on non-human power, rather than that provided by sweating slaves. This change of heart was a direct result of an implicit theological assumption of the worth of humanity, but it did not of course prevent the church from cashing in.

The first actual record of a windmill in Europe is an oblique one from 1180, in which an abbey in Normandy receives the gift of a piece of land at Montmartin en Graine, which is described as being 'near a windmill'.[101] The first English mill seems to have been one built at Bury St. Edmunds in 1191, but demolished again almost immediately by order of the lord of the manor, an abbot who held the local monopoly on grinding. There was never an act of parliament that made this practice legal, but by custom known as the Milling Soke – hence the expression to be 'soked' or 'soaked' by taxes – all tenant farmers were forced to grind their grain at the lord's mill, giving him one sixteenth of the proceeds.[502]

The oldest surviving windmill is a seventeenth-century construction at Bourn in Cambridgeshire, so our knowledge of developments in the four centuries before that, comes entirely from medieval art. From illustrations in psalters and from church carvings. There is a boss in the cloisters of Norwich Cathedral from about 1325, showing a woman riding toward a windmill with a sack of corn balanced on her back to relieve the horse of this additional weight. The same joke is repeated in brass at St. Margaret's Church in Kings Lynn, and on a bench end at Lyng in Somerset.

These illustrations show that the early mills were all 'postmills', built to pivot in their entirety around a central column, so that the whole structure could be turned to meet the wind. This daring piece of carpentry, held together by mortice and tenon joints, hardwood trenails and sheer craftsmanship, was eventually replaced in the fourteenth century by more substantial 'towermills', in which only the cap and sails moved round. The oldest known illustrations of both types are in the Pierpont Morgan Library in New York.

The first mills were rigged with reed-covered blades, but by the Middle Ages they all used canvas sails tied across the stocks in a variety of ways. As the wind speed increased, each sail could be reduced in area by reefing it into shapes known as 'dagger points' or 'sword points', which left canvas exposed only on the ends of the arms.

In 1772 Andrew Meikle, a Scottish millwright, invented the 'springsail', which consisted of hinged wooden shutters connected to a bar running the length of the stock, which could be opened and closed together like venetian blinds. In the earliest versions, the mill had to be stopped to make such adjustments, but later models were equipped with springs that opened under tension to spill the air and slow the sail if the wind became too strong. And finally, in 1807, William Cubbitt designed a patent mechanism which linked all the sailboards to a series of levers and chains that made it possible for them to be adjusted simultaneously without stopping the mill or going outdoors.[28]

When he did stop his mill, a good miller took care to leave it in a position likely to sustain the least strain, but in time a whole language grew up around particular settings. If the four arms of a classic mill pointed straight up and down and across, it meant the miller was home, or would be back soon. If they pointed diagonally, it usually meant that the mill was closed for the day. Other settings indicated sickness, celebration, or simply the miller's need for the services of a good hired hand.

Because mills were nearly always built on hill tops in full view of the surrounding countryside, they were inevitably used, by day or night, to send signals only the locals would understand – to give advance warning of the arrival of enemies or taxmen. And when a miller died, 'all the twenty boards in the arms of the mill were taken out and the mill stood motionless for a given time, as if in grief over the loss of an owner'.[134] A miller's wife rated nineteen boards, a child thirteen, his parents eleven, while the child of a cousin was no more than a one-board bereavement.

At the height of the Wind Age, there were probably 100,000 windmills working across Europe, scattered like a vast fleet of ships in sail across the countryside. Only some of these were actually milling – grinding grain or extracting oil. The rest were pumping water or driving machinery for stamping, cutting and sawing, but they were all called 'mills'. They were manned by millers, by an army of men who must have become as wind-sensitive as Polynesians, alert for the first sign of a breeze, watching the sails of other mills in the distance to see which way the air was moving, careful to adjust their own to the strength and direction of the blow, conscious of any slight change in the rhythm of sail or stone.

When off watch, waiting through a calm, many had their own lands to work, but they could always be distinguished from other men by the blue marks on the backs of their left hands caused by hundreds of tiny fragments of steel embedded in the skin from tools used for regular roughening or 'dressing' of the stones.

Today there may be 3000 mills in all left standing, some converted into homes, a few lovingly restored along with all their machinery, but most in urgent need of attention. And the army who worked them have long dispersed and are remembered only in the common names of Miller, Mills, Milne and Milner – or their equivalents in other tongues.

Germany once had 18,000 windmills; Holland and England 10,000 each; Denmark about 7000; Portugal and Spain another 7000 between them; France 6000; and a scattering across the lowlands of Poland, Russia and the Balkans – and those parts of the United States that were settled by immigrants from these countries.

For a long time windmills were the only complex machines in daily use in most places. Each village had a mill before its church tower could boast a clock. They were machines that worked well, serving their communities, but always only at the will of the wind, which was in the end just not consistent enough to compete with the new touch-of-a-button heat engines.

Today only 5 per cent of the mills survive in any form, and a fraction of these heroic old structures with their expressive arms and their great cogs and gears, are in the care of dedicated windmill societies who see them as something more than just museum pieces.

123

They are more. They are, with the possible exception of sailing ships, the nearest we have yet come to perpetual motion. They are the only ethically satisfying machines we have ever devised, and the art and craft that went into building and using them ought to be rigorously preserved wherever it can be found.

It may yet be of service to us.

New Winds

Wind is moving air, with mass and energy.

To tap it as a resource, all we need to do is slow a part of it down. The energy lost by the wind is then gained, more or less efficiently, by the wind-device that stood in its way. In the case of sail, the energy is turned into movement. And when the sails are on a mill, through movement into useful work.

Until the development of the steam engine, windmills ranked second only to wood as an energy source, and continued even after the Industrial Revolution to provide a large proportion of our energy needs. This proportion rose dramatically in 1850 with the development by Daniel Halliday of the famous multibladed American farm windmill. Six million of these ubiquitous machines clanked and spun, pumping water up from below the ground and making life possible in many of the drier parts of the United States. They are still the only things that stand between ranchers and ruin in many areas of Mexico, Brazil, Argentina, South Africa and Australia.

And it was these homely devices, that seem to have inspired the first serious attempts to transform wind and movement into a form of energy that could be used elsewhere, or stored for use at another time.

There are no waterfalls or coal deposits in Denmark, but by

1890, there were 7000 windmills in operation across the nation, supplying a quarter of the country's direct power needs. A far-sighted government decided to make themselves even less dependent on foreign energy supplies by developing large-scale wind-powered electric generators, and by the turn of the century seventy-two huge machines, the first of their kind, had been built.

These consisted of towers twenty-five metres tall, supporting rotors with four blades twenty-two metres in diameter, that were connected by drive-train to generators on the ground. Power from these machines, and their successors, continued to be fed into the national grid until 1968, when the last one was closed down because the cost of electricity supplied by local wind was then twice that of hydroelectric energy imported from Sweden.[123]

In the United States, early development was of a more private nature, confined largely to the manufacture and sale of 'wind chargers' – two or three bladed propellers about four metres in diameter that delivered just enough for a farmhouse to power its radio, one or two lights for reading, and perhaps an ice-maker or a wringer washing machine. Half a million of these machines were scattered across the midwest until the Rural Electrification Administration finally succeeded in 1960 in supplying most country homes with powerlines.[164]

Meanwhile in Europe, a number of inventors were trying to find a more efficient way of tapping the energy in the air.

The most eye-catching solutions were those based on the 'Magnus Effect' – the discovery by a German physicist that a cylinder rotating in an airstream produces a lateral thrust. Anton Flettner in Germany built two such cylinders on the deck of a vessel he called a 'rotor ship' and succeeded in crossing the Atlantic with it in 1925. Another engineer tried to harness the effect by putting rotating cylinders twenty-seven metres high and eight metres wide on rail cars, siphoning off the power that was generated by the wheels of this strange train as it ran round and round a one-kilometre circular track.[138]

The Finnish inventor Sigurd Savonius in 1923 sliced his rotating cylinder in half from top to bottom and pulled the two pieces apart slightly, producing something that looked a little like a cup anemometer. His principle is now embodied in a generation of 'S' shaped wind rotors which have been very successful in ventilators.[422]

In France, the power engineer Georges-Jean-Marie Darrieus produced a range of wonderfully elegant airfoil designs, the best known of which looks like an old-fashioned egg-beater. These vertical rotors continue to show promise in performance, but are handicapped by the fact that most need a starter motor to get them going.[372]

On balance, the most effective machines seem still to be those based on the classic 'Dutch' propeller, so it was to this design that the United States returned for its first major windpower project in 1939.

Palmer Putnam, the engineer in charge, decided to build the world's largest rotor, two blades with a rotary diameter of fifty-three metres, and to put it on a windswept hill known as Grandpa's Knob in Vermont. The device was completed in the autumn of 1941 and began almost immediately to feed 1500 kilowatts into the wartime grid. But in March 1945 one of the two blades tore itself loose.

> The erection foreman, who was a powerful man, was on duty aloft. Suddenly he found himself on his face on the floor, jammed against one wall of the control room. He got to his knees and was straightening up to start for the control panel, when he was again thrown to the floor. He collected himself, got off the floor, hurled his solid 225 pounds over the rotating mainshaft, reached the controls, and brought the unit to a full stop in about ten seconds by rapidly feathering what was found to be the remaining blade of the turbine.

The other eight-ton stainless steel blade was found crumpled against the hillside 240 metres away, and the project was abandoned.[258]

Since then, the Americans have tried again with another large blade rotor in Ohio, one in New Mexico, and are opening the world's largest wind generator – with a 100-metre turbine – in Hawaii in 1985. The Russians have built several big turbines near the Black Sea, including one that turns to meet the wind like an old postmill. The French have tested a number of units near Paris and at St. Rémy in the south. The Germans are producing some advanced variable-pitch propellers on the banks of the Elbe. The Danes have brought at least one of their old machines back into use. And the British have put a turbine blade with articulated tips into their power-generating network at a station in Wales, and are planning a new generation of vertical axis machines that have the advantage of not having to turn to meet the wind.

The energy crisis of the 1970s has started everyone thinking again about alternatives to fossil fuels, but all windpower projects seem to falter on the questions of constancy and cost. Approximately 2 per cent of all radiant energy that falls as sunlight on Earth is converted into the kinetic energy of wind. There is enough power in circulation in the wind at any time, day or night, to provide for all the world's energy needs, but the problem is that it is so thinly distributed. No country, no town even, can afford to rely only on seasonal shifts and changes, and nobody has yet produced a cheap and effective way of storing the excess energy which accumulates when winds do blow.[445]

There is no question that wind-generation can and should provide at least a part of all domestic energy needs, and that small scale windpower systems have a big future, particularly in rural areas and developing countries. Modified versions of the delightful wood and canvas windsails that decorate the plains of Lassithi in eastern Crete, can be built anywhere out of local materials and are already pumping water in Ethiopia and Gambia. But we are still a long way from turning wind energy into a major source of urban or industrial power.[497]

The best hope at the moment seems to lie in 'wind farms'. In concentrations of state-of-the-art generators at specially windy places, where each giant turbine, with blades as big as the wings on a jumbo jet, can be expected to produce enough electricity to supply 1000 average homes. At this pitch of performance, which is high by present standards, it will still take 1000 such monsters to equal the output of a single nuclear reactor. And, in order not to steal each other's wind, these huge propellers would have to be spread out over an area of 750 square kilometres. A solution may be to locate such wind farms at sea, on barges moored to the continental shelf in line with the prevailing winds and not too far from population centres.[137]

A better solution might be to combine wind turbines on such platforms with a second system designed to take advantage of the rise and fall of wind-generated waves, which continue to roll in long after the breeze drops. In 1920, the French inventor Bouchaux Praceique designed and built a turbine driven by wave motion and used it to light his house by the sea. Several such systems are now being tested in which the rising waves compress air in a tank and turn a turbine. Some have a two-way pulse system, so that when the wave falls, air pressure falls with it, opening a valve and pulling new air into the tank to turn the turbine once again.[330]

But perhaps the most exciting windpower idea of all, goes

directly to the source and exploits the jet stream itself. A combination of balloons and kites could lift an aero-generator system to the necessary altitude of 10,000 metres, and the speed and regularity of the stream would provide a constant lifting force and a steady power output. The anchor lines could of course act simultaneously as conductor cables. At the moment this is science fiction, but the technology is not beyond us. It is a concept of such breathtaking simplicity and purity that it might well work. It is probably the only realistic chance we have of harnessing the wind in a way that meets our major power needs.

That torrent of thin air that howls perpetually around Earth just below the stratosphere is, like all winds, a by-product of solar radiation. So, naturally, it has occurred to someone to produce smaller, more manageable, winds in the same way. Experiments are now in progress in several countries to create a captive artificial wind by heating air in large plastic greenhouses, and then funnelling and accelerating this rising draft past delicately balanced bicycle-wheel turbines.

Some artificial winds now hurtle past test beds in special tunnels at speeds even greater than the jet stream. At the Langley Research Centre in Virginia, NASA has built a transonic wind generator that uses 135,000 horsepower fans to produce a frigid whirlwind of nitrogen at $-160°$ Centigrade that howls past scale models of supersonic aircraft at a velocity of 1470 kilometres an hour. By firing models explosively into such a stream, it is possible even to simulate winds blowing at the inconceivable speed of 230,000 kilometres an hour.[416]

These are captive winds, imprisoned behind bulwarks of steel, but at the University of Clermont in France, they have been experimenting with feral whirlwinds. To do this, they have constructed a diabolical machine they call a 'meteotron', which consists of two huge pumps that drive a ton of fuel a minute to 100 gas burners. When these are spread out over 15,000 square metres of the Lannemezan plateau in the Pyrenees and fired simultaneously, they produce a towering column of thick black smoke that generates a cumulus cloud. And if the winds over the plateau are just right, the base of the column begins to turn and within fifteen minutes *voilà*! – an artificial whirlwind exists, sucking dust out of a growing hole in the ground.

On one occasion, when the air overhead was particularly unstable, the group succeeded in spawning a perfect twisting little column with a tube a metre in diameter and 600 metres high, that persisted for four minutes after the burners were extinguished. And they have announced plans to 'provoke the formation of artificial tornadoes by using near the burners a kind of large churn to initiate a gyratory motion'.[102] The reactions of their neighbours are not on record.

On the other side of the Atlantic, an equally wild project led in 1947 to dropping forty kilograms of dry ice into the eye of a hurricane 'just to see what happened'. The storm at the time was 600 kilometres off the Florida coast and heading out to sea. What happened was that 700 square kilometres of stratus cloud on the wall of the eye were converted to snow, the hurricane stalled, changed course abruptly and forty-three hours later wrecked a large part of Georgia.[446]

The results may have been disastrous, but the response was so dramatic that a more carefully controlled study was launched by the United States Weather Bureau in 1960 under the name of Project Stormfury. They had to wait three years for a suitable storm to practise on, but finally found one called Beulah who, in August 1963, was far enough out in the open Atlantic to present no threat to land no matter which way she might decide to turn. Two aircraft equipped with silver iodide generators flew into the eye of the hurricane and deposited a dense sheet of crystals right across the wall, while ten other planes made direct observations and took regular measurements of wind, pressure, temperature, humidity and cloud structure.

The effect of the seed crystals is to raise the temperature of the wall and therefore lower its pressure and, theoretically, reduce the gradient between wall and eye that produces hurricane-force winds. Within minutes of seeding, supercooled cumulus clouds exploded over 5000 metres up through the storm front and Beulah's eye broke up. There was an overall reduction in wind speed of 14 per cent and the storm centre expanded considerably, but it then drew in and reformed fifteen kilometres east of its original track. The experiment had succeeded at least in giving the storm pause for thought, and deflecting it in a predictable direction.[154]

The next lady accosted by Project Stormfury was Debbie, in 1969, who was waylaid and seeded five times within eight hours. The result was that maximum wind speeds decreased by 30 per cent, from 180 kilometres to 125 kilometres per hour. By displacing the walls of the eye, pushing them out and redistributing their energy, it does seem possible to lower overall wind speeds in a tropical storm. But within twenty-four hours of the first ambush, Debbie was back on her beat, perhaps even a little more forceful than before, and further seeding succeeded only in producing a temporary reduction in wind speeds of 15 per cent.[155]

Project Stormfury has now stalled for lack of further suitably remote hurricanes, but seemed anyway to be reaching a consensus that, while it was possible to modify a hurricane in some small way, the process was expensive and the effects too shortlived to be really convincing or useful.

A number of other countries have since been involved in similar deliberate attempts to modify wind and weather and, following the signing of the Law of the Sea in 1982, concern is now being expressed over the need for a similar international mechanism for atmospheric management, for a Law of the Air.

There is good reason to be concerned about proposals for drastic geographical engineering – such as the creation of 'thermal mountains' by painting or staining large areas of the desert black in order to increase convection and the possibility of rainfall. Or the dusting of ice caps in Greenland and the Arctic in order to increase heat absorption and melt enough ice to raise the ocean level. Or blocking the Bering Straits as a first step in draining the Arctic Ocean. Or putting a carpet of metallic needles into orbit round Earth to contain radiant energy that would otherwise be lost to space. All these things are possible and any one of them could start a chain reaction over which we have no possible control.

It may be that we will one day be able to 'fine tune' wind and weather in order to offset general warming or cooling trends that are likely to prove disruptive. But we are a long way from having anything like the necessary knowledge or expertise.

Right now any attempt at interference is a little like leaving the repair of a priceless antique clock to a five-year-old with a hammer.

Free Winds

There is no such thing as still air.

All air is composed of molecules in constant and rapid motion at speeds approaching those of the fastest jet aircraft. Any object in air is bombarded by millions of such molecules, and if neither the object nor the mass of air is moving, this bombardment is equal on all sides. But as soon as an object moves relative to the air, then wind blows and all the rules change.

The effect is most evident if the object is an airfoil – if it has the property of splitting up an air mass unequally, as a wing or a sail can do. All such objects are shaped to persuade the air to flow more quickly on one side and slow it down on the other, with the result that air is less dense, and contains fewer molecules, on the side with the fastest flow. And the cumulative effect of such an unbalanced bombardment by unequal numbers of molecules, is that force is applied to the inside curve, the side with the slowest densest flow, providing lateral thrust to a sailboat or uplift to a wing.

This ability to create and sustain unequal density in the air is the secret of all free flight on the wind.

The earliest artificial airfoil was probably the *boomerang*. We know it by the name used in South Australia, but the same principle appeared in use in South India, Egypt, West Africa, and amongst the Hopi in North America. Traditionally it was made of hardwood, often *Acacia*, with two arms at an angle of anything between 70 and 120 degrees. One arm was often slightly longer than the other, and the tips were always bent a little upwards, and at angles in opposite directions to each other. But the most important and invariable characteristic of all designs, is that the lower surface was flat, while the upper was curved to provide a difference in flow.

Boomerangs were used in hunting, mainly for birds, who seem to see them well enough, but fail to recognise one as a threat until it is too late. They are thrown by holding near the tip, flat side towards the thrower, with a strong overarm sweep, flicking the wrist down at the last moment to produce rapid spin. Most travel forward, spinning like a wheel for all of their effective range of about fifty metres, and then, if nothing is struck, flatten out, swing to the left, rise high in the air and volplane back. During all of the return phase, the boomerang acts and behaves like a flying saucer or 'Frisbee' with air under increased pressure trapped beneath the spinning disc.

The next development, historically, seems to have been a tethered airfoil – the *kite* – which dates back to the first production of the necessary restraining silk thread in China, perhaps as early as 2600 BC.

The earliest actual reference is in a text from the fourth century BC, which credits the engineer Kungshu Phan with the invention of a 'wooden bird' that flew for three days without descending. The Greek mathematician, Archytas of Tarentum, is said to have independently developed his own model of a flying bird in the same century.

By the second century BC, kite design was so well advanced in China that large models were being used to airlift food at night into a besieged city. And by the time the invention reached Japan with Buddhist missionaries in the seventh century AD, it had become necessary, in order to forestall aerial invasion, to forbid the construction of any kite large enough to carry a man. Legend has it that the samurai warrior Tametomo, forced into exile on the island of Hachijo with his son, built a kite there that carried the boy safely to freedom on the mainland.[378]

When Marco Polo reached China in 1282, kites were being routinely used to carry bricks and tiles to workmen on pagoda towers, and he gives a graphic description of another less pleasant practice that followed the launching of a new ship.

> The men of the ship find some fool or drunkard and bind him on a hurdle, since no one in his right mind or with his wits about him would expose himself to that peril. And this is done when a strong wind prevails . . . which lifts it and carries it up into the sky, while the men hold on by a long rope . . . and if the hurdle going straight up makes for the sky, they say that the ship will have a quick and prosperous voyage. But if the hurdle has not been able to go up, that ship stays in port that year.[384]

From China, kiting spread to Burma, Thailand, India, Korea and to the islands of the Pacific, in each area evolving its own design and significance. In monsoon lands, kites were flown when the winds blew, to encourage them to keep on doing so. In Korea, a kite inscribed with the name of a newborn son was released and carried away with it any evil spirits that might have been attracted to the birth. In the Solomon Islands, kites of *Pandanus* palm leaf were used to carry fishing lines out far beyond the reef. The Samoans used them for pulling canoes. And in India, as more recently in Nagasaki, a whole subculture grew up around the sport of kite-fighting between rivals who try to cut each other's lines with the aid of ground glass glued to their own.

Kites seem to have appeared in Europe only in the sixteenth century, probably with the return of the first traders from the East Indies, and by the eighteenth century, provided an enormously popular pastime for children. The first recorded adult interest is an experiment performed by Alexander Wilson of Camlachie in Scotland, who in 1749 used a train of kites to carry several thermometers up to a height of almost 1000 metres. It was three years later that Benjamin Franklin flew his famous, and highly dangerous, kite and key in a storm.

Franklin later used a kite to propel himself, floating on his back, across a pond. This seems to have started a craze for kite propulsion that led in the early nineteenth century to the *char-volant*, a light carriage pulled along at thirty kilometres an hour; to a cable strung by kites across Niagara Falls; and to a series of devices for rescue at sea designed to take advantage of the fact that a kite could be flown toward a lee shore by the same wind that put a ship on its rocks. But despite the early Chinese and Japanese success, it was only toward the end of the nineteenth century that combinations of kites, including one called 'The Levitator' designed by Boy Scout founder Baden-Powell, managed to lift a western man.

Most of the load-carrying kites were 'man-lifters', with the passenger suspended on a long line some distance below the kite, but between 1894 and 1896 Otto Lilienthal in Germany and Charles Lamson in the United States, flew proper 'man-carriers' – biplane designs which look remarkably like the powered machine that was to take briefly to the air at Kitty Hawk some seven years later. The Wright brothers in fact began their career in 1900 with a tethered glider of their own, which Orville, like Lamson and Lilienthal before him, controlled by shifting his weight from side to side.

We tend to forget that the glider predates the powered airplane by nearly half a century. In 1857, Jean-Marie Le Bris, a French sailing captain, built a machine based on his observations of albatross at sea and had it launched from a horse-drawn cart. The horse bolted and the towline broke, wrapping itself around the unfortunate coachman and destroying the artificial bird, but not before Le Bris, the plane and the coachman, were all briefly airborne.

In the 1890s, Otto Lilienthal jumped off a man-made hill near

Berlin with a series of fixed wing-gliders based on his study of seagulls. On one of them at least he succeeded in floating for over 230 metres, before crashing to his death in 1896. His work was carried on in the United States by Octave Chanute, who designed the first successful soaring biplane and contributed materially to the Wright brothers' designs. After their successful powered flight, which lasted just twelve seconds, Orville Wright went on to establish the world record for soaring without power with a controlled glide of nine minutes and forty-five seconds in 1911.

By the terms of the Treaty of Versailles which ended World War One, Germans were denied the use of powered aircraft. So they turned instead to gliders and in 1921 at a meeting in the Rhön Mountains, Wolfgang Klemperer 'discovered' clear-air thermals and made a flight of thirteen minutes and three seconds. Within a year, flights of two and three hours became common and today the best production sailplanes in the world, which in the right conditions can stay aloft almost indefinitely, are still either German or based on German designs.

The popularity of soaring, as opposed to simple gliding down off a slope, has led to a great deal of new meteorological knowledge about the way in which air behaves in clouds or along weather fronts, and to the discovery of standing waves such as the famous 'Bishop Wave' in the lee of the Sierra Nevada in California.

The debt that powered flight owes to gliders and tethered kites is reflected in the fact that, for a long time, all the early free-flying aircraft were fondly known as 'kites', but the two streams of development soon diverged, with the man-carriers of the last century almost forgotten until the space program began. It was Francis Rogallo, an aeronautical engineer with NASA, who revived an interest in hang-gliding and produced the present boom with his invention of the soft inflatable kite and later, the delta wing-glider for gentle, accurate, controlled landing of anything from a space capsule to a solitary enthusiast.

Rogallo's designs are pure and clean, reflecting his belief that form should follow wind flow rather than fighting against it, and have directly inspired the most recent and the most impressive innovation in the history of the airfoil – fellow American Domina Jalbert's invention of the parafoil in 1963.

This is a lifting surface made entirely of fabric, which maintains its form by the action of wind entering openings on its leading edge and setting up an internal pressure. It is a combined balloon, parachute and kite – the lightest, most efficient and economical lifting surface ever devised, a realisation at last of the proverbial 'sky hook'.

It is the parafoil which will make it possible, if anything can, for us to tap the jet stream. Its stability and load-bearing capacity are extraordinary. When used as a parachute, it gives ejecting pilots the ability to fly up to five times as far as they fall, and the control necessary to perform elaborate evasive manoeuvres and pick the best landing spot on the way down. And its potential as a sail is only just beginning to be explored.

At the University of Leiden in Holland, they are working on the 'bottomless yacht', the ultimate sailboat which is all sail and no hull. It is steered by someone suspended from a complex parafoil and connected to the sea surface only by an underwater kite, a paravane, which provides just enough drag to keep the wing aloft. With maximum efficiency through both air and water, performance never dreamed of by naval architects, suddenly becomes possible, and costs no more than the price of putting the craft into the best available wind.

Meanwhile, without benefit of the high mathematics that seems to be necessary to get the best out of low technology, more and more people are finding that wind can be fun.

Body surfers and boardsurfers enjoy the by-products of the wind; while windsurfers and sailors go directly for the breezes. Skate-sailers, iceboaters and land yachts pick up on the energy in moving air over land; and skysurfers, hang-gliders, kiters, parachuters, sailplaners and balloonists take off wherever and whenever the wind allows. Each approach has its devotees and its centres of activity, but all share in a conscious awareness of wind which is something new for city dwellers. Even if they come from centrally heated homes, and travel to and fro in air-conditioned vehicles, everyone involved in a windgame has taken a huge step beyond urban confines, a move away from the alienation and monotony of climate control, which deadens the natural senses.

To live, even for a few off-duty hours, at the pleasure of the wind, is to recapture the thrill of being a part of something larger.

Like a prodigal child acknowledging the parentage of Gaia, and being made very welcome.

PART THREE
Wind and Life

Without wind, most of Earth would be uninhabitable.

This is obvious, but there is an effect of the wind on life which is less evident and more exciting.

It has to do with dispersal and begins in the northern spring of 1918. Woodrow Wilson had just proposed the establishment of a League of Nations, but long before the Treaty of Versailles could be signed, the world was united in another and more fundamental way. Half the human population and a large number of other species, were involved in a pandemic of influenza that spread in a strange way and when it stopped, left more than thirty million people dead.

There were local areas in which illness progressed slowly as though it were being passed from hand to hand, but it also appeared simultaneously, apparently out of thin air, in widely separated parts of the world. It was detected in Boston and Bombay on the same day, but took three weeks to travel from Boston to New York, despite the volume of traffic between the two American cities. And it needed four weeks to get from Chicago to Joliet, just sixty kilometres away.

When the onslaught ended a year later, influenza had killed more people than the World War. In the Punjab, 'streets were strewn with the dead bodies of the victims, and at railway stations trains had to be continually cleared of dead or dying passengers'.[247] Towns and even cities were wiped out and not even the most remote places were spared. Over 1000 Inuit died in the winter of 1918 in Alaskan outposts accessible only with difficulty by dogsled; and Kodiak, off the western coast, was struck despite the fact that all connections with the island had been severed by severe winter conditions. Australia, on the other hand, though visited almost daily by ships, many of whom reported outbreaks amongst their passengers and crew en route, had no local cases of influenza until February of 1919.

Influenza has been known since Hippocrates described something like it in 412 BC. The name comes from the Latin *influentia*, meaning 'influence', because whenever it appears it does so suddenly and disastrously, as though it were under remote control – possibly, it was thought, from some astrological source. The first well-documented pandemic took place in 1580, at a time of great religious discord. It started in Asia and spread rapidly through Europe, augmenting the priestly massacres with thousands of extra deaths every day. Then, just as suddenly as it had started, it stopped, 'as if it had been prohibited'.[35]

There were further pandemics in 1732, 1781, 1800, 1830, 1847, 1857 and 1889. And since the great outbreak of 1918, there have been others in 1933 (a case of the 'English sweats'), 1946 (a homegrown Australian version), 1957 ('Asian flu'), 1968 ('Hong Kong flu'), and 1976 (the 'New Jersey bug'). Each followed roughly the same pattern, moving rapidly and randomly out from a centre of origin or recognition, flaring to a peak and then disappearing as quickly as it came. And each time pandemic influenza appeared, it did as a result of a new virus, one never before recorded, for which there was no natural immunity.

Classic disease theory requires that there be a reservoir for such pathogens, in either human or animal populations. And that, once an illness breaks out of this natural store, it spreads by individual animal-to-person or person-to-person contact. But the problem with each new influenza virus is that it attacks humans, pigs, horses, chickens, ducks, donkeys, sheep, ferrets, baboons, bison, elk and wild birds with almost equal severity. And that it spreads in ways which ignore the normal patterns of animal migration and the technology of human travel.

When maps of the onset of an influenza virus are produced, it is possible to draw flow lines representing the spread of a pandemic in space and time. The results are either hopelessly complex, with arrows going in all directions, apparently at random, leaving inexplicable gaps; or they are ridiculously simple, showing the virus travelling across a nation state by state, county by county, as though air and bus and train routes did not exist.[221]

There is only one template that can be laid on these designs to produce a reasonable fit, only one pattern that makes sense of both the general simplicity and the particular complexities. And that is a chart of all the world's winds.

In the broad sweep of jet streams and zonal wind systems, and in the fine detail of storms and local turbulence, is everything necessary to make sense of the patchy global distribution of virus infection. Influenza has to be windborne.

But where does it come from?

To be influenced by both stratospheric streams and ground level eddies, viruses have to be either boosted from the troposphere, or else fall down through the two levels from a point perhaps higher than either of them.

Gravity, and our own difficulty in shrugging it off sufficiently to reach escape velocity, make the first alternative seem unlikely, except perhaps during volcanic eruptions. And almost all of us have conceptual difficulties with the second alternative. Space as a place seems inhospitable and remote. Astronomer Sir Fred Hoyle, however, has a nice way of dealing with this alienation. 'The vertical distance to the space above our heads,' he says, 'is not more than an hour's drive in a car.'[223] Put that way, the transfer in either direction seems more reasonable, even inevitable. And, given the fact that gravity works *with* a falling object, assisting the downward movement of anything in our atmosphere, the connection from thin air to Earth is easy.

It may be the key to an understanding, not only of influenza, but of all the plagues and epidemics that break out at random intervals with devastating results. Most of these are produced by recognisable pathogens, are clearly transmitted from person to person and draw on identifiable natural reservoirs, but there are problems in the geography and epidemiology of nearly all of them. Sudden outbreaks of cholera, seasonal fluctuations of measles, and widely separated instances of plague, are difficult to fit into the usual theories of disease propagation. But they become much more amenable if we assume that their origins are at least atmospheric.

At the heart of this, and many other dilemmas, is our tendency to think of Earth as a closed system. It is not. We do not live in a sealed spacecraft, isolated from the environment in a convenient bubble of air. We travel rapidly through space and time with our windows open, constantly exposed to the complex ecology of the galaxy and all it contains. We are windblown. And, while one of the consequences of this openness may be bouts of influenza, another could be the very existence of life itself.

There is a wisdom that clamours to be heard in the Latin root *anima*, meaning both 'wind' and 'spirit' – which leads to *animus* the 'soul', *animare* to fill with 'breath', and ultimately 'animal'. And the root *spirare* to 'breathe', from which come 'spirit', 'aspire' and, in the end, 'inspiration'.

In most languages there are similar spirals, racial reminders that life and breath and spirit are one with the air, and that wind is its divinity.

The multitude of plants and animals that now constitute our biosphere, share between them a total of about one million different genes. These could have been acquired as a result of random mutation through the ages, but the range of variation and the speed with which evolutionary changes are often made, suggest another mechanism. We seem to need, and to be provided with, substantial new sources of variation. New genes, like those of the latest kind of influenza, that come floating down like manna from heaven.

Gaia is a galactic creature, breathing cosmic air, and feeding on the fruits of the universe. The winds are her system of circulation, passing on energy and information, and we are one of the results. A recent flowering that gives her awareness both of the universe and of herself.

In the words of the metaphysical poet George Herbert, 'The wind in one's face, makes one wise.'

5. THE BIOLOGY OF WIND

Air is full of surprises.

In 1959, a scientist in New Zealand collected a number of samples of fresh snow, catching it as it fell well above the timber-line on mountain peaks 2500 metres high. In addition to traces of the usual and expected air baggage such as dust and pollen grains – in New Zealand there were very few of either, he discovered that all the samples also contained dissolved organic matter. And, most intriguing of all, the majority of this material was in the form of soluble albumins, the proteins that occur in egg white, milk and blood. As though the air itself was scarred by life being vigorously lived.[525]

But where?

New Zealand lies in the westerly air stream of the southern hemisphere, downwind of over 2000 kilometres of Tasman Sea. Could the sea be the source? The bulk of the ocean contains organic matter in concentrations sixty times lower than the sampled snow, but the surface of the sea is a better bet. It is very much more lively.

The thin sunlit zone swarms with bacteria, yeasts, diatoms, algae, fungi and all the host of tiny plants and animals that enrich a soup of plankton. Most of the debris from this complex ecology falls to the ocean floor as silt, but the lighter products concentrate in the uppermost few millimetres where they are whipped by wind and wave action into frothy, still somewhat mysterious, substance we call sea foam.

When this is disturbed, bubbles burst at the interface between water and air, and each one throws up a small rich droplet. The haze produced by large numbers of such laden particles is clearly visible over whitecaps at sea, or above beach surf on a sunny day, and part at least of it is carried off on the wind.

In New Zealand it has long been known that farms near the coast carry more sheep than the poor soil theoretically allows. This effect decreases with distance from the sea and peters out altogether fifteen kilometres inland. It seems that the coasts are fertilised by prevailing winds which drop most of their nitrogenous load near the shore. Recent measurements suggest that two or three grams of organic matter are in fact deposited every year on each square metre of coastal pasture, but sea breezes also carry enough of their nourishment on to salt the snows a long way inland.[526] And such generosity is by no means peculiar to the Antipodes.

A study of spring rains in Missouri, shows that they are astonishingly rich in vitamin B12. This is cobalamine, an organic compound that is vital for successful cell division. It is extracted from liver or fish and fed to livestock to enhance red blood cell formation and subsequent growth. And its presence in waters, both fresh and marine, is essential for the bountiful green algae that lie at the base of most food pyramids. There is no evidence that B12 is carried into the air on dust particles. It seems to be an integral part of the atmosphere, brewed up perhaps by local winds, but deposited in sufficient concentration to produce a visible bloom of algae on ponds and lakes after just one thunderstorm.[373]

The more we learn about 'thin air', the more substantial it becomes. In some places it is almost thick enough to plough and, in others, it is primed by the wind with so much organic matter that it comes close to being a living tissue in its own right.

Windseeds

Frogs, dogs, pomegranates and people have a great deal in common.

They are all tunes played on the same instrument.

It is in essence a very simple instrument with just four basic notes, or nucleotides. Variety is produced by combining these notes, three at a time, into chords. But of all possible combinations, only twenty are actually used in any composition. And in every performance, patterns of more than a hundred chords are strung together into long fixed sequences. Each of these complex musical phrases is known as a gene, and it is an arrangement of such phrases that forms a particular tune, or species.

Closely related species, such as lions and tigers, have very similar tunes which may differ only in the construction or position of a single phrase. But there are also astonishing similarities between the tunes of species as distinct as cats and baboons, which have been shown to carry some of the same genetic material and which seem to be playing from the same master score, following the same hidden melody, but interpreting it in different ways.[31]

There are, on the other hand, some surprising differences between the tunes of organisms that ought to belong to the same species. An examination of the influenza virus that first appeared in New Jersey in 1976, showed that by the time this pathogen reached Asia, it had changed its tune. There were alterations in as many as sixty of the component phrases, giving almost every patient in Hong Kong a sneeze in a different key.[355]

The genetic difference between the various forms of this one virus are greater than those which separate humans and chimpanzees. It would not be stretching the analogy too far to compare the infection in this influenza epidemic to attacks by, for example, raccoons at one extreme, and by rats or rabbits at the other.

The inability of viruses to hold a tune, their lack of genetic stability, is due largely to their tendency to make mistakes when reproducing themselves. They not only breed faster, but change and mutate more often and with greater facility than any other organism. They have a natural ability to edit the genetic material, to tack loose bits of gene, odd phrases, together to form new compositions. This mechanism is a complex and possibly dangerous one, for which most other organisms have no need. There is really only one natural situation in which genes break often enough to require a full time repair service, and that is exposure to strong ultraviolet light.

Thanks to the ozone layer in our atmosphere, most of the Sun's potentially lethal short waves are filtered out before they reach us. So there is normally no risk of exposure to them and no reason for any organism on Earth to develop a specific defence against radiation with a wavelength around 2500 Angstroms. But that is precisely what viruses do have. And, given their genetic volatility, it is highly unlikely that this one talent should have survived unchanged from a time, perhaps 3000 million years ago, when Earth may not have had an ultraviolet screen. Evolution is notably parsimonious and seldom develops structures or abilities without good and recent reason.

The conclusion seems inevitable. These viruses possess efficient ultraviolet repair mechanisms because they need them, because they have been recently and strongly exposed to the ravages of ultraviolet light. And there is nowhere that this could happen to them except above the ozone layer, at an altitude of at least thirty kilometres.

Somewhere up there, it seems, there is a breeding ground.

In 1677, the janitor of the City Hall at Delft in Holland discovered creatures too small to see with the naked eye. Anton van Leeuwenhoek's hobby was grinding lenses and in one of them he saw the first recorded protozoans, which he called 'animalcules'. Six years later, right at the limit of resolution of his most powerful microscope, he also got a glimpse of tiny structures that, from his description, could only have been bacteria. No one else was to see them again for over a century, but the science of microbiology was born and for 200 years one of its principal concerns was the question of the origin of microscopic life.

It was no longer believed that maggots materialised by magic out of ripe meat or that mice came from dirty wheat, but there was widespread agreement that microbes probably arose from non-living material by a process of 'spontaneous generation'.

The first to question this conclusion was Louis Pasteur, who rescued a vital industry in 1854 by showing that fermentation involves living organisms, yeasts, and that maturing wine or beer could be prevented from going sour if the wrong species were killed by heating the product gently to 50° Centigrade before corking.

His opponents suggested that the failure of the microbes to appear in such sealed containers was due to nothing more than the destruction of a 'vital principle', something in the air, which was responsible for the generation of life. So Louis was forced to find a 'pasteurisation' process that did not involve heat.

He boiled meat extract in a flask and then left it exposed to the air, but only by way of a long narrow neck with an 'S' bend in it, a gooseneck that curved down and then up again before it opened. This allowed unheated air to reach the broth, but only after it had been deprived of dust, which settled to the bottom of the outside curve. The soup did not spoil, there was no decay, and no trace of micro-organisms. Pasteur announced his results to a gala meeting at the Sorbonne in 1864 and added 'sterilisation' to his string of discoveries, which was later to include the 'germ' theory of disease.

Spontaneous generation was routed. Everyone accepted the reality of infection and contamination and, after a struggle, even the medical profession began to boil instruments and steam bandages before they used them. It was agreed that air was dirty, that there was something alive or dormant in dust all the time. But despite this consensus, another possibility still lurked beneath the surface of science, an ancient, almost alchemical urge to squeeze life out of the non-living. And in 1953, almost a century later, spontaneous generation once again became an issue.

Suppose, they said, you kept a sterile solution for longer than Pasteur allowed it to stand? What if you had an ocean of solution and left it for 1000 million years? And what effect would it have if the air above that solution were of a different composition?

A young graduate student at the University of Chicago was sent to find out.

Stanley Miller took sterilised water and put it into an atmosphere of hydrogen, methane and ammonia. This magic mix was chosen because it provided all the necessary chemical ingredients and because it matched the gases believed to exist on Jupiter. *Pioneer 11* has since confirmed that these are the principal ones in the Jovian atmosphere, but there is no good reason to assume that Earth was ever like that.

However, Miller imitated the storms that may have raged in any early atmosphere by energising his model with an electrical discharge. He left it sparking for a week and, at the end of that time, analysed his little ocean. He found that it actually contained organic compounds, including even a few simple amino acids – the basic components of all living proteins.[336]

The experiment was hailed as a classic. 'After all,' they said, 'if in one week and in one small set-up, Miller could get amino acids, how much could be done in a billion years.' Other and more sophisticated tests followed and succeeded in synthesising sugars, complex amino acids, and even some nucleic acids – the direct precursors of genetic material.

Life, or something very close to it, *can* be created out of non-living matter, and spontaneous generation rides again. All you need is water, a mixture of special gases and a few thunderstorms. Then you can pluck life seeds out of thin air and plant them in a nourishing 'primaeval soup', where they could, given time, grow to maturity.

But, says astronomer Fred Hoyle, no Miller-type experiment has yet produced a living organism. 'The proof of the soup is in the eating.' He does not believe there ever was such a brew. And he objects strongly to the implicit, almost pre-Copernican, assumption that life had to begin right here, where we live, on Earth's surface.

Apart from the highly questionable prerequisite of a strongly hydrogen-dominated primaeval atmosphere, there are other unanswered questions. Could enough energising radiation penetrate to give large enough yields of all the biochemicals? Could all the basic biochemicals form with the same starting substances and under the same set of conditions?[222]

Hoyle and his colleague Chandra Wickramasinghe at the University of Cardiff are amongst the few who doubt the geocentric explanation; who reject the rigid orthodoxy of the soup-sayers; who question the articles of scientific faith that draw hard and sharp limits to life less than twenty kilometres over our heads.

They are, however, in good company.

The Swedish chemist Svante Arrhenius made a habit of challenging accepted opinion. His 1884 doctoral thesis on ionisation was so radical that one of his examiners rejected it outright and the others awarded him the lowest possible grade. Twenty years later, the same study earned him a Nobel Prize in chemistry.

Something similar happened to him in 1908 when he turned his attention to the larger mysteries and published his theory of *panspermia*. This proposed that life on Earth began with living spores which reached it through space. And he suggested that the spores, which could resist harsh conditions for long periods, were carried here by the solar wind.[10]

They laughed at him. But astronomer Carl Sagan has recently estimated the radiation pressure necessary to move particles of various sizes through space, and calculates that all viruses, some bacteria and many spores could be blown along by solar wind quite effectively. At peak velocity they would travel as fast as spacecraft, moving from Earth orbit to Mars in a few weeks, to Jupiter in a matter of months, and as far as Neptune and Pluto in no more than three or four years.[417]

None of these voyages would be impossible for terrestrial microbes, such as the bacterium *Bacillus subtilis* and the mould *Aspergillus fumigatus*, that have shown in the laboratory that they can withstand exposure to ultra-high vacuum.[387] Nor would it be beyond the capacity of micro-organisms such as the common intestinal *Bacterium coli* which seems to be able to survive even quite high experimental doses of ultraviolet light.[444] Several species, including the virus responsible for polio, have now been sent 150 kilometres up by rocket and returned apparently none the worse for the experience.[214] Arrhenius could be right.

A major objection to panspermia has always been that it avoids the question of the origin of life, which must after all have taken place somewhere. Fred Hoyle believes that he knows where, and his argument is simple and persuasive.

The gaps between the stars are punctuated by clouds of gas and dust. This is the stuff of which new stars are made. Radio 'fingerprints' from these regions show that an astonishing amount of this raw interstellar material is already organic, consisting of carbon monoxide, cyanogen and even complex compounds such as formic acid and ethanol. In some areas, the dust particles have become so coated with sticky organic tar that they clump together into larger masses in which still more lifelike chemical reactions take place.

Infrared radiation from these sources strongly suggests that they consist mainly of cellulose, the substance of all plant-cell walls and natural fibres, and still the most abundant organic material on Earth. 'It seems possible,' says Hoyle, 'that all the essential biochemical ingredients of life existed already in large quantities within the interstellar cloud of gas and dust from which our solar system was born.'[223]

There is no need to strain credulity with early hydrogenous atmospheres and hypothetical soups. The existence of distant cosmic cradles with all the equipment necessary for the origin of life, lets Arrhenius off the hook. Life could have begun almost anywhere in space and then been distributed later, as panspermia suggests, to other suitable habitats, like Earth. It could have been cast adrift at the mercy of solar wind, but Hoyle has even been thoughtful enough to provide a suitable protective vehicle.

About ten times each year. our solar system is visited by celestial outlaws. These luminous partial vacuums – 'the nearest thing to nothing that anything can be and still be something' – swing through the cosmos, mopping up interstellar material like giant sponges. Gradually they take up definite shapes, with most of their matter concentrated into a nucleus a few kilometres across, trailing a diffuse tail that may be more than a million kilometres long. These are, of course, comets and their core or coma consists largely of ammonia, methane and water ices dusted with debris, like a large and dirty snowball.

Some comets flash across our skies when perturbed by passing stars, that tilt their paths in our direction. Others have an elliptical orbit that brings them back at regular intervals like wild satellites of the Sun. Either way, they travel far enough into the spaces between the stars to pick up the seeds of life and nurture them. Conditions inside their cores are still largely unknown, but the amount of matter there must give individual particles of incipient life some protection from ultraviolet radiation. And it is possible that the alternate heating and cooling as a comet moves towards or away from a star, provides exactly the sort of fickle environment that seems to be necessary for the encouragement of adaptation and evolution.

As a comet moves, volatile material is stripped off its nucleus and swept out into the dramatic tail. Particles are scattered across millions of kilometres of space or swept into windrows by solar radiation, dusting the air that Gaia breathes, infecting her with new and radical ideas. One of these may have been life itself, but diffusion through space is probably an ongoing process and it seems likely that Earth is still being spiced with cosmic seed, picking up galactic gossip and the occasional bout of influenza.

It becomes difficult to deny that there could be some kind of community out there, even if it is only the most rudimentary of nurseries. And difficult too to avoid the logical implication of panspermia, which is that Earth may not merely be a customer for life seeds, but also a supplier. Answers, if any are to be had, will be found one day on the frontier, in the no-mans-land between Earth and space.

Boundaries here are hazy. At a height of perhaps 80,000 kilometres, which is about six Earth diameters out, our atmosphere merges into that of the Sun. In this area it is almost impossible to define an atmosphere, and terms such as wind and temperature have little meaning. Charged particles from the Sun flow through in a variable stream that blusters like a wind, blowing sometimes hard and sometimes falling away into an ominous calm. It is a tenuous flow, with a density of just five ions per cubic centimetre, compared to the twenty-seven million million million in an equal volume of air at sea level. Without this refinement, the temperature would be devastating – around 160,000° Centigrade.

The solar stream does not pass our planet unimpeded. It sweeps by like the flow around a rock in a river, building up a bow wave on the Sunward side and meeting again in turbulence downstream. As it passes, there is an interaction with Earth's own field and the result is a concentration of particles into two standing waves arranged in broad belts about Earth's waist at about 4000 and 20,000 kilometres. These are the Van Allen belts, named after the American physicist who pioneered cosmic ray studies. They are regions of high-energy radiation where gases begin to break the normal rules. Gravity no longer holds them in thrall, and molecules are so widely scattered that the lighter ones can travel progressively outward, eventually darting off into space at speeds of up to 40,000 kilometres an hour.

Earth seems to have its own cometary tail. Our passage through space is marked by a trail of debris, by bits and pieces shed like flakes of skin. Most of this wake consists of stray ions of air, but it may well include particles of dust and possibly even an occasional microbe.

The most concentrated volcanic blasts, like those of Tambora in 1815 and El Chichon in 1982, throw plumes of dust and ash up right through the stratosphere. Krakatoa – 'the biggest bang in history' – may even have rung our atmospheric bell, hitting the top of the mesosphere eighty kilometres high. The big thermonuclear devices, especially those tested in the atmosphere, have created their own puncture wounds, sending fireballs and mushroom clouds of water and debris up fifty kilometres. All these eruptions, both natural and man-made, have left us with a legacy of lighter particles in suspense high above the weather. In recent years, a number of rockets fitted with special collecting devices have been fired into this frontier zone and all have returned with samples of something.[58] It may be that we have discovered nothing more than a very expensive way of picking up our own garbage, but the evidence suggests a more exciting conclusion.

In 1974, a series of six Russian rockets probed the upper edges of the mesosphere over the deserts of Kazakhstan. At the peak of their trajectory, between fifty-eight and seventy-seven kilometres, each nose cone opened briefly to expose a film coated with a sticky meat broth and then parachuted this sealed sample back to Earth. After fourteen days in a thermostat, the packages were opened to show a total of thirty-one microbial colonies growing on the nutrient medium.[232]

Six species were involved. Two of them were bacteria known to be parasitic on humans. *Micrococcus albus*, which belongs to a group of pathogens responsible for meningitis and gonorrhea; and *Mycobacterium luteum*, whose relatives cause tuberculosis and leprosy. The remaining four were microscopic fungi, including two common moulds – *Penicillium notatum*, the same species which led Alexander Fleming to his discovery of antibiotics, and *Aspergillus niger*.

All the fungi collected are species that occur in a wide variety of habitats, in marine and fresh water, in soil and on plant and animal remains. They are opportunists and it is fascinating that all four of them produce black spores. Pigment is a rarity in fungi, but it is essential for any microbe that is likely to need protection from the strong ultraviolet light at altitude. It can be no coincidence that these species, which are pioneers of their kind, should have acquired this particular adaptation. It suggests that flights above the stratosphere have been a normal part of the experience of at least these species for some time.

How long? How many others are involved? And just how high can they fly?

None of these answers are available. We know that terrestrial dust particles heavier than any of these microbes have been recovered from altitudes of 100 kilometres, but the upper air has never been systematically explored.

There was a Bioprobe program organised by the Goddard Space Flight Centre for NASA, which involved shooting an Aerobee 150 rocket carrying sterile film to be exposed at altitudes above 150 kilometres. *Bioprobe 1* returned in 1966 and results were promised after several additional flights had been performed. These have still not been published. The Gemini space flights were scheduled to collect material during extravehicular activity in orbit, but there is nothing in the mission reports on such studies. One wonders why. Did they find something too awful to tell us about?

Aerobiology is still in its infancy, but already we have enough hints to suggest that we are dealing at high altitude not only with accidental intrusions carried there passively by freak conditions, but with an ecology adapted to the edges of space as a normal habitat. One third of all bacteria collected in the upper air are able to withstand exposure for forty-eight hours to temperatures as low as −26° Centigrade.[390]

It is no longer reasonable to regard air simply as a means of transport, as an occasional and temporary medium for the dispersal of some plants and animals.

To go from dust to dust may be the inevitable fate of all organisms trapped in the usual terrestrial cycle, but evidence grows for the existence of another kind of ecology which is predominantly aerial. For a community of the air that reproduces itself, possibly even living permanently at great heights, rising and falling with the atmospheric tides.[309]

An aeroplankton.

Windspores

The notion that all matter consists of tiny particles is one we owe to the Greek philosophers Democritus and Epicurus.

They held that these fragments were infinitesimally small and therefore indivisible, or a-tomic. Beside these, only the void, the space between atoms, existed. They did not realise that air had weight, Aristotle thought that it tended rather to have 'levity'; but as far as they were concerned, air was already more than a space between things, it had some kind of substance. It was one of the active principles, along with water, earth and fire.

It was left to the Roman philosopher Lucretius to suggest that some of the observed properties of different kinds of air might be due, not to its own substance, but to something that came with it.

He noticed the scintillation of dust motes in a beam of sunlight playing across the still air in a darkened room, and concluded that their movement must result from bombardment by invisible Greek 'atoms' that kept them in suspense.

This started him thinking about the origin and dispersal of pestilence. It was, he decided, in the nature of things that 'baleful particles' carried in clouds by the wind would settle on wheat or the lining of a lung. And with this concept, almost 2000 years before Pasteur, Lucretius set the scene for all modern studies of pathology, allergy, and infection.

The impetus for most work on invisible particles in the air continues to come from medicine or agriculture and their joint concern about airborne disease, but this obsession with 'germs' conceals the fact that only a tiny fraction of everything blowing in the wind is inimical. The truth is that air supports an unseen ecology that is every bit as rich and varied as a rain forest or a coral reef.

The census of microscopic species in the air has only just begun, but already includes a host of viruses; close to 1000 varieties of bacteria; 40,000 species of fungi; hundreds of different kinds of algae, mosses, liverworts, ferns and protozoans in at least some stage of their life cycle; and more than 10,000 species of flowering plants that are wind-pollinated.

Many of these species are only a temporary part of the aerial community, present as relatively inactive cells or small groups of protected cells called *spores*.

Some bacteria, many fungi and most mosses, ferns and flowering plants use the air purely as a means of dispersal for these seeds of new life elsewhere, but a large number of species take to the air as active adults.

The abundance of any wind-living creature at any stage of its life depends very much on local conditions. Broad ecological boundaries of the various air habitats are drawn by thermal layers of the atmosphere and by the flow of global wind systems, with greatest population densities in the troposphere and usually closest to the ground. In addition, there are other, more biological, limits that have become the concern of micrometeorologists – who are not little weathermen, but scientists interested in microclimates.

Within all the recognised climatic zones, and almost independent of variations in the weather, there are smaller geographical units of air that experience quite different conditions.

The influence of Earth itself on the atmosphere is effectively limited to the lowest ten kilometres, to the troposphere, but on a daily basis this influence is further restricted to a very much shallower zone known as the *planetary boundary layer*. In it, there is constant turbulence produced by the drag of wind across rough terrain and by the bubbling up of parcels of air from the heated surface. There is a tendency for the layer to wax and wane in rhythm with the Sun. By day, it may be between 1000 and 2000 metres deep, but at night when the surface cools, it can shrink to a thickness of less than 100 metres.[366]

Within the boundary layer, there are several other zones controlled by surface features. In immediate contact with Earth and all projections from it, there is a microscopic layer in which air is held by atomic forces. Except for molecular movement, this area is still and completely windless.

Directly above it is the *laminary layer*. This again is thin, usually no more than a millimetre or two deep, but the air here is in motion, sliding in parallel streams along the ground or over the surface of a leaf. It is a viscous layer, a dust trap, anything that arrives here usually stays. But, in contrast to the relatively stable conditions a metre or more above the ground, the climate in this laminary layer can be violently changeable. 'A seedling pushing its first leaf through the soil may, even in a temperate latitude, experience in a single day a range of temperature that for man would correspond to commuting daily between the tropics and the Arctic.'[471]

At all heights above a few millimetres, air is subject to turbulence. Winds blowing over rough surfaces produce eddies which stand as stationary waves, or which break and tumble away through the air in the wake of an obstacle. The higher the wind speed, the greater the turbulence and the thicker this layer becomes. This is the zone we know best, it is the one in which our heads, and the Stevenson screens that hold most of our weather-recording instruments, lie. On hot, windy days it may reach a height of 150 metres, but on clear calm nights this turbulent layer is usually no more than a few metres deep.

Above the level of major turbulence, is a transitional zone in which normal friction effects gradually disappear, ending altogether at a ceiling between 500 and 1000 metres high, which is often visible as a distinct dust horizon. This is the top of the planetary layer and the boundary for the vast majority of living and other windborne things, but it is a line that is frequently breached.

Air rises more rapidly than usual in thermals over bare rock,

155

roads, cities and ripe crops of cereal, climbing at three metres a
second to heights of more than three kilometres. And at any one
time there are approximately 1800 thunderstorms in progress
somewhere over Earth, lifting masses of air, along with most of
their contents, right up to the edge of the troposphere. As a rule,
these pockets of laden air take just an hour to reach the top of a
thunderhead, but it might be twenty days before they sink back
again into the boundary layer, by which time it is possible that they
will have gone right around the globe.

In microclimatology, everything above the planetary boundary
is known as the 'free atmosphere', where movement is largely a
product of Earth's rotation. It is fed by convection with life and
food from the lower levels, but it is in every other way a distinct
habitat with its own characteristic species.

Within the boundary layer, the sheer quantity of living things is
phenomenal. Over every square centimetre of land surface, there
must be an average of something like 10,000 microscopic organisms. That is fifty to a hundred million above every square metre
or, if you like, a tower of roughly thirty million living things
balanced on your shoulders in a precarious column 1000 metres
high.

The smallest and lightest of creatures are represented in this
ancient circus act, but not many of them. Very few viruses are
specifically adapted for life in the open air. Arguments still rage
about whether or not they are even truly alive. The fact is that they
have some of the properties of obviously living things, while
lacking others, and need to be intimately related to organisms
whose status is not so much in doubt. Viruses range in size from
0.000015 to 0.00045 of a millimetre and are obliged, it seems, to
hijack larger cells and live inside them, taking over and adapting
the local machinery to their own ends.

Migration to a new vehicle, the infection of another organism,
usually takes place by direct contact or with the help of a middle-
man, who is most often a biting insect. It is claimed that psittacosis
in humans, foot and mouth disease in cattle, and some viral
infections of poultry, are airborne in their own right. This may well
be true – as it seems to be for influenza – but one of the problems
is that particles as small as viruses, are so strongly attracted to any
surface, that it is difficult in the first place to dislodge them directly
into the air, and almost impossible to prevent one that has become
airborne from sticking to the first solid obstacle it encounters.

It is possible, of course, that this might be a dust particle or even another organism which could carry it piggy-back to a suitable host. The neatest arrangement would be for a virus to become hereditary, to attach itself to a spore, sperm or pollen grain and move directly from generation to generation like a hidden Greek being carried into Troy by the Trojans themselves. There is at least one virus parasitic on mushrooms which does precisely that, hitch-hiking along inside the travelling spore.[424]

On the next rung of the evolutionary ladder are a group of larger creatures – cylindrical, spherical, spiral or rod-shaped – that range in size from 0.0003 to 0.015 of a millimetre. Which means that you could still put 1000 of the largest ones side by side across a fingernail. Linnaeus had his doubts about them and called them *specia dubia*. Their taxonomy remains contentious, but time and better microscopes have left little doubt about their reality and vitality. The air everywhere seethes with bacteria.

The best study of free living bacteria is one made in Paris during the last quarter of the nineteenth century.[339] Pierre Miquel sampled the air at several selected sites every day for thirty-four years and always found bacteria there. He discovered seasonal cycles, with higher counts in the summer months, but recorded an annual average of 290 bacteria in each cubic metre of air in the Parc Montsouris five kilometres from the city, 5555 in the quiet Passage St. Pierre, and 7480 in each cubic metre near the busy Hôtel de Ville in the centre of St. Germain. His counts were even higher inside buildings and highest of all in a crowded hospital, but fell to a surprising 3835 in the main sewer beneath Boulevard Sebastopol.

Bacteria are simple cells that live in a variety of habitats where they can feed on organic material, usually either living or decaying plant and animal tissue. Humans make ideal hosts and support flourishing populations which are distributed liberally wherever we go. With each 100 spoken words, particularly those with explosive consonants like 't' and 'p', we put 250 tiny droplets into the air. Forty per cent of these contain one or more bacteria, usually of the *Streptococcus* or *Staphylococcus* types.[110] A single cough is worth 2000 words, wheeling out 5000 potentially dangerous droplets. But a sneeze is the Rolls Royce of bacterial vehicles. With acceleration from a standing start to 400 metres per second, almost the speed of sound, a sneeze generates as many as a million drops of various sizes, most of them infected.[45] This biological tornado burdens the air with as many bacteria as would normally be dispersed by someone speaking 400,000 words – which would mean talking non-stop for fifty-five hours or reading *War and Peace* out loud.

Some people always spray more vigorously than others, usually due to idiosyncrasies of speech or dentistry, and there are a few who seem to be almost living aerosols. On one occasion all the newborn infants in the nursery of a hospital became rapidly and heavily infected with *Staphylococcus aureus* and the epidemic was traced to a child that was putting out such a concentrated stream of bacteria that it became known as the 'cloud baby'.[120] The reason for this effusion remains unclear, but it seems to have been produced in part by a viral infection that narrowed the air passages and increased the velocity and the bacterial loading of every breath.

It is possible that there is a symbiotic relationship in this case between virus and bacteria, both of whom benefit by rapid and effective dispersal. It even begins to seem likely that a sneeze, which can after all be a violent and painful experience, frequently damaging the nasal membrane, is something imposed on us by the virus for its own nefarious ends. It is a pattern of behaviour with high survival value for the parasite, but offers little to the host. Which may be why people everywhere look askance at sneezing, believing that it bares the soul, and greet each personal explosion with an automatic chorus of ritual blessings. Perhaps these talismanic comments are the product of a deep-seated recognition of possession by, and surrender to, an alien authority?

It is not always necessary for bacteria to make their own way out into the open air. Large numbers take advantage instead of the stream of vehicles which pour off the production line at the surface of our skin. Every touch, every breath of wind, carries away thousands of epidermal skin scales that are light enough to float free. Each time we move, we leave behind us a smoky vapour trail of flakes, all just 0.008 of a millimetre across. And every one of these little dry discards is a life raft, a vessel capable of carrying whole colonies of bacterial passengers. Convoys of these rafts, several thousand strong, are commonplace in every office, factory and home, and they become almost visible, reaching concentrations as high as 390,000 per cubic metre, in the air over a freshly made bed.[175]

It is likely that these samples over-emphasise the abundance of bacteria in cities. The high concentrations in urban air are not so much a reflection on our personal hygiene, only a fraction of bacteria are pathogenic anyway, as a consequence of disturbance. Cities are busy places and the constant activity keeps dust and bacteria airborne. The actual number of bacteria resident in the country is far higher.

It is estimated that just a gram of fertile soil could house as many as 1000 million individual bacteria. But the majority of these are locked into the soil cycle, where they are responsible for the formation of humus and for the breakdown of organic matter into essential nutrients. Only a relatively small number of bacteria hold a watching brief, waiting for the occasional windfall that brings them a ripe apple or a dead elephant to dispose of. These are the spore-bearers, the ones at the surface which most often become airborne and which, like their city cousins, frequently do so on rafts of soil or vegetation.

On hot dry days convection alone is enough to lift material off the ground, but most take-off is wind assisted. It has to be, to get micro-organisms out of the thin film of still air at the surface and through the laminary layer of thick slowly moving air just above the ground. *Dust devils* whirl millions of bacteria high into the air and, once there, they are at the mercy of every stray breeze. A large bacterial cell in a gentle but steady wind of just ten kilometres an hour, will travel over 10,000 kilometres before falling 100 metres back to Earth.

Air masses moving from land out over the ocean start off with average concentrations of around 500 bacteria per cubic metre. This falls rapidly to an average of less than fifty at a distance of 160 kilometres from the coast, and by that time part of the number consists anyway of marine species picked up by the land breeze as it passes through the spray zone above beach surf. Counts over the open ocean average less than one bacterium per cubic metre, which provides an interesting contrast with the land – and with the figure of about 100 for air in the wheelhouse of an ocean liner, and over 1000 in its air-conditioned saloon. Every ship, it seems dilutes the possible therapeutic effects of an ocean cruise in clean air, by travelling in a cloud of self-contamination.[150]

Similar problems face polar explorers. They find that wounds suffered by themselves or their dogs very quickly turn septic. Perhaps because the lack of other suitable habitats attracts the undivided attention of the few bacteria that are around. Some of these are undoubtedly carried in by the expeditions themselves, but the local wildlife seems to have almost as much trouble. The skin injuries inflicted on each other by rival male seals and sea elephants in beach colonies, are commonly infected with bacteria. And seal hunters know from experience that the skins of their prey have a habit of biting back long after the original owners are dead.[92]

The immediate source of many of these potential pathogens lies in the air and on high-level winds. It seems that nothing anywhere is immune from seeding by the wind.

As a general rule, there are more organisms in the lower than the upper air. But there are notable exceptions to this rule, produced mostly by local weather conditions. Clouds are probably the greatest concentrators of airborne micro-life. Rising bubbles of air in convection cells bring collections of organisms to heights where some of them are frozen and thrown to Earth as hail, but others simply get washed down to lower levels where rain evaporates, allowing them to be carried up once more and eventually heaped up in greater and greater numbers at certain fixed heights.

It seems likely, in the tropics at least, that whole populations get caught up on this active roller coaster in cumulus cloud, become concentrated in certain areas, and are held there long enough for them to be considered almost as permanent residents. Even a few hours in the life of bacterium is a long time – enough for several generations. A spore liberated from a cloud at 1000 metres in completely still air would take over a day to reach the ground. But no air is ever that still and the average time spent aloft seems to be something more like ten days. Air that rises over the equator, often spends weeks or even months at altitude before descending with its load at an entirely different latitude.

There is, it seems, good reason to accept that a distinct 'biological zone' with an indigenous fauna and flora, exists and persists in parts of the upper air. Much of this community is passive and dormant, simply carried up into the air by the wind, but parts at least are active. Bacteria have been seen to repair themselves, making good damages done by ultraviolet light.[104] And if they can do this, there is no reason why they should not feed and reproduce. How the rest of the community behaves up there, remains to be discovered.

Fungi probably do pretty well. There are certainly enough of them.

At last count there were 50,000 different mushrooms, moulds, rusts and yeasts scavenging their way round the world. Most textbooks call them plants, but none have chlorophyll. They live more like animals, descending on and attacking their food with powerful digestive juices. Although some are quite large, none are multicellular, consisting instead of a mass of protoplasm which flows along inside a threadlike cover that invades the body substance of their chosen prey. Parts of this may break off and waft away to new feeding grounds, but between meals most species take to the air in the form of minute spores, just twenty-five times larger than the average bacterium, which are produced on fruiting structures that rise above the dead air near the ground, and are usually far more colourful and conspicuous than the rest of the organism. Mushrooms, toadstools and puffballs are good examples of the elaborate launch pads devised by these wind-conscious creatures.

Mushrooms rely largely on gravity, growing the characteristic cap that protects spores in narrow gill slits until they are ready to drop into whatever turbulence may be available beneath the umbrella. But they make up for this casual arrangement by liberating an astonishing half a million spores a minute – and by going on doing so for as long as a week. Even an ordinary edible mushroom such as *Agaricus campestris* can release up to 3667 million spores.

The shaggy mane mushroom *Coprinus comatus* is even more profligate, producing an estimated 5240 million. Woody bracket fungi such as *Ganodema applanatum*, which cling like orange flanges to the trunks of beech trees, can churn out 3600 thousand million spores during a six-month growing season. But the production record belongs to the giant puffball *Lycoperdon giganteum* that holds as many as 7000 thousand million tiny missiles.[61] Puffballs grow globular fruiting bodies, some as big as footballs, whose skin is thin, waterproof and flexible. Each time a raindrop strikes it, the ball dents inward and, acting as a bellows, forces a jet of air laden with spores out through an aperture. During a rainstorm, the forest floor comes wonderfully to life, throwing up little puffs of coloured cloud, as though war had broken out between Lilliputian forces armed with antique cannon.[174]

Other fungi are more selective, producing just eight large spores. But they give each of these a fighting chance by actually firing it from a gun. The spores are contained in a long thin cell which becomes more and more taut as it grows, until the whole structure is under considerable hydrostatic pressure. Eventually it ruptures at the tip, the elastic wall contracts violently, and the spores are shot out in rapid succession. In the species with the longest range *Podospora fimicola*, they travel up to fifty centimetres. Which may not sound much, but is equivalent in human terms to shooting a large cannonball twenty-five kilometres. And this is something no artillery weapon was able to do until the closing stages of the First World War, when the German Army brought out fabled 'Big Bertha' to join in the long range bombardment of Paris.

Even this ballistic prowess, however, takes second place to one of the yeasts, which seems to have invented rocketry. The spore in this species is attached to a gas-filled bubble which explodes, sending it off into fungal space like a miniature jet.[367]

There are seasonal influences on most species of fungi, with a tendency for the dominant moulds to flower and fill the air in late summer when grasses and crops are ripening. It is then that 'clouds' of fungal spores float through the air like gritty lower banks of fairweather cumulus, at times so numerous they become clearly visible. To get some idea of normal density, the air over an arable field in southern England was sampled every day during an entire summer.[176] The numbers of fungi varied enormously, but it was not unusual to find 120,000 in each cubic metre of air. Which means that, walking through farmland on a clear summer's day anywhere in the British countryside, you are likely to find yourself inhaling, amongst all the other airborne nourishment, at least sixty microscopic mushrooms with each lungful of air.

This is an alarming prospect, particularly as some fungi are known to cause asthma and alveolitis, but it is also true that many of our most useful antibiotics are extracted from moulds. It is possible that country air owes its healthy reputation, not to a freedom from smoke and industrial pollutants, but to contamination with a flourishing population of fungi.

Once in the air, fungi are at the mercy of the winds. Viable spores are as common high over the ocean as they are over land, and the number in any air mass depends not so much on its height or location, as on its origin. The leaf rust *Hemileia vastatrix* grows only on coffee trees and has always been a problem for planters in Asia and Africa. Trees imported to Brazil were free of the fungus until April 1970, when it appeared suddenly and simultaneously on several plantations over 1000 kilometres apart.[46] It seems significant that between January and April, southeast trade winds blow directly across the Atlantic ocean from Angola, at speeds which would make it possible for spores to travel 6000 kilometres from this infected area in less than a week.

All winds are laden with fungi, but tropical air which passes over the agricultural lands of North America seems to pick up the heaviest loads, carrying as many as 2800 spores per cubic metre right up into the arctic.[369] The majority of these are always the ubiquitous summer mould *Cladosporium herbarum*, but recent surveys in the Antarctic have found high numbers of other species, forty of which seem to thrive there despite the harsh conditions.[470] Balloon flights into the stratosphere have even collected samples of fungi at an altitude of thirty kilometres, where their density is down to one organism per 300 cubic metres.[58] Nobody knows the upper limit. It may well be possible to distinguish Earth's path through space from that of all the other planets by the presence on the wayside of an occasional hopeful mould.

Fungi have an important role to play in Gaian ecology, but they and all the rest of us would be unemployed without the food-makers, the ones that contain chlorophyll and eat the Sun. Dominant amongst these are the algae, which are aquatic rather than aerial, though a surprising number of them live in environments that are barely moist. Wherever there is vegetation of any sort, there are algae, covering leaves and bark in a fine felt or powder, clinging to the cracks in stones on the desert floor or on to the bare faces of buildings. And from time to time, some of these become airborne.[326]

Skeletons of at least one group keep dropping out of the air. Charles Darwin was surprised to find the decks of the *Beagle* covered in fine dust as she passed north of the Cape Verde islands, and delighted to discover that his sample of dust contained large numbers of the sculptural little pill-box algal cells called diatoms.[94] These were, of course, the fruits of the *harmattan*, gathered and dispersed from the ancient marine bed of the Sahara. The similarity of diatoms everywhere suggests that, between the ocean currents and the winds that move them, global populations are extraordinarily well-mixed.

A similar situation exists amongst the simplest of all animals – the protozoans or one-celled species. Every freshwater pool in the world tends to contain the same freeform amoebae, no matter whether it is tropical or polar, deep or shallow, fed by volcanic springs or melted snows. Given the geographical separation and the short lifespan of most seasonal ponds, they can only be seeded with these species from the air.[313]

When pools dry up, the inhabitants enclose themselves in sheaths or cysts which get caught up in the wind and blown around the world. Falling finally into water or a suitably moist environment, the cocoons crack and the creatures emerge again, 'some pulsing hexahedrons, some shaped like spaghetti, honey combs, doughnuts, pretzels, hourglasses, seahorses, roulette wheels, pyramids and boomerangs – everything and more than the molecules they are made of'.[351]

No count has found concentrations in the air greater than three tiny protozoan cysts per cubic metre, but apparently this is enough.

Moss and fern spores are relatively heavy and equally sparsely distributed in the air. There are local concentrations as high as 300 per cubic metre, but the average is usually less than five. Ferns have a sort of slingshot which fires as the spore-containers contract in dry weather. And some mosses possess a magnificent 'air gun', which consists of a capsule containing a bubble of air. As this dries, the gas is compressed until at a pressure of about five atmospheres, a circle of weak cells breaks, blowing off a lid and shooting the spores fifteen centimetres into the air.

Other moss relatives, the delicate little liverworts, depend more on random currents of wind and water for their dispersal and may produce as many as a million spores.[234]

Although spores from any of these plants are not common in air samples, their systems of dispersal must be efficient. South American ferns have successfully travelled in the westerlies to colonise Tristan da Cunha 8800 kilometres away. And mosses are amongst the first plants to appear on new ground anywhere – many of the same species occurring on islands right across the Pacific.

Which leaves just one major contribution to the micro fauna and flora of the wind – and that is a considerable one, amounting at times almost to a tidal wave.

The source, of course, is pollen.

Windlovers

Incest is probably the only naturally dirty word.

There is as much resistance to inbreeding amongst plants as there is amongst humans and other animals. With only a few exceptions – violets and pansies are outrageous slatterns – flowers go to great lengths to avoid self-fertilisation.

The problem for plants is that they are largely sedentary and enjoy sex only by proxy. They need a middle man, an outside agent, who can transfer male sex cells to female receptors, and to meet this need they have evolved two sexual strategies.

The first and flashiest is called *zoophily* and involves the use of beetles, birds, bats, wasps, flies, butterflies and moths as pollen carriers. These surrogates are enticed into the business by showy advertisements of large and colourful blooms, and by offers of sweets in the form of nectar. Moving directly from flower to flower, such 'animal lovers' are very efficient, and the plants which use them need to produce only a small quantity of pollen.

The second strategy is that of the 'wind lovers'. *Anemophilous* plants are more austere. They eschew sexual frills and have only small, green, unscented, inconspicuous flowers. This represents a considerable saving in effort and expense, but these species have to compensate by producing enormous quantities of pollen. With the result that every square metre of Earth surface is dusted each year with an estimated ten grams, or 100 million grains of pollen. This is clearly visible on the surface of ponds, on snow at high altitude and on arctic floes. Pollen falls not only provide food for a whole community of animals, but are dense enough at times to absorb extra heat and even melt the snow or ice on which they lie.[97]

There are two annual pollen peaks in each hemisphere. One is in the spring, when temperate forest trees and shrubs flower at the beginning of the growing season, many of them even before the new leaves appear. The second surge comes later in the summer, when annual herbs are full grown and beginning to think about putting something by for the future. These are two quite different approaches to reproduction, but it is significant that almost all the species in both groups of wind-loving plants are temperate forms, living in the zone of regular westerly winds.

The minimum wind velocity necessary to lift most pollen is about ten kilometres per hour. Below this, pollen either fails to become airborne at all, or falls around the parent plant. At twenty kilometres an hour, an average pollen grain can reach a height of ten metres within a hundred metres of its source, and once into turbulent air, the sky is the limit.

Juniper pollen has been collected 480 kilometres from its source, larch a distance of 700 kilometres, spruce and birch 1000 kilometres, and several species of fir over 1200 kilometres from the nearest parent trees.[389] There are records of clouds of pine pollen from Scandinavia travelling at least 1200 kilometres in upper airflows before pouring down like rain on to the icefields of West Spitzbergen in the arctic.[385] There is at least one report of pollen from Australian pines landing 1800 kilometres away on the Tasman Glacier in New Zealand.[340] And a study of peat from the mid Atlantic island of Tristan da Cunha, has found pollen from the southern beech *Nothofagus*, 4500 kilometres from its nearest source in Argentina.[190]

Over most of the Earth surface, the ecologically dominant plants now belong to wind-pollinated species. Temperate and tropical grasslands, coniferous and deciduous forests, and nearly all semi-desert lands, are dominated by species which reach up to shed pollen into turbulent air. Only rain forest, where wind is largely excluded, relies on animal-pollinators. Just ten per cent of the world plant species are wind lovers, but they account for ninety per cent of all individual plants. And paramount amongst these for the last fifty million years have been the grasses, the cereals which have come to mean so much to so many species, setting the stage for the evolution of the mammals and the emergence of human beings.

The grasses appeared so suddenly and spread so fast that Darwin described them as 'an abominable mystery'. Looking back at their emergence, Loren Eiseley said, 'In that moment, the golden towers of man, his swarming millions, his turning wheels, the vast learning of his packed libraries, glimmered dimly there in the first grain of wheat.'[121]

The almost negligible weight of a handful of new grass pollen changed the face of the world and made it ours. And the success of all the grains ever since has been due in very large measure to the strategy these first ones borrowed from the pines, to an ability to harness the restless air, which can shake and bend and break the biggest trees, and make of it instead a gentle lover.

The notion of wind as an inseminator is not unique to botany. The Christian account of creation begins with a spirit like the wind that 'moved upon the face of the waters'. In the Babylonian epic, life begins with 'raging winds filling the belly' of a primaeval dragon. The Hindu gods come into being when primordial waters are inflated by wind 'as a blacksmith blows up a fire with his bellows'. The Algonquin hero Hiawatha is born after his mother becomes pregnant by the breath of the west wind. In Egyptian mythology, vultures and the mares of Lusitania are fertilised by the wind.

Everywhere there is a sense of wind as a creative spirit, an invisible force exercising, in the modern idiom, a genetic influence.

It is not a random one. If it were, genes would be dispersed along an even gradient around their source. Breeding would take place most often between close relatives. But all studies on the effects of wind distribution show a strange double-humped curve. Breeding does occur between close relatives, but it also takes place just as often between distant relatives, though seldom with those in between.[20]

What this means is that wind, while securing a firm base for any population, is also a potent force in favour of genetic novelty.

It shapes new forms, probes new habitats, stirs up unrest and foments evolution.

'A fine wind', said D. H. Lawrence in one of his less pessimistic moods, 'is blowing the new direction of Time.'

6. THE SOCIOLOGY OF WIND

Take a breath. Look at it under a microscope.

Even the cleanest air, in the centre of the south Pacific or somewhere over Antarctica, has 200,000 assorted bits and pieces of flotsam in every lungful. And this count can rise to as high as 375 million per gasp on a six-lane highway during the rush hour.

But avoiding these extremes, consider instead an average breath, taken with pleasure after a late summer shower of rain, in the streets of a quiet suburb, not far from a coastal town, one with a little light industry, lying on the edge of farmland which bears a crop of ripening grain. This half-litre of cool evening air will still, under magnification, become an astonishing gruel of twenty-five million particles and inclusions, each smaller than the diameter of a single human hair, all stirred up by the tireless spoon of wind.

Most of the material, perhaps 99 per cent, will be minute particles of salt, clay, ash from the smoke of forest, industrial or domestic fires, and some of the thousands of terpenes or unsaturated hydrocarbons that are produced every day by natural vegetation. These are the *cloud condensation nuclei*, which form blue and grey hazes on dry days, and concentrate water vapour into mist and rain when the humidity is high.

A far smaller fraction of the flotsam consists of larger, angular mineral fragments, either blown up from the ground, carried in from distant volcanic eruptions, or falling down from that frontier where meteorites burn themselves out on contact with our atmosphere. Or of particles of paint and asphalt and rubber produced by friction on our roads. These are the *ice nuclei*, which come to lie like secret seeds at the heart of every pearl of hail.

And growing in, or growing on, or simply being carried along by, this fertile soil, is a zoological garden of local and exotic fauna and flora.

In that same lungful there are likely to be – a few stray viruses in transit between hosts; four or five common bacteria; fifty or sixty fungi, including several rusts and moulds; one or two minute algae drifting in from the coast; anything from two to twenty pollen grains, depending on the season; and possibly a fern or moss spore, or an encysted protozoan.

All this is inevitable.

It is the stuff of life and air, and the fact that we never see any of it, unless light should strike the microscopic particles just right, changes nothing.

We live in and breathe a rich and potent brew.

And we share this environment with more than a million macroscopic and highly visible species – an astonishing number of which take to the air and travel on the wind.

Dispersal

Zephyrus, the west wind, was a savage and baleful influence, living in the mountains of Thrace and taking pleasure in brewing storms.

Until, it is said, he was diverted by the gracious Chloris, a nymph by whom he had a son. Thereafter, he became a sweet-scented gentle breeze whose breath fanned Elysium and coaxed the spring into bloom. When Chloris died, she turned into one of those early flowers, a pure white one with five delicate sepals. She became the 'windflower' – *Anemone*, which continues both to be fertilised by the wind and to entrust its soft white plumed fruit to the breeze.

Today there are species of anemone growing on mountain sides and open pastures throughout Europe, Asia and North America, and even one *Anemone thompsonii*, on the upper slopes of Mount Kilimanjaro, south of the equator.

The direct effects of wind on vegetation are obvious. Trees growing on cliffs and windy shores are permanently swept to leeward, 'as pines keep the shape of the wind, even when the wind has fled and is no longer there'.[430] The extent to which such 'flag trees' have been wind brushed, can even be used to assess the prevailing wind direction and its average speed. The extreme condition, of course, occurring on mountain peaks, where trees prostrate themselves before the wind in the totally submissive posture known as *Krummholz*, or else change their whole lifestyle and become an 'elfin wood'.

Indirect social effects are less obvious.

Vegetation, the total shape and look of plant cover in an area, is strongly influenced by wind. Although the species involved in any community are drawn from a wide variety of families, they tend to share a common strategy.

The closed canopy of a tropical rain forest protects thin soil from sun and rain, but it also excludes wind and by doing so, forces most of the species involved to adopt alternative strategies, such as insect-pollination and animal-dispersal. But in all other societies, wind is a vital, often dominant, factor.

In hot dry climates with sparse vegetation and isolated showers of rain, wind carries the fruits and seeds of plants that have managed to flower, to other areas where they might be rained on again. And in temperate and polar climates, both wet and dry, the predominantly westerly winds spread seed as widely as possible during the short summer season.

Adaptations to the wind are as varied as the species involved and range from movement of the entire plant, to the development of seeds so small that they get carried as far and as fast as pollen.

On deserts, steppes and prairies, there are several species so mobile they have all earned the name of 'tumbleweeds'. The most abundant and conspicuous of these is *Salsola kali*, of the spinach family, which originated in Asia and has somehow been introduced into North America, Africa and Australia. This stiff spreading herb with prickly leaves grows in an almost perfect sphere and as it dries, detaches itself and bowls along in the wind, scattering seed as it goes. Sometimes several get tangled together, 'until at length they form a ball as big as a cartload of hay. Such balls or steppe-witches have been lifted up by whirlwinds and driven bounding over the plain till stopped by a fence or other obstacle.'[399]

Another striking tumbler is the Rose of Jericho, *Anastatica hierochuntica*, a delicate wild mustard plant of normal form until it dries, when the branches all close in to form a snaky ball, like hair on the head of a gorgon. This rolls across the desert floors of the Middle East and North Africa until it gets wet. Then the branches recurve, the ball opens and the seeds drop out and germinate.

Such total, whole-plant response to the needs of wind dispersal seems to be confined to desert species, who both need to be migratory, and happen to live in areas where there are open wind arenas. Species in temperate climates tend to concentrate instead on mechanisms which detach the fruits and seeds and keep these in the air for as long as possible. They have accomplished this by the invention of the sailplane and the parachute.

Gliding, which is essentially a way of delaying an inevitable surrender to gravity, is something which has evolved mainly amongst tall trees, in which fruits and seeds already have the advantage of height. The only notable exceptions are the orchids, whose epiphytic way of life allows them to take advantage of someone else's height.

All 15,000 species of orchid produce minute seeds in which the coat is thin and loose and drawn out into some kind of delicate wing. And, casting their fortune to the winds in this way, they hedge their bets in abundance, a single plant sending out as many as 374 million emissaries. Half a million of the lightest seeds weigh just one gram, so it is not surprising that some at least of these airborne embryos land on suitable substrates a very long way from their parent plants.

There are orchids growing today on a number of remote oceanic islands, including Agalega, 650 kilometres downwind of Malagasy; Bermuda, 930 kilometres from the coast of North Carolina; the Galapagos, 960 kilometres from South America; Kermadec, 1120 kilometres off New Zealand; the Seychelles, 1500 kilometres east of Africa; and the Sandwich Islands, which are perhaps 3000 kilometres from their nearest source of supply.

The seeds of trees on which orchids grow, are far larger, but their coats have been spread into wings in much the same way. The simplest development has been an irregular, more or less circular, outgrowth that allows the seed to drift away like a dry leaf, but there are a number of refinements that give the seeds of some species distinct aerodynamic advantages.

One of the most effective devices is the growth of a single or oblique wing attached to only one side of the seed, so that it rotates rapidly as it falls, spinning in the wind, putting off the moment of touchdown. The seeds of all pines are good examples of this technique, falling in a fair wind as far as 800 metres away from the tree that bears the cone. There is a record of at least one seed from the Scotch Pine *Pinus sylvestris* travelling seven kilometres in a storm. The giant feather-winged seeds of the Malayan climber *Macrozanonia macrocarpa* take to the air on windy days like swarms of butterflies and have even been known to fall on the decks of ships several kilometres out at sea.[496]

In a number of other species, the whole fruit is winged. Maple and sycamore trees produce a pair of winged seeds, that can helicopter for fifty or sixty metres from a parent tree. The Norwegian maple *Acer platanoides* has a slight curve in each wing that makes the fruit rotate so rapidly that it can, given a decent breeze, even gain altitude and fly for 100 metres or more. Logical developments of this system are found in the European hornbeam *Carpinus betulus*, which has three wings, the climbing verrain *Congea velutina* with four wings, and the Malayan shrub *Ancistrocladus extensus*, which has a five-bladed fruit that looks exactly like an old-fashioned propeller.[399]

Other variations on the theme are provided by trees producing fruits which are basically spherical, but have thin stiff wings running longitudinally around them. African species of *Combretum* and *Terminalia* have four or five wings, while Malayan trees of the genus *Pentace* vary, according to the species, from four to as many as ten fins. It is interesting that the most widely distributed and abundant species in all these groups are those whose fruits have fewer, larger wings. But, with the exception of the orchids, no plant which relies on the development of wings of any kind to disperse its seeds or fruit, no matter how large or efficient these appendages may be, has ever managed to move more than a few kilometres in this way. There are no such species on distant oceanic islands.

Taking advantage of passing winds presents a very different problem for plants that grow close to the ground. They have to rely on a totally different strategy, which involves the production of fruits and seeds decorated with fine silky plumes that function like parachutes.

Typical amongst these are composite flowers, including many species of *Carduus* or thistledown, which produce a fluffy mass of seeds. Each seed is suspended beneath a plume of white floss that slows its descent in still air for as long as eight seconds a metre. All it takes is a wind of just three kilometres an hour to keep such a seed permanently airborne, at least in dry air. Though the plume contains an oil that sheds water, rain tends to bring most of these seeds fairly quickly down to damp ground, which is precisely what they need for germination.

Other species with plumed fruits or seeds have similar capabilities. The common groundsel *Senecio vulgaris* has seeds which fall at 3.5 seconds per metre. The umbrellas of *Taraxacum*, the dandelion, fall at a rate of four seconds a metre in still air. Thoroughwort daisies of the genus *Eupatorium* take 4.3 seconds to fall a metre, while some of the *Salix* willows prolong the process for six seconds or more. At times the air near flowering willows is so full of pale hairy seeds that it looks as though a snowstorm is raging. But the most buoyant of all plumed seeds seem to be those of *Typha latifolia*, the common bullrush or cattail, which subsides with considerable reluctance, even in still air, falling at a rate of ten seconds or more per metre.

Bullrushes and the common reed *Phragmites communis* are amongst the most abundant of all plants on Earth. They, a number of other grasses, and a variety of daisies and thistles, have spread through the air to colonise places as remote as Juan Fernandez, 640 kilometres from Chile; the Cape Verde islands, 800 kilometres off the coast of Senegal; St. Helena, 1820 kilometres from Angola; and even Tristan da Cunha, where a little creeping composite called *Chevreulia stolonifera* now grows, over 8000 kilometres downwind of the similarly sandy hills of Uruguay.

A number of plants without special aerial adaptations or apparatus benefit from prevailing winds by being blown by brute force along the surface. The dry fruits of *Mesembryanthemum*, the South African 'everlasting', roll erratically across the flat Karroo, scattering their seed as they go. The seeds of many rock and wall plants are carried into new cliff crevices by stray drafts. And there is even a small arctic daisy *Ambrosia trifida* whose fruits stand up on several smooth feet that carry it for kilometres across the ice like tiny skis.

Ordinary everyday winds put a large number of plants, at all stages of their life cycles, to flight. Collections taken from the summer air over Louisiana found the seeds of four species of daisy or thistle, five different grasses and a cottonwood, all lurking at a height of 1500 metres, just waiting to descend on some new and likely habitat.[161] Powerful movements of air, the real winds of occasion, play a sporadic but perhaps even more influential role.

Hurricanes, tornadoes and dust devils undoubtedly pick up and redistribute masses of material, which includes whole plants and quantities of seed.

On July 15th, 1822, a rain of seeds from *Galium*, the common bedstraw plant, fell on Marienwerden in northern Germany. Two days later, another similar mass fell hundreds of kilometres away on the banks of the Oder in Silesia, and the rest of the load was dumped unceremoniously over what is now Poznan in central Poland. The progression of this vegetable cloud suggests that it was being carried slowly eastward at high altitude in a westerly airstream and stripped of its burden by a series of storms.[142]

Other deliveries have been somewhat more idiosyncratic and suggest that some species, no matter how far they have been carried or how long they may be airborne, actually benefit from the time spent aloft.

One day in the summer of 1897, a number of small blood-coloured clouds gathered over the Adriatic coast of Italy. When the storm broke, it brought down showers of seeds that carpeted the ground around several small towns near Macerata. These seeds proved, on examination, to be those of a legume, the Judas Tree *Cercis siliquastrum* native to central Africa, and had probably been carried aloft, along with a mass of fine red dust, by an updraft starting somewhere south of the Sahara. The most fascinating thing about this particular cargo is that, by the time it fell, 'a great number of the seeds were in the first stage of germination'.[141] They had begun, like well-behaved passengers, to prepare themselves for landing at their final destination.

Mass movements and random falls of this kind are rare, and probably even more rarely successful. But it does not take freak weather or unusual emigration of this order to make any plant species a good colonist.

Most plumed seeds under ordinary circumstances in fair winds stand a good chance of travelling forty kilometres in a single flight. They settle in a suitably protected spot, wait out the winter, produce a new plant, and the following year make a similar flight on prevailing winds in the same direction. Leaping along in this way, a species can travel 40,000 kilometres, which is a complete circumnavigation of Earth, in 1000 years. Or all the way from London to Peking in less time than it took a group of Bronze-Age Britons to build Stonehenge.

Trees, which take a long time to develop from seed until they are old enough to fruit, move much more slowly and the wide distribution of any species is a better indication of its antiquity.

But wind everywhere is so potent a force for dispersal that it accounts on its own for the present geography of an enormous number of plants that have been blown, swept, lifted, rolled, tumbled, skated, or even sailed, to almost every corner of this air-conditioned planet.

Migration

We tend to underestimate the power of wind.

The problem is that our experience of it is that of a large animal, exposed for the most part to moderate winds, blowing horizontally not too far from sea level. We can only guess what it must be like for a small animal facing high winds that can on occasion blow straight up from the ground.

The difference is one of relative body surface. The surface area of any animal is a product of two dimensions and varies as the square of either. Which means that the exposure of skin on a common shrew just five centimetres long, is roughly forty times greater for its weight, than that of a giant shrew which, for some inconceivable reason, happened to grow to the size of an adult human. As a result, the problems which winds pose for tiny shrews can be forty times greater than those faced by human beings in similar situations. And as wind speeds increase, the problems multiply alarmingly.

The pressure exerted by a wind varies with the square of its velocity. So, when a moderate breeze of just twenty-five kilometres an hour, which is barely enough to raise dust, increases to the top speed of a full gale at 100 kilometres an hour, its pressure multiplies, not just four times, but sixteen times. Which means that a gale is sixteen times as bad as a breeze for human beings, but it becomes 640 times – that is 40 × 16 times – worse for hapless shrews.[93]

A wind of twenty-five kilometres per hour produces a dynamic pressure of about half a gram per square centimetre. Which is already enough to bowl over any five-gram shrew stupid enough to stand broadside on to such a breeze. Few shrews are that daft, but accidents do happen. In the playground of a Yorkshire school, a child weighing twenty-five kilograms was lifted by a freak gust of wind so that a teacher standing on a balcony could see her hanging there at eye level six metres above the ground, 'with her arms extended, and her skirts blown out like a balloon'.[6] Heaven knows what such a gust could do to a careless shrew.

By rights the air on really windy days ought to be filled with a flurry of unwary voles, mice and lemmings. It isn't, at least not obviously so, but enough wildlife does get wafted about to make it necessary to take wind seriously as a force to be reckoned with in zoogeography.

Many biologists find it hard to accept that wind can lift and carry anything larger than a pollen grain. There was a time, not too long ago, when a similar faction found it equally difficult to believe that a heavier-than-air machine could fly. But given the right conditions, both of these things become possible, and evolution has not been slow, either to take advantage of accidental dispersal by wind, or to see the survival value in becoming deliberately airborne.

Ecologists recognise more or less natural boundaries to plant and animal communities, drawing, for instance, a 'timberline' around the world at the highest altitudes and latitudes where forest can survive. On Mexican volcanoes and African peaks near the equator, this may be as high as 4250 metres, in the Himalayas it is generally around 4100 metres, and on the edge of the arctic circle, it descends all the way to sea level.

Everything beyond this line, which roughly corresponds with the highest mean monthly temperature of just 10° Centigrade, is known as the alpine or polar zone. It is inhabited by dwarf plants, cushiony lichens and a variety of animals, most of whom creep into warmer moister microclimates in convenient crevices. The far boundary of this zone is the permanent 'snowline', where mean temperatures never rise above 0° Centigrade, and where flowering plants, as well as the life they support, come to an end. This line was thought, apart from a few accidental strays, to mark the limits of life on Earth, but in this century an entirely new life zone has been discovered.[474]

Tiny dark-coloured three-pronged bristletails, and two-pronged springtails, perhaps the oldest and most primitive of all insects, have been found living, quite successfully, on snowfields and glaciers. And a number of jumping spiders have been seen, in apparently robust good health, bouncing across the ice at 6700 metres on Mount Everest, where 'they were not only in proud position of being the highest permanent inhabitants of the earth, but seemed to be alone in their isolation. No other living thing has been found to share their loneliness. There is nothing but rock, snow, and ice. What they get to feed on is a mystery.'[423]

The secret of their success, is that they have their food sent in. Close examination shows that snowfields and glacier surfaces are littered with a choice variety of refrigerated food, airlifted from the plains.

In the warmer months, updrafts build up around all mountain ranges, particularly in the afternoons, pushing parcels of air upwards at ten metres a second or more, carrying along with them anything light enough to be lifted. Pollen, spores, seeds and fragments of plants are levitated and, as this elevator slows toward sunset, the cargo is scattered over the higher slopes, some of it on snow. Bacteria and tiny algae thrive on the organic deposits, turning the snowfields bright yellow or red and green. But what the carnivorous spiders and springtails are waiting for is an appetising variety of aphids, gnats, midges, beetles, butterflies, moths, grasshoppers, ants and mites that are cast out on to the snow like seasoning. They are waiting to be fed by the wind.[315]

Lawrence Swan of the San Francisco State College was the first to recognise this wind habitat and called it the *aeolian zone* in honour of Aeolus, Greek god of the winds.[473]

The quantity of dead creatures is considerable. In just twenty minutes of one afternoon, 400 specimens were deposited on ten square metres of the slope of Pir Panjal in the Himalayas.[315] An average of five or six insects fell on each square metre of snow each day in California's Sierra Nevada, producing seasonal totals as high as 1376 specimens per square metre.[371]

During migration of butterflies or locusts, the fallout can be even more impressive. And this living rain goes on continuously through the summer, through all the summers of history, with the result that a whole community of beetles and scavenging flies have come to congregate around the melt point of snowfields and glaciers, waiting for the thaw of deep-frozen meals, some of them kept on ice for centuries. 'Grasshopper Glacier' on Mount Cook in Montana is studded with whole swarms of Rocky Mountain locusts which fell around its source between 350 and 600 years ago.[187]

Much, but by no means all, of this faunal fallout is killed, presumably by prolonged exposure to low temperatures at high altitudes. But some of the species involved are merely moribund, while others are still unquestionably alive and active. The fact that as many as eleven specimens per square metre have been found on flat arctic snowfields without the benefit of mountain updrafts, suggests that at least some of these creatures arrive there of their own volition, attracted perhaps by ultraviolet reflections from bright patches of new snow – in precisely the same way that insects can be lured to a light trap at night. It is certainly significant that dirty snow patches usually have very few or no insects on their surface.[117]

Some biologists are still reluctant, even in the face of this evidence, to accept that fallout in the aeolian zone involves anything more than a relatively small number of creatures who happen, by accident, to be carried away from their normal haunts. But it becomes very difficult to ignore the implications of these studies, which point to the existence of a distinct aeroplankton, a windborne ecology, whose strays and discards alone land up on distant snowfields.

An argument in favour of such a wind zoo, is the fact that a significant portion of the air fauna is neither winged, and therefore already airborne, nor light enough to be blown away under normal circumstances. There is at least one group of invertebrates who are willing aeronauts, deliberately constructing their own magnificent flying machines.

Every once in a while, when conditions are just right, usually in the spring or the autumn with a gentle breeze or updraft in otherwise still air, the largest and lightest canopy in the world is erected in California. It is almost 2000 metres wide and it soars into a diaphanous, but mighty arch 1200 metres high over Sentinel Rock in Yosemite National Park. Sometimes there is another, slightly smaller awning stretched right across Bridal Veil Falls. And both of these are made in a single day by millions of little vagabond spiders.[218]

Prompted by subtle environmental cues, involving temperature, humidity and the degree of crowding, great numbers of spiderlings at certain seasons get a common urge to travel. They climb, each one, to the top of the nearest vantage point, which may be a rock, a tree or a fence post, and stand there on tip-toe with their abdomens in the air. Then their spinnerets begin to work and one or more threads appear. 'Occasionally the spider will attach a small flocculent mass to these threads which increases the force of the current of air upon them. This spinning process is continued until the friction of the air upon the silk is sufficient to buoy up the spider. It then lets go its hold with its feet and is carried off by the wind.'[79]

Normally each young aeronaut would drift off on its own, but if wind comes up, the threads are blown across to other rocks, trees or furrows, weaving together to form a sheet of silk that is almost invisible, unless it is touched by light and dew, when it ripples 'like the wake of the moon on slightly disturbed water'. And if the wind drops at the critical moment, and the ground beneath the communal web is warm enough to produce a little lift, the whole incredible canopy floats high over places like Yosemite for a few magic moments.

It is windblown fragments of such webs that have come to be called 'gossamer', from the French *gaze à Marie* – the 'gauze of Mary', which is said to have fallen from her shroud at the Assumption. Gossamer has puzzled people for centuries. Chaucer considered it amongst 'the unsolved riddles of the universe'. Spenser attributed it to 'scorched dew'. James Thomson in *The Seasons* said: 'How still the breeze! save what the filmy threads of dew evaporate brushes from the plain.' But Gilbert White, the ever-reliable naturalist of Selborne, knew what they were. He described a 'prodigious shower' of gossamer that fell on September 21st 1741, blinding his dogs, which struggled for hours to rid themselves of the sticky webs.

The French naturalist Jean Henri Fabre, in his usual elegant way, experimented with young orb-weaving spiders to find out just how much updraft they needed to achieve successful lift-off. Bringing a nest into his study, he found that even with doors and windows closed, one shaft of sunlight was sufficient to create a column of rising air, and it was not long before the first spider found it.

With the aid of a sail of thread so fine as to be invisible, she makes her ascent. She ambles with her eight legs through the air; she mounts gently swaying. The others, in ever-increasing numbers follow by the same road. Anyone who did not possess the secret would stand amazed at this magic ascent without a ladder. In a few minutes most of them are up, clinging to the ceiling.

On another occasion, watching crab spiders disperse from their nest, Fabre described

a continuous spray of starters, who shoot up like microscopic projectiles and mount in a spreading cluster. In the end, it is like the bouquet at the finish of a pyrotechnic display, the sheaf of rockets fired simultaneously. The comparison is correct down to the dazzling light itself. Flaming in the sun like so many gleaming points, the little spiders are the sparks of that living firework. What a glorious send-off! What an entrance into the world! Clutching its aeronautic thread, the minute creature mounts in an apotheosis.[130]

Charles Darwin, watching a spider on a fence post in South America, noticed the behaviour of the separate threads. His specimen

darted forth four or five threads from its spinners. These, glittering in the sunshine, might be compared to diverging rays of light; they were not, however, straight, but in undulations like films of silk blown by the wind. They were more than a yard in length, and diverged in an ascending direction from the orifices. The spider then suddenly let go its hold of the post, and was quickly borne out of sight.

He suggested that the lines diverged and were kept separate because they carried similar and mutually repulsive electric charges – which is a suggestion still worth pursuing.[95]

Once airborne, spiders can travel almost anywhere. Aristotle watched spiderlings wafting over a summer meadow. The early English naturalist Martin Lister climbed to the summit of York Minster in 1670 and saw them sailing by above him. An American balloonist reported groups travelling along 1000 metres high over Texas in 1874. Darwin found thousands of small red spiders in the rigging of the *Beagle* when his ship was 100 kilometres off the coast of Argentina.[87]

The literature is full of progress reports on the travel and travails of spider aeronauts, but the first quantitative study was one made from an antique biplane over the southern United States and northern Mexico. Beginning in 1926, this succeeded in catching 1500 of the tiny balloonists belonging to more than forty-five species, operating at a variety of altitudes.

Most were taken during the winter months, possibly as a result of autumn movements, but even as late as December the air was filled with gossamer and 'often when landing the struts would be almost white with the silken webs wrapped around them'.[161]

Spiders were caught at every altitude from 10 to 4500 metres, with concentrations at 60 and 300 metres which may have been due to a natural weather or thermal ceiling at that level on those particular days. The loftiest catch of all was a sheet-web weaver of the genus *Linyphia* taken at over 4600 metres, but given the fact that spiders jump about on the Himalayas at 6700 metres, it is likely that there are other eight-legged astronauts floating undetected in the air a great deal higher still.

This pioneer study was supplemented in 1975 by a year-round survey using a suction trap just thirteen metres above ground level on the roof of a building in Texas. This device caught 3400 spiders in the act of leaving the ground, most of them on still days in spring and autumn. The majority were small sheet-web builders and crab or lynx spiders.[418]

Most spiders in flight are, predictably, young ones off on their first great migration, but all collections include a number of specimens that are fully adult. Which suggests that taking to the air may be something that spiders do more than once in their lifetimes, and usually by design.

In theory, these 'arachnauts' ought to end up wherever the wind takes them, which could involve random landings after several circumnavigations of Earth on the skirts of a jet stream. In fact, the distribution of even the most wayward spiders is incredibly precise. Each species has its favoured habitat and tends to be found only in certain areas, on certain kinds of vegetation. Like even the most sedentary creatures, they are rare here and common there, with large gaps in their distribution. It is likely that many of these blank spaces remain so, because spiders which land there and find conditions unsuitable, take off again.[484] But this may not be the only explanation.

The definitive study of spiders in flight remains to be made, but the little evidence we have suggests that the creature suspended beneath its gossamer balloon may be very much in control.[363] By reeling in one or more of the parachute threads, a spider can regulate its descent and might, with the help of such a spider spinnaker, even be able to steer. 'If the wind rises he can reef in his thread to "shorten sails". If the wind drops he can pay it out again like a yachtsman in a tight race. His own body is his craft and crew.'[351] And there is nothing short of hunger to stop each sailor voyaging on for thousands of kilometres, even across whole continents or seas, remaining aloft for weeks.

There is a tendency, even amongst biologists, to belittle spiders as 'drifters' and to label their travels accidental or inadvertent, but this does them an injustice. The winds may determine their general direction of travel, but few if any of them are transported passively like leaves or dust.

It makes no more sense to accuse them of aimlessness than it does to describe human passengers in commercial aircraft on scheduled routes as arriving at their destinations by chance. Spiders and humans, under normal circumstances, only become airborne in the first place as a result of a deliberate act that launches them from the ground, and everything we know suggests that, once on the wind, spiders have very much more control over their destiny than human passengers in the belly of a 747.

Locusts have had an equally bad press ever since they 'covered the face of the whole earth, so that the land was darkened' in Egypt. The plague that beset Pharaoh is described as being brought by the east wind and then taken away by a 'mighty strong' west wind that cast them into the Red Sea.

That may not have been the end of them. There is at least one report of a swarm carried out to sea from the Barbary coast of North Africa by a trade wind, which settled at night on the surface of the water, drowning at least half its number. But at dawn, the survivors rose 'in a mass which glittered in the sun' and made it on the second day to the Canary Islands, where they laid waste to Tenerife.[230]

Entomologists down through the years have continued to link the movements of the desert locust to 'accidents' of this kind or to convenient and prevailing winds. But a recent study casts grave doubts on these old and comfortable notions.

Desert locusts are nothing more than ordinary short-horned grasshoppers with a Dracula complex. Under certain conditions of overcrowding, they react to each other by growing vivid red cloak-like wings and take off in coherent swarms, thousands of millions strong, intent on devouring 'every herb of the land'. The transformation of ordinary green grasshoppers into fearsome red and brown hordes is so complete that even their name changes, from *Schistocerca solitaria* to *Schistocerca gregaria*.

The swarms travel at about fifteen kilometres an hour, following a roughly circular route in a clockwise direction around the Sahara, with some groups making side trips to breeding grounds on the Persian Gulf and the great lakes of East Africa. What they need is recent rain, enough to wet the ground sufficiently to keep the eggs from drying out, and to supply new green vegetation for their offspring. They find the rains along the doldrums or intertropical convergence zone, which drifts north and south with the seasons, but there is no way they can do so by merely following local winds. Low pressure zones bubble up all along the wind equator, creating a confusion of short-lived squalls and calms which would soon lead passive drifters astray.

Close observation of locust swarms shows that they have short-term tendencies to head for high ground, which is more likely to attract rain, and that they will fly headlong into the wind to get to areas where they can see or smell rain already falling. In the absence of either of these direct cues, however, they maintain station on their broad migration circuit by following preferred compass directions of travel, regardless of the wind. Which brings the swarms to, and keeps them in, areas where the doldrum rains are most likely to fall, even if this does not happen until several weeks after their arrival, or on occasion fail to take place at all.[503]

The moving swarm is so clearly in active control of its destiny, that it even adopts two distinct strategies according to wind direction. If the wind blows in the preferred direction, the swarm takes advantage of it, rising like a thundercloud to heights of 3000 metres and bowling along as a cohesive mass. Individual insects in this column gain ground in the wind at height, and then drop off the leading edge of the swarm to feed while the rest of the mob rolls by.

When forced to fly against the wind, the tactics are entirely different. The swarm becomes diffuse and hugs the ground, keeping as far out of the turbulence as possible. There is clearly nothing accidental about such procedures. The behaviour in each case is highly adaptive and designed to give the species the best chance of breeding in the most favourable habitat.

The grasshopper gambit differs from the spider solution only in the form of its specific adaptation to the wind – which boils down in the end to a question of aerodynamics.

Flight

Getting anything that is heavier than air off the ground involves the creation of lift. Which, in essence, means an ability to be pushed up into the air by wind.

The first creatures to try anything like this deliberately were almost certainly insects living in Silurian times, probably as a means both of escaping from predatory relatives and of reaching new and more desirable habitats. The first flights may have been no more than exaggerated hops in the manner of the Wright brothers' earliest attempts, but in the 300 or 400 million years since then, every imaginable way of becoming airborne seems to have been exploited. Some of these are extraordinary, but all share the common problem of creating sufficient aerodynamic lift to counteract body weight.

There are two basic approaches to the problem. The first assumes a quantity of air in relatively rapid motion past a still body, and the second requires the reasonably rapid movement of that body through still air. In other words, either you wait for a decent wind, or else you make your own wind by jumping off a high place. It is hard to say which was the earliest approach, but the first method is the one still used by dandelions, by all other plants with plumed seeds, and by spiders.

Spider aeronauts are sometimes described as ballooning, which implies being lighter than air, which they are not. A better description of their system of transport would be parachuting. This involves some kind of wing or canopy which develops a drag force that resists weight which, by definition, acts vertically downwards. A human dangling beneath a silk chute has only enough drag to slow inevitable descent to a safe speed. But a lighter spider can, by increasing the size of its silk canopy, produce a drag force greater than its weight and rise from, rather than fall to, the ground. There are, however, few other natural parachutists. A more common approach is to become a glider.

All the winged seeds are gliders. So too are most aquatic creatures who soar effortlessly with assistance from the greater density of water, but there are several who exercise similar talents above the surface.

A small blue species of squid will, when alarmed, jet itself singly or in small groups out of the water, rising to a height of four metres or more and then glide, with fins spread, in perfect formation like a flight of javelin back to the sea up to fifty metres away. This is the flying squid *Ommastrephes pteropus*.

Another species, the hooked squid *Onychoteuthis banksi* has, in addition to the fins on its mantle, broad membranes on the third pair of arms to give it extra lift. This is the animal which landed many nights on the decks of Thor Heyerdahl's raft *Kon-Tiki*, and which has been reported, in windy conditions, to smash into the bridge and superstructure of merchant vessels at sea.

There is, of course, a whole family of fish – the *Exocoetidae* – whose lower tail fin is elongated into a propeller that beats eighty or a hundred times a second at the surface, hurling the animal out into the air. Once airborne, facing always into the wind, flying fish erect their broad winglike pectoral fins and glide for fifty or sixty metres before falling low enough for the tail to beat the water into a froth once again, sending it on the second or third bounce back through the surface perhaps 200 metres away from whatever it was that originally put the fish to flight.

The principle of gliding differs from that of parachuting in that aerodynamic lift is never vertical. It is created at right angles to the airflow past a body that slips sideways down through the air from high places to lower ones. Weight and gravity still act directly downwards, but they are held in abeyance for a while by the airflow, by an induced wind passing a body stretched to make it as broad and as flat as possible.

There are examples of gliders in every class of vertebrate animals, including the frog *Rhacophorus dulitensis* which evades its predators by leaping out of tall forest trees in Kalimantan and floating down to lower levels on sails created by spreading the broad webs between its toes. Reptiles are represented by *Draco volans*, the Indonesian gliding lizard that keeps all four of its limbs free for use and soars instead on a pair of frilled and folding paravanes attached to its sides on elongations of the ribs.

Amongst mammals, gliding has been made possible by the

independent evolution of an identical technique in four separate groups of animals. Each has grown a fold of skin, called a *patagium*, which is stretched between the arms and the legs by spreading these out as wide as they can go. *Glaucomys* in North America, *Pteromys* in Europe, *Eupetaurus* in India, *Petaurista* in China, and *Petaurillus* in Indonesia are all true squirrels, most of them nocturnal, that extend their leaps from tree to tree by gliding.

In Africa, their niche is occupied by Anomalurids, a strange mouselike family which takes to the air in precisely the same way, but differs in that its members have scaly tails which act as climbing irons, clamping on to tree trunks to help get them back once again to gliding height. In Australia, gliders are of course marsupial. *Petaurus* phalangers and *Schoinobates* opossums, that can travel more than fifty metres between tall and relatively isolated eucalyptus trees. And finally there is, in the Philippines, a unique cat-sized creature with large eyes called colugo *Cynocephalus variegatus*, which is the best of all the mammalian gliders, swooping in long smooth flights as far as seventy or eighty metres at a time.

Gliding, as a rule, works well only over short distances downhill, but there are some creatures that have parlayed it into a successful long-distance technique.

Albatross have discovered that life above the ocean in the southern westerlies can be all downhill. They glide downwind on a gentle slope towards the surface of the sea, picking up speed as they go. Then, just above the waves, the great birds wheel to windward and rise again. Normally, this would result in a sudden loss of speed, but in the open ocean there is a dramatic wind-shear effect in which the lower layers of air are slowed by friction with water and wind only reaches its full velocity perhaps fifty metres up. So, as the albatross climbs, it passes into air that is moving more and more quickly, and the net result is that its momentum is maintained and it arrives back at its original height with little or no loss of energy. And it can go on looping in this way for as long as the wind blows, which in its home waters, is practically without respite all year round.

Birds like vultures that soar over land, use a different technique. This was first analysed by a British chemist working in India before the First World War, who noticed that cheel kites *Milvus govinda* with one-metre wing spans, took to the air each morning an hour or more before black vultures *Otogyps calvus* with twice their span and wing loading – and that adjutant storks *Leptoptilus dubius* could only put their huge three-metre wings into glide position over grass fires or on the warmest of summer after-noons.[193]

He noticed also that birds soaring with ease would sometimes be poised directly above others forced to flap madly just to maintain height, and questioned the conventional wisdom that tells of tall and continuous thermal columns of rising air.

He was right. In 1953, two meteorologists finally proved that warm air lifts from Earth's surface in great buoyant bubbles, exactly like the vapour that rises from the bottom of a kettle of gently-boiling water.[427] These bubbles or 'thermal shells' are self-contained meteorological systems with their own internal turbulence. If they were visible, they would look like giant rotating smoke rings. And it is inside these that the larger birds with broad wings come to soar so effortlessly over land, that they seem, in the words of Keats, 'to sleep wing-wide upon the air'. They are carried aloft as passengers inside parcels of air that rise faster than the birds' gliding flight causes them to fall, and never reach the bottom of such a bubble, because the air inside it is spinning fast enough to keep them flying horizontally.

This bubble-stream system of convection is clearly the most effective way of transporting heat upwards without too great a loss. Once they reach a certain height and size, the bubbles drift with the wind, redistributing warmth and, with it, the creatures held captive inside the temperature shell. Swallows have been seen soaring and feeding inside such bubbles at heights of up to 2000 metres and can, in an average high altitude wind of around fifty kilometres an hour, be carried 1000 kilometres away in a single day.

Without wind-shear over the ocean or thermal bubbles over the land, soaring is impossible and gliding becomes a problem. The main difficulty with them as ways of getting about, is that there is always some drag. Gliding flight can never be horizontal without drastic loss of speed. To fly on the level successfully there has to be thrust, there has to be power behind the flight. It is not enough just to have aerodynamic wings. You also have to flap them.

Efficient powered flight required three evolutionary changes. First, the wings of gliders had to be lengthened, to increase the area swept with each beat, and therefore the ability to support weight in the air. Second, the muscles attached to these wings had to be modified for the transmission of power on a long-term basis. And third, the body had to be made as streamlined as possible to resist the frictional effects of drag.

The insect solution was two pairs of wings. No creatures except the chimaera that decorate Chinese imperial tombs, or some stained glass seraphim at Chartres, have ever had more than this.

There are more than 700,000 species of insect and 90 per cent of these fly. All seem to have derived their wings from the same ancestral form with eight longitudinal veins, but there are now countless variations, ranging from a single pair of stiff small gliding foils, to elaborate pairs of huge interlocking sails. In termites, both pairs are alike, while flies have concentrated on the forewings, reducing the second pair to tiny balancers. Beetles use only their hind wings, folding and covering them between flights in rigid casings developed from the pair in front.

There are lacewings, netwings, wings like gossamer, horn and leather; wings that are fringed, grooved, painted, stained and stencilled; wings that are metallic, prismatic, polished, opaque, pellucid or totally transparent. And all, except for the dragonflies, whose fixed wings vibrate on the ends of special muscles, beat indirectly by flexing bands of muscle that raise, lower and rotate the airfoils by changing the whole shape of the insect's body.

Insects with four wings, beat them all in unison, sometimes locking fore and aft wings together with a clasp or yoke that makes the coupling more efficient. Very few have a straight up-and-down stroke. The tendency is for them to trace an undulating figure of eight path, moving back on the up and forward on the down-stroke, but the wing surfaces are never completely flat, being thrown into waves, frills and reciprocating curves that adjust to the stress of air pressure at every part of the pattern.

Most butterflies beat their wings at a rate of eight or nine times a second, hawkmoths at about seventy-two, large wasps at 110, solitary bees at 190, horse flies at 240, and house flies at speeds which may reach 330 beats a second. All in the process produce vibrations, hums or buzzes that are characteristic of each species and can be used to distinguish one from another.

In the Amazon, where there are more than 600 species of mosquito, the blood-sucking females of each have their own slightly different frequency of whine, which attracts males with sensitive plumed antennae to mates only of their own kind. But it is a recognition system that has to be used with caution. Angry wasps seem to change their tone, and a honey bee going about its routine business normally vibrates its wings 435 times a second, producing a note that corresponds to A above middle C – but a tired worker of the same species slows a little and drops to an exhausted E.

Insects had the air and the airwaves almost to themselves for 150 million years. The first vertebrate to take off was probably the pterosaur or pterodactyl, a piratical reptile which, 175 million years ago, produced an immensely elongated fourth finger supporting a web of skin that made a fairly efficient wing. It suffered, however, the same fate as the rest of the ruling reptiles, and vanished with them more than seventy million years ago.[51]

It was replaced in the air by bats and birds. Bats improved on the pterodactyl arrangement by supporting their flying membranes on three rather than one extended finger, and the flexibility of this hand-wing gives them an unbelievable degree of manoeuvrability in low-speed flight. But the innermost panel of the wing still has to be supported and tensioned by the hind leg which has, as a result, been rotated at the thigh so that the knees point backwards. This makes it almost impossible for bats to walk on a flat surface, and forces them to take off by dropping from elevated perches. Bats are nevertheless highly successful, but without this handicap, and with a good grasping hand, they might have become a very direct challenge and a viable alternative to the emerging primates.

Bird wings are very different, and can afford to be, because of the unique advantages provided by feathers. These are as thin and light as membranes, but stiff enough to be supported at one end only, instead of having to be stretched between two bones. Which virtually gives all birds two entirely independent locomotor systems, freeing the hind limbs for walking, perching, feeding, fighting or swimming.

In both birds and bats, power in flight is provided by the downstroke of the wing produced by a huge depressor muscle on each side that forms as much as 20 per cent of the whole body weight. Similar development in a man would give him pectorals each as large as a fullgrown Christmas turkey. But there is a definite upper limit to the size to which such muscles can grow, effectively limiting flight to creatures weighing no more than about twenty kilograms. The largest flier ever was probably a super-condor called *Teratornis* that lived in North America about a million years ago.

In species that hover, and therefore have to work particularly hard to keep airborne, there is comparable development of a wing elevator muscle which also lies below the wing, but functions via a pulley arrangement over the shoulder joint, like hoisting sail from the deck of a vessel.

Large birds generally beat their wings relatively slowly, but rates increase in smaller species to reach what seems to be the limit for skeletal muscle in hummingbirds, whose wings blur into invisibility at 100 beats per second. The design, balance and energy consumption of all bird muscle is so good that the most efficient fliers, such as the swifts, remain airborne for most of their lives, landing only for the purpose of mating and breeding.

In the course of 130 million years of evolution, bird wings have taken on a wide variety of shapes, but these are all broadly based on four functional models.

The blunt rounded wings of partridge are designed for manoeuvrability, for rapid short-distance flying at close quarters in dense vegetation. Grouse, quail, woodcock, rail and limpkin all share this configuration, fluttering away when flushed, only to settle again as soon as possible. And jays, cardinals, bulbuls, thrush, flycatchers, darting predators and woodland owls seem, for similar reasons, to have made the same choice.

As the need to cover longer distances takes over from short-range manoeuvre, the primary feathers lengthen and vary, and the wings become more acute. Thus is the pattern of orioles, crows, pigeons, doves and tern, lengthening with each increase in speed, and reaching scimitar perfection amongst martins, swallows and swifts. And on a side branch of their own, the miniature hummingbirds, whose narrow little wings are practically propellor blades.

Equally tapered wings, but ones that have become long and narrow, stiff and streamlined, are the gliding aerofoils of petrel, shearwater and albatross. These are perfectly adapted to minimise drag and create only the smallest possible spillage of air at the tips. Which is a very different approach to that adopted by soaring landbirds, whose solution has been a broad square load-bearing wing with feathers at the tip drawn out into a fringe of fingers. Widely separated pinions on the ends of stork and vulture wings curve upwards in flight, forming an aerodynamic fence that cuts drag by keeping air from flowing round the tip.

During the course of evolution these basic forms have been taken each by its species and bent to special ends. Each has become unique by subtle blends of genes and needs, combining beating and floating in ways that give every species its own characteristic rhythm – so that there is no mistaking the flappet of a lark, the twitch of a hummingbird, the undulating line of a goldfinch, or the dip and swoop of a woodpecker. 'The air', said that literate naturalist W. H. Hudson, 'is their element; they float on it and are borne by it, abandoned to it, effortless, even as a ball of thistledown is borne . . . to dwell again and float upon an upper current.'[224]

Whatever their style, the distance flown by birds or bats is determined in the end by fuel consumption. Amongst migrant species, as much as 50 per cent of the body weight may be consumed in this way, giving a small perching bird with a cruising speed of thirty-five kilometres an hour, a maximum range of about 2000 kilometres before it has to refuel. Many species do make passages this long, crossing over deserts and oceans. The longest is said to be the 3700-kilometre crossing of the Pacific from Alaska to Hawaii by the golden plover *Pluvialis dominica*. But the majority break their voyages down into shorter legs with more frequent feeding stops.

Few migrations, however, take place in still air, so the distance, duration and direction of all flights naturally depends to a very great extent on wind.

When wind is too strong, migration stops altogether. But ducks will take to the air in gales greater than fifty kilometres an hour if the winds are blowing in the right direction. Smaller perching birds prefer speeds of less than twenty kilometres per hour, while some of the soaring species, such as hawks and eagles, find that any wind at all destroys the thermals on which they depend for gliding altitude.[398]

The 'right direction' is not necessarily the same as the preferred compass route, though there is good evidence to show that patterns of migration are set in the first place to take advantage of seasonal wind conditions. The first cyclonic storm of the autumn takes place in North America any time between the beginning and the 20th of August, but when it does, it triggers the departure of swifts and nighthawks from all cool northwestern areas. What they seem to need and like is turbulence, rather than a following wind.[3]

Most birds are uncomfortable in tail winds, which tend to reduce their relative movement through the air dangerously close to stalling speed. The optimal conditions for many species seem to be provided by steady winds that blow at an angle to the chosen path, from behind.

There is a clear tendency for birds to pick those altitudes where the winds suit their particular flight pattern best. By climbing on to the westerly circumpolar flow, giant petrels from Australia are able to travel completely around Antarctica. Slender-billed shearwater from Tasmania use winds in a similar way to carry them much of the way on their formidable loop migration up into the North

Pacific. But no matter how attractive winds might be, most birds avoid them altogether on cloudy nights when they run the risk of becoming displaced by wind, without sun or stars to help them make appropriate course adjustments. Radar studies show that hardly any nocturnal migration takes place on nights which are both windy and overcast.[29]

Despite all precautions, however, it is not uncommon for migrants to be caught unawares and to be blown a long way off course.

Shearwaters and petrels, frigates and tropic-birds get caught up unawares in the eye of hurricanes, trapped there as the storms form and become condemned to travel with them wherever they may go. Tropical Atlantic species are sometimes carried all the way up to Maine and Nova Scotia and dumped on arctic shores. There is one report of a Brazilian hummingbird being deposited on the island of Dominca. The Cape Sable sparrow *Ammospiza mirabilis* is found only on one salt marsh on the southwest tip of Florida, and is probably a hurricane refugee. It was discovered there in 1918 and almost wiped out again by the next severe storms of 1935. Smooth-billed ani *Crotophaga ani* from Haiti, and Bahama honeycreeper *Coereba bahamensis*, are other recent windborne immigrants to the American mainland.

Onshore winds during the autumn migration of oceanic petrels from the Arctic to areas south of Iceland, occasionally throw them exhausted on to European shores. In 1952 a five-day gale 'wrecked' 6700 of Leach's petrel *Oceanodroma leucorrhoa* on the shores of Britain, Holland and France, leaving some dying as far inland as Germany and Switzerland. And the same thing happened again thirty years later in 1982.

Strong offshore winds sometimes carry small migrants bodily away, despite everything they can do, and variable winds can have the effect of confusing even the most responsive species, who compensate constantly for changes in direction. An ornithologist travelling on Cunard liners on 100 voyages across the North Atlantic, found a total of over 300 land birds taking refuge on the ships in mid ocean. Most of these, 260 birds of fifty-seven species, were American land birds – including some migrant warblers and strong fliers such as red-shouldered hawks *Buteo lineatus* and osprey.[113]

Only thirty-nine birds representing nine species were of European origin, showing quite clearly that most of the strays were being carried to sea by predominantly westerly winds and anticyclones in the strong circumpolar flow. Many species reach Europe in this way without hitchhiking passing ships, and some at least do so in sufficient numbers to become new residents. After several wind-enforced emigrations, the American spotted sandpiper finally gave up the battle and in 1975 a pair are reported to have settled down and nested in Scotland.[524]

On occasion, there are unusual airflows which allow crossings in the opposite direction. In 1929, more than 1000 lapwing plover *Vanellus vanellus* were blown from Ireland to Newfoundland in a single day. And in recent years enough small thrushlike fieldfare have been carried off from Scandinavia to start a flourishing breeding population in southern Greenland.[419]

There is no question that winds have played, and continue to play, a large part in distributing birds about the world. The same may well be true of bats. Freetail bats *Tadarida brasiliensis* from central America arrive in Texas each spring simultaneously with southerly winds, and leave again in the autumn when the cold northerlies begin. The presence of bats on remote oceanic islands, where they may be the only resident mammals, suggests that some have been blown a long way astray.

But when it comes to mass movements and far-reaching effects, there is nothing to compare with the influence of wind on flying insects. This is a patronage that leaves all the rest of us living, somewhat uneasily, at the bottom of a vault of insect-laden air.

Sometimes it is obvious.

In the early days of settlement along the Mississippi River, buffalo gnats or *Simulid* black flies bred in vast numbers in the swamp lands, and when the west wind blew strongly, it carried clouds of the bloodthirsty insects into the streets of Memphis, Tennessee. 'At such times so great were their numbers and voracity that they were known to stop the street cars by killing the mules that pulled them dead in their tracks.'[512]

Charles Darwin, off the shore of northern Patagonia, found the *Beagle* surrounded by vast numbers of butterflies, 'in bands or flocks of countless myriads, extending as far as the eye could range. Even by the aid of a telescope it was not possible to see a space free from butterflies. The seamen cried out "it was snowing butterflies", and such in fact was the appearance.'[95]

W. H. Hudson watched *Vanessa* butterfly migrate across the pampas in Argentina and calculated that 65,000 insects passed over every 100 metres during the flight, 'which lasted from nine o'clock in the morning to a little after five in the afternoon. The breadth of the column was about three miles. On the following day they continued for about seven or eight hours passing in the same number, then the numbers began to decline, and on the third day the whole migration finished.' He decided that he had watched the passage of seventy-five million individuals.[224]

William Beebe, in a break from bathysphering, stood in the narrow Portachuelo Pass in Venezuela and tried to count the butterflies pouring through it from the coastal plain. He estimated a minimum of 'a thousand a second going past in the face of a gentle breeze. In the narrow trail above the gorge it was necessary to put on glasses, so dense were the crowds impinging upon our faces . . . For many days this continued, millions upon millions coming from some unknown source, travelling due south to an equally mysterious destination.' He found that there were at least 245 species of butterflies involved, along with another fifty-two kinds of day-flying moths.[27]

Every year millions of monarch butterflies *Danaus plexippus* travel over 3000 kilometres from their northern breeding grounds in Canada to traditional overwintering sites in California and Mexico. During September, a moving carpet of the tawny insects flows down across fields and city suburbs, piling up at night around carefully selected trees along the route. And the horde keeps relentlessly on, heading south by southwest, regardless of wind direction, stopping only to let the most severe cold fronts pass by.[523]

A migration of the tropical white butterfly *Belenois mesentina* in East Africa was estimated to involve thirty-six million individuals crossing a line more than a kilometre wide each day.[257] A cast of painted ladies *Vanessa cardui* 3000 million strong fluttered across a California stage in full twelve-hour performances on three consecutive days.[303] But even in these spectacular displays, the individual butterflies were still two or three metres apart and the actual density was only about one insect to every sixty or eighty cubic metres of air. This number has been known to rise to one per cubic metre in some small pieridine species such as the southern white *Ascia monuste*, which flows in particularly tight streams just a metre deep, but it is still small compared to the density in which some less obvious insects occur.[361]

Thrips which congregate around the flower heads of daisies are almost invisible and too small to be collected in most traps, but there are at times two or more of these tiny insects with their elegantly fringed wings in every cubic metre of warm summer air.

And aphids, the little pear-shaped plant lice that bedevil farmers and gardeners everywhere, accumulate in even more bewildering numbers. At peak density, there are 1000 million rose-grain aphid *Metopolophium dirhodum* in every hectare of wheat, and a well-attended bean crop of the same area can produce a further twenty-five million new black bean aphid *Aphis fabae* every day.[285]

The winged adults take off in mid afternoon on calm days, putting an estimated six or eight million aphids per hectare into the air every hour. Actual measurements with a suction trap above an English bean field one summer afternoon produced counts of from three to six aphids in every cubic metre of monitored air. Which may not sound much, but if this density prevailed only as high as the planetary boundary layer at about 1000 metres, it would mean a population of 6000 million bean aphid over every square kilometre of land.[243]

And there may, of course, be more than one species involved. A gauze cover placed over the air intake on a research station at New Brunswick in Canada, trapped 31,482 aphids of over 150 species.[2] So, if all species are taken into account, the density of aphid swarms can rise as high as fourteen insects per cubic metre. Which is higher than that of desert locust in the most ravenous and compact downwind swarm ever seen, but because of this insect's smaller size, aphid flights pass almost unnoticed. At least until winds concentrate them even further and they reach plague proportions, blanking out car windscreens and aircraft canopies, getting tangled in people's hair and drowning in the drinks at pavement cafes.

The normal existence of untold millions of insects in the air around us, is something of quite recent discovery. Even entomologists are only now just beginning to realise what this vast wind-burden means in terms of dispersal and insect-borne disease. Consciousness of the continuous rain of insects on all parts of Earth all the time, has been very slow to grow.

Gilbert White, alert as ever, recorded an aerial invasion of black aphid over Selborne after an east wind in 1785. Sir William Parry, the arctic explorer, found aphids scattered over ice floes in the Polar Sea in 1827. Once lightships with permanent crews were established at sea, reports began to accumulate about delicate slow-moving land insects, such as caddis fly being attracted to the lights at night. And progress reports on the reinvasion of the bare slopes of Krakatoa demonstrated that at least some of the waifs were capable of establishing themselves after their long ocean voyages.[74]

Just a month before Archduke Ferdinand was assassinated in 1914, a certain Dr. Everling of Halle in Germany wrote to the journal *Scientific American*, commenting on a butterfly he had seen from his balloon and suggesting that other aeronauts might make observations of insects at altitude. But it was not until the 1920s, when the European gipsy moth *Lymantria dispar* began to spread inexorably across the United States, destroying forest at an alarming rate, that there was any sort of systematic approach to the study of insects in the air. The results were surprising.[131]

Everyone accepted that an occasional passive insect could be blown away to distant places – the ecologist Charles Elton had just published his classic record of spruce aphid *Cinara picea* that drifted 1300 kilometres over the Arctic Ocean from Russia to Spitzbergen.[124] Nobody dreamed, however, that there were whole populations of insects, an aeroplankton of species, arranged into definite and discrete layers in the wind, going to great altitudes, rising and falling at will, treating the air as a natural habitat. But this is precisely what the first detailed studies discovered.

These began from moving cars and trains, but took to the air when mechanics working on World War One fighters noticed large numbers of insects sticking to the engines and air frames. The first attempts to collect aloft were very casual, involving little more than holding a net or a piece of sticky flypaper out at arm's length. Charles Lindbergh did this while flying across the Atlantic, but in France one summer, Lucien Berland dragged proper nets behind a monoplane to heights of 2300 metres near Paris.

Berland identified a *zone terrestre* below 300 metres, containing most of the larger flying insects, and a *zone planktonique* above 500 metres, filled with smaller weaker fliers.[33] In postwar England, Alister Hardy made a similar study, more economically, by tow-netting from kites.[195] But it was entomologist Perry Glick in the United States who provided quantitative data which is still the best available.[161]

Glick designed a screen trap that could be fixed to the wires between the wings of an antique biplane in which he flew a total of 1456 flights, exposing his traps for 1119 hours and capturing 33,934 specimens at a variety of altitudes up to 4800 metres.[162] From the density of his catch, it was calculated that in mid-summer on most days there were normally around fourteen million insects hovering below 4300 metres over every square kilometre of Louisiana.[163]

He was able to identify over 700 species from 216 different families, with flies and mosquitoes being the most numerous, followed by beetles, wasps and bees, with the hopper and aphid family in fourth place. Beetles are in fact the most abundant insects on Earth, outnumbering all other orders by two to one, but these proportions do not show up during the day. In the relatively small number of hours Glick flew and collected by night, beetles resumed their rightful place at the head of the league table.

Information provided by these studies on the presence or absence of certain species at particular levels, is only marginally useful. Some species do seem to exercise a preference, but a great deal depends on seasonal factors, on weather conditions and on the local geography.

Most insects take to the air by day, in warm weather under high barometric pressures, but where winds are slowed down by convergence, insects can 'pile up' in a narrow zone along a weather front. Cold fronts often coincide with influxes of certain species such as dragonflies, which W. H. Hudson described as flying before a line squall on the pampas so fast that 'one could not tell what those swift creatures were that came flashing and rushing past one's face'.[224]

There is also a tendency for many species to congregate along windbreaks and hedges where there are marked changes in airflow, or simply to congregate over conspicuous landmarks like church steeples or standing stones that may have nothing to do with air movement.

But the one thing that is abundantly clear from all studies, is that air everywhere is normally heavily populated up to great heights, both by individuals and by social groups of an enormous number of insects of every possible size.

The early researchers were impressed by the numbers of small species they found at high altitudes, and concluded that these insects were the most likely to be carried aloft unwillingly. But recent studies show that, even at ground level, only 2 per cent of all insects captured by powerful suction traps are larger than the common house fly anyway. The high proportion of small insects in aerial plankton is due simply to the fact that most insects *are* very small, and it seems that even the smallest have the ability to decide when, and for how long, they are going to be airborne.[244]

The latest attempts to track insects in flight are being made with the help of needle-fine radar beams that can track a single aphid up to 500 metres away and even decide what species it is from the frequency of its wing-beat. With equipment of this kind, it has been possible to show that groups of aphid regularly rise to over 1500 metres, travel widely, and descend on a target field of their favourite plants despite adverse winds and rising drafts of air. Experiments with single potted plants of alfalfa set out in the centre of a California desert, have successfully attracted aphids from as far as 135 kilometres away. And it seems that, failing to find what they need, even supposedly diurnal insects can and will go on flying by starlight.[103]

We are apparently justified in assuming that an insect is carried by the wind against its will, only when it is not flapping its wings. Even then there is no certainty, because the radars also show that aphids, and most other species, glide whenever possible.

The gipsy moth which catalysed major quantitative studies in the United States, has in the end proved interesting in itself. The caterpillars of this drab little moth are very hairy and light enough to be lifted like plumed seeds and blown fifty kilometres downwind from their hatching sites. Which may well have been how the whole process began, giving an impetus, an experience of wide dispersal, which pushed evolution in the right direction, into the arms of the wind.[76]

There are some wingless mealy bugs that still float on the wind with the aid of long silky filaments on their backs, but somewhere between 300 and 400 million years ago, an insect exchanged these rudimentary structures for the familiar net-veined wing that, in one guise or another, is now used by all flying insects.

Wings in insects are today confined to mature adult animals, which implies that they arose as supplementary aids to more effective breeding. Some species, such as termites, still only have mating flights, which mark the one occasion on which they ever take to the air, producing a wonderful explosive scattering of individuals and genes. But for the majority of species, flying is now a complete way of life; a way of ensuring successful reproduction, while at the same time providing food, shelter and protection from predation and the problems of overcrowding.

Without assuming that dispersal to a variety of distant habitats must have taken place on the wind, it is impossible to make sense of the present distribution of insects and many other animals across formidable ocean barriers.

Geological evidence shows that most Pacific islands have risen directly from the ocean floor as a result of volcanic activity, and have never had continental connections. Remote islands tend to have no mammals other than bats, few land birds, amphibians or snakes, and only a handful of the sort of lizards, mainly skinks and geckos, likely to have been rafted there on floating fronds of vegetation. But the older ones have quite large insect faunas.

Generally speaking, the older they are, the richer they tend to be, and apart from those species that have been carried by humans or could, as in the case of boring beetles, have survived an ocean voyage inside floating logs, most must have arrived by air. And, while it is possible for some to have hitchhiked on birds – a duck shot over the Sahara 160 kilometres from the nearest source of water, was found to have live mollusc larvae attached to its feet – it seems likely that wind is the principal carrier.[158]

Between 1957 and 1966, the Bishop Museum in Hawaii used a number of devices to collect insects from the air above ships sailing across the Pacific. They took over 13,000 specimens and found that Hemipterans – plant bugs and leaf hoppers and aphids – were the most numerous, with the Dipteran order of flies and mosquitoes a close second. The vast majority of all species collected were of Asian rather than American origin, which fits well with what is known of the island faunas. Even Hawaii, the Marquesas and Easter Island in the eastern Pacific, have close affinities with the Oriental and Australian regions and very few similarities with North or South America.[212]

Distribution has quite clearly taken place from west to east, following the flow of wind and weather in both hemispheres, and it is highly significant that the proportion of each insect order represented in the shipboard catch, corresponds directly to the abundance of each of these orders on the islands. In other words, species such as aphids, plant hoppers, cicadas, fruit flies, weevils and wood-boring beetles, which are most often caught over the open ocean, are also the most common insects on the distant islands. And the more distant an island is from Australasia, the higher the proportion of these windblown species it is likely to have.[178]

The older islands, such as Tahiti and Hawaii, have of course had time to grow new species of their own. The entire list of over 5000 species of insect recorded in the Hawaiian islands, can be traced back to radiation from less than 254 original immigrants. It has been argued that the chances of any windborne migrants ever finding remote islands, which represent less than 3 per cent of the total Pacific area, is vanishingly small. But, given the age of Hawaii, it would take only one successful landing every twenty thousand years to account for the present fauna.[539]

A similar situation exists in the Caribbean, where there are strong similarities between the insect fauna, particularly ground beetles, found on some West Indian islands and those on the coast of Africa 5000 kilometres away. There are northeasterly trade winds that could carry such insects part of the way out over the Atlantic, along with Saharan sand, but it is possible that they travel the rest of the way on hurricanes. The tracks of tropical disturbances in the North Atlantic, connect the Caribbean with the doldrums several times each year, like runaway trains bearing huge loads of miscellaneous freight.[93]

There are about forty such storms across the world each year and they must, throughout history, have played their part in stirring up plant and animal geography. There is, for instance, a small leaf frog *Oreophryne annulata* which is restricted to cool cloud forests in the high mountains of New Guinea, Sulawesi, Kalimantan and the Malay Peninsula, despite the fact that there is not, and never has been, a range connecting these peaks. They, or their eggs, can only have been carried leapfrog across 4500 kilometres of inhospitable valleys and seas on the backs of passing monsoon or typhoon winds.[233]

The whole surface of Earth, everywhere that winds blow, is subject to an incessant and largely unseen bombardment. A constant rain of millions of creatures of all kinds comes pouring down over land and sea. The problem really is to explain why insects in particular are not even more widespread and more homogenous than they seem to be. Part of the answer must be that some are damaged or killed by long periods in the air, but a larger part is that many wind wanderers are either lost at sea or else land in areas where they cannot survive for very long or ever live at all.

After hurricanes in 1947 and 1950, Puerto Rican water striders *Rheumatobates minutus* were seen in temporary fresh-water pools in the Florida Keys, but they died when the pools dried up a few days later.[205]

Polar regions get their share of fallout, but there are only fifty species of insect known from the whole continent of Antarctica, and the majority of these are chewing or sucking lice that attach themselves to penguins and seals. The few that have managed to survive on their own, are all relatives of the springtails that live also on glaciers and snowfields around mountain tops in lower latitudes.[179]

The community of the wind is worldwide and works best when it serves those species that are winged, light, have some natural resistance to low temperatures and high radiation, and perhaps already live in open windy habitats where take-off is easy. But it touches all the rest of us all the time. We all depend to a greater or lesser extent on windblown water, soil, nutrient and seed, and none of us can ignore the constant flow of information.

The air is alive with news. Each breath of wind carries tales of fruit and flower, of moose and moth in search of mates, of old dust, new rain and the constant ebb and flow of busy genes. The chances are that one of the viruses or bacteria you inhaled today is already looking for ways of fomenting revolution, spreading the message, and carrying on its craft of social and genetic engineering.

Even the trees, it seems, are listening.

Botanists working in woods near Seattle have just discovered that willows and alders warn each other when they are being attacked by leaf-eating insects. The natural defence of these plants to attacks by tent caterpillars or webworm is to produce alkaloids in their leaves that make them unpalatable. And when a tree is infected it sends out an airborne warning cry in the form of a chemical, probably one of the terpenes, that starts other trees some distance away preparing to meet the onslaught. Sugar maple and poplar saplings do something similar when mechanically damaged by having their leaves plucked or torn.

Willow talk? Indeed.

And, as always, the answer is blowing in the wind.

PART FOUR
Wind and Body

We are air conditioned.

Every adult has around two square metres of skin in contact with surrounding air. And, in turn, surrounds a roomful of it each day with eighty square metres of tissue that line our lungs.

We can survive for twenty-four hours on just fifteen cubic metres of air in a small closed room, but it would be very uncomfortable for at least half of that time. In practice, because of our own motion and the movement of air past us on the wind, we come each day into contact with at least enough air to fill a large auditorium. And, even without travelling at all, the chances are that on a day with a pleasant breeze, some of this air will be from over 300 kilometres away, or will have come tumbling in on a downdraught from somewhere close to the stratosphere.

We live in the stream of Gaian breath and what touches her, touches us and changes us.

Awareness of this fact is nothing new. Five thousand years ago in the reign of Emperor Huang Ti, the art of *feng shui*, which involves the sacred principles of wind and water, was already being practised in China. Houses, gardens, beds and tombs were located and aligned in a landscape so as to enjoy the maximum benefits provided by the quality of air and the shape of the land. For, it was believed, human nature was strongly influenced by such surroundings.

Two thousand years later, Susruta – the first recorded Indian physician – compiled his *Ayur Veda*, a compendium of the effects of the seasons on humans and animals, recommending a retreat in summer to cooler, drier climates in the hills.

His concern is echoed still further west by the Greek poet Hesiod, who quoted rural precepts which spoke in praise of the south, west and northerly winds, but had nothing good to say about the east wind which 'bursts upon the misty force of the open sea, bringing heavy distress to mortal men'. He admonishes his brother Perses to 'Beat this weather!' and 'Get on homeward before the darkening cloud from the sky can gather about you.'

In the sixth century BC, the philosopher Anaximenes suggested that air was the foundation of everything in the universe and affected all living things directly by its concentration or dilution. A century later, Empedocles extended this stark chemistry to include water, earth and fire, creating a four-square cosmology which continues to influence the way we think. We speak still of 'raging elements', when what we mean is air and water being lashed into fury by storm.

The Greeks recognised Asclepios, 'of the clear bright radiant air after a storm', as the god of healing and established temples or Asclepeions in which priest-physicians practised their arts. One of these early clinics at Knidos concentrated on 'treating the disease' and is the spiritual source of much modern medicine and its preoccupation with killing germs. But there was another clinic on the island of Kos which believed in 'treating the patient', and it was this one that, in the fifth century BC, produced Hippocrates.

We know little about the man, except that his family were members of an hereditary guild of magicians and healers, and that more than fifty books are attributed to him. Most of these were brought together centuries later in the library at Alexandria and described as the 'Hippocratic Collection', but there is one that is indisputably his own. It is a splendid piece of dogma, without any sort of evidence to support its considerable conclusions, but there is at the same time a curious restraint and a dignity about it that are compelling.

It is called *Airs, Waters, Places* and begins:

Whoever wishes to pursue properly the science of medicine must proceed thus. First he ought to consider what effects each season of the year can produce; for the seasons are not at all alike, but differ widely both in themselves and at their changes. The next point is the hot winds and the cold, especially those that are universal, but also those that are peculiar to each particular region . . . On arrival at a town with which he is unfamiliar, a physician should examine its position with respect to the winds.[210]

As a free citizen in a republic, Hippocrates was properly contemptuous of those who were not their own masters. 'Asiatics are feeble,' he said, 'being governed by kings.' And he was chauvinistic enough to suggest that it was only lands, like his own, that were 'bare, waterless, rough, oppressed by winter's storms and burnt by the sun', that could produce men who were 'hard, lean, vigilant, energetic, independent, well-articulated, well-braced and hairy'. But most of his observations were less radical and are still well worth reading.

Dry years were on the whole, he felt, healthier than wet ones and resulted in lower mortality. He associated wet seasons with fevers, gangrene and tonsillitis; and suggested that drier times, though on the whole more healthy, nevertheless carried the risk of arthritis, dysentery and inflammations of the eye. He connected cold winds with coughs, sore throats and constipation; and warm winds with weakness, headaches and vertigo.

He was astute enough also to realise that some people, because of their particular build and temperament, were likely to react quite differently. His most important contribution, however, was the suggestion that all living things are, to a greater or lesser extent, weather-sensitive and not only respond to changes in the air, but are able very often to anticipate them.

In the two thousand years after Hippocrates, little was added. Theophrastus, in the third century BC, collected some interesting wind and weather omens. Celsus, the Roman encyclopaedist, advocated long sea voyages, and the physician Galen sent his patients to the mountains near Rome to convalesce. The Islamic physician Avicenna in the tenth century, wrote on the influence of the wind on human temperament. Francis Bacon in 1653 noted that, 'Every wind has its weather.' And a contemporary of his, the physician Thomas Sydenham, described 'noxious effluvia' in the air.

It was not until the eighteenth century that winds and weather, health and illness became once again matters of special concern, and it was three remarkable and weather-sensitive men who made it so.

The first was Voltaire, who suffered not only from constant persecution for challenging accepted belief, but also from chronic ill health, describing himself in a classic piece of diagnosis as 'susceptible to the east wind'.[140]

The second was the German poet Goethe, who built up a network of meteorological observations and corresponded for years with the dramatist Schiller about the effects of wind and temperature on their mental efficiency.

And the third, a century later, was the philosopher Nietzsche who noted, 'A weather that is unusual and unpredictable makes people distrust each other; they become obsessed with the new, because they have to give up their habits. That is why despots like all those areas where the weather is moral.'[362]

These literary giants were followed by a number of physicians, physicists, anthropologists and hygienists, most of them German, who during the nineteenth and early twentieth centuries began to quantify the intuitions of Hippocrates. Together, they founded a new science, now known as 'bioclimatology' or 'biometeorology', which encompasses all the effects of the atmosphere, both natural and artificial, on plants, animals and humans.

Biometeorology is still in its infancy and remains very largely a matter of European concern, in the hands of the heirs of Kos rather than Knidos. It has, until very recently, been ignored by germ-killers in Britain and the United States, who seem to feel that there is something not quite respectable, a little too mystic, about the notion that wind and weather may in themselves determine the onset and etiology of disease. But times and attitudes are changing with the weather and realisation grows of just how much we are all influenced by 'something in the air'.

The key to wind wisdom lies in the fact that air is much more volatile, more eccentric, than water, and all we creatures who crawled out of ancient swamps to live on the shores of the atmosphere, have been caught up in its caprice.

We are touched by the wind and have, as a result, grown more quickly and into greater sensitivity than would otherwise have been possible. And it should not surprise us if we now find ourselves moved by emotions we cannot properly understand – by extraordinary and simple things such as autumn leaves swaying, 'rusty with talk',[485] or a winter gale 'howling like a tyrant in the valley'.[421]

7. THE PHYSIOLOGY OF WIND

All people are not equally sensitive to wind and weather.

Hippocrates expressed his feelings about such individuality in a botanical analogy. 'The things that grow in the earth,' he said, 'assimilate themselves to the earth.' But, he pointed out, the seasons change and 'it is natural to realise that generation too varies in the coagulation of the seed, and is not the same for the same seed in summer as in winter nor in rain as in drought.'

He equated 'coagulation of the seed' directly with formation of the foetus, and concluded that changes in the 'airs and waters' during growth, had a direct effect on the shape and nature of both plant and human adults.

People are different. And do seem to respond differently to the weather. But if Hippocrates was right, then individuals with the same genes and experience ought to respond in similar ways. William Petersen, an American pathologist whose work deserves to be better known, proved that they do.

In the summer of 1940, Petersen took three medical students

who were also identical triplets, through six weeks of exhaustive testing. He kept them separately, under identical conditions on the same diet, and every day took a series of thirty-six measurements – including blood pressure, pH value, cell count and sedimentation rate; body temperature, weight, strength and rate of fatigue; and ammonia, uric acid, protein and haemoglobic output.[382]

At the end of the test period, he compared the results and found that Harold, Eli and Seymour Cohen showed wide fluctuations in all the parameters being measured, but that the triplets, although physically separate, remained physiologically linked. Their bodies were somehow synchronised, responding in the same way at the same time to common cues.

June that year was dry and colder than normal, while July was a little warmer than usual. In other words, the weather was changeable. During the six test weeks, fourteen cold fronts and seven warm fronts passed over the laboratories in Chicago, presenting Petersen's subjects with what he called 'unusual demands on the involuntary adjustment mechanisms'.

He kept careful track of air movement, temperature, pressure and humidity, and discovered that there were dramatic and convincing correlations with the students' changes in biochemistry.

Each time a cold front passed, bringing wind, a sharp rise in pressure and equally sharp drops in temperature and humidity, their blood pressure fell, their body weight and blood acidity rose, and their white blood cell count increased. These things seldom happened precisely at the time a front was overhead. The brothers slurred their reactions, sometimes delaying for as much as a day, but Harold, Eli and Seymour kept pace with each other by producing the same responses to the weather at the same time.

They marched in step to the sound of the same distant drummer, to the rhythm of Gaia's pulse beat being carried on the wind.

Physical Effects

Wind is like massage, stimulating ten million nerve ends on the surface of our skin. We experience it as 'a form of vasomotor gymnastics for the superficial blood vessels',[241] and it feels good, up to a point.

A fresh breeze can be exhilarating, clearing the air, blowing sometimes, it seems, clean through us, carrying away obstructions that lie in the way of perfect freedom of mind. But anything more than a strong breeze becomes annoying, swirling up dust that irritates the membranes of nose and throat, causing acute discomfort to the eyes. We have a surprisingly low threshold to wind – almost as though at a certain speed, the skin begins to transmit warning signals to the brain, making us feel uneasy, perhaps a little anxious and irritable, even prey to dread.

There are good, purely physical, reasons for a vertical animal poised somewhat precariously on just 2 per cent of its skin surface, to feel insecure. For wind, which is nothing but a lot of air wrapped around some microscopic inclusions, packs a surprising punch and strikes some sneaky blows.

Puget Sound in Washington State was bridged in 1940 by a ribbon of road suspended between two tall and elegant steel towers. The bridge seemed in every way to represent a triumph of engineering. It was designed to withstand tornadoes, but just four months after it was completed, the Tacoma Narrows Bridge was totally demolished by a wind that never once rose above moderate speed.

The process began on the morning of November 8th with a wind of sixty kilometres an hour, just strong enough to blow foam off the tops of waves and make the local fishermen think twice about going out. But this was sufficient to get under the roadway of the central section of the bridge and start it oscillating up and down. Before long, it began to 'gallop'. Standing waves were set up, with as many as four crests at a time humping their way across the Narrows, sending frightened drivers fleeing on foot from their abandoned, but by no means stationary, cars.

This lengthwise deformation continued for two hours, until the wind increased slightly to sixty-seven kilometres and introduced a new factor into the equation. The bridge began to twist, turning the roadway through forty-five degrees like a corkscrew, spilling cars down into the Sound. The structure had developed a 'Karman Vortex', which is the engineering equivalent of hysteria.

Quite independently of the wind, the bridge began to excite itself, until the wavelength of movement was exactly equal to the width of the roadway, setting up a resonance that coincided with the structure's own natural frequency. Film taken by members of the crowd which had time to gather at either end, shows the entire central section of the giant bridge twisting and flopping like noodles strung between the hands of a Chinese chef until, inevitably, the deformation became too great and the fabric parted.

Three hours after the wind and movement began, the bridge collapsed into the waters far below.[330]

It was not the first to do so. The Dryburgh Abbey suspension bridge in Scotland in 1818, the Chain Pier Bridge at Brighton in 1836, and the bridge over the Firth of Tay in 1879, all failed for the same reason – an unwillingness on the part of engineers at the time to concede that winds are dynamic, lively forces quite different from the steady pressures exerted by static weights and loads.

Every stalk of grass 'knows' this and makes appropriate allowances. The weight of a wheat flower is precisely the quantity necessary to act as the most efficient momentum absorber for a stem of its length and diameter, preventing the plant from falling victim to the sort of runaway oscillations that destroy bridges. The shape of the stalk and tapering leaves of a rush, produce complex movements that avoid the risk of simple harmonic resonance in the wind. Palm trees have split leaves that spill high winds with a minimum of resistance and, under stress, the whole trunk leans just far enough to let excess wind out from under the crown.[314]

A whole new discipline is growing up around a study of the way in which different plants wave in the wind. It is, perhaps appropriately for a concern with such artistic overtones, centred in Japan where it is known as *honami*.[236]

Humans have similar problems.

We have the ability to get out of the wind, but while in it, we might as well be wheat stalks anchored to the ground. At any wind speed, the air directly on the surface around the soles of our feet, is motionless. It has been braked, by friction and its own viscosity, to a standstill. But air above the ground is stratified, with every layer moving a little faster, reaching official velocity only at ten metres, which is the height at which wind speed is traditionally measured.

On a day when the weather bureau warns of storm-force winds blowing at ninety kilometres per hour, the velocity at human head height on open ground will be about sixty kilometres an hour. At waist height, it is likely to be around twenty-five kilometres per hour, at knee level ten, around the ankles less than three, and about the toes it will be practically stationary. So anyone standing in the open in such a storm, will have their feet in a calm, their knees in a gentle breeze, a strong breeze blustering around their middles and a gale about their ears. And, in the face of such an unequal onslaught, they will find it difficult to stand upright.[153]

We all face these mechanical problems, and profit by them. Wind is an important training factor in our growth and development, working against the body like an isotonic machine, increasing our muscular strength and resistance, improving and exercising our sense of balance. But there is a point, a wind speed or force, at which we become particularly vulnerable.

Meteorology has come a long way since the days when wind speeds were assessed according to their ability to pluck the feathers from a live chicken.[501] Anemometers in common use today range from the familiar cups mounted vertically on a rotating wheel, through tiny propellers or pressure tubes oriented into the wind by vanes, to delicate sonic instruments that measure the local disturbance experienced by a pulse of ultrasound. It is perfectly possible with a hot-wire anemometer that detects a loss of heat due to air movement around a fine platinum thread, to record turbulence so slight it lasts less than one hundredth of a second. But sensitivity of this order is of little use in biometeorology.

What we need to know about a wind in human terms is its general ability to make life more or less pleasant for us. And it would be useful to have some common terms of reference that enable us to communicate our assessment of this ability to others.

Sailors were the first to quantify the wind, starting with descriptive terms such as 'calm', 'breeze', 'gale' and 'storm'. This led in the seventeenth century to a scale worked out by one Charles Tomlinson, who rated winds from 1 – or just enough to provide steerage in a fishing smack, to 8 – at which point it seemed prudent to make for the nearest harbour. But it was the English naval commander Francis Beaufort, later Sir Francis, Rear Admiral and Hydrographer to the Royal Navy, who in 1805 devised the scale that we still use today.[255]

Beaufort chose a scale with thirteen divisions and used the ship with which he was most familiar, the English sailing frigate or man-of-war, as his pointer, his wind probe. The scale began at 0, which was a wind speed too low to move the ship at all, and ended at number 12, which indicated a wind speed too high for the ship to carry any canvas. For a warship, this meant wind of any speed greater than about 120 kilometres per hour.

There are of course greater winds, both at sea and on land, and when antarctic expeditions encountered such blizzards during the early part of this century, they simply extended the scale to winds of up to 320 kilometres per hour at number 18. And when it came to rating the exceptional winds of tornadoes, other extensions were proposed, beginning with new sets of numbers going up to 'incredible', 'inconceivable', 'intensely devastating' and 'super tornado' at Fujita[149] and Torro,[328] forces 10 to 12.

Perhaps because we are bilateral animals, with two eyes and two hands, we find it easiest to compare quantities of things if one is either half or double the other. Most people, just by eye, can pick the midpoint on a line, but very few can divide anything accurately into thirds or fifths. The Beaufort scale makes superb use of this innate preference for multiples of two. The midpoint on the scale at force 6, is taken as zero by the British Meteorological Office. The next notch up, force 7, marks the point at which humans on land begin to find walking difficult and is the lowest number at which it is considered necessary to issue a warning to shipping on the coast. Force 9, which is the midpoint between 6 and the top of the normal scale, is considered the lowest number in regulations of the British Board of Trade which permits a captain to plead 'stress of weather' in case of casualty to his craft. Maritime warnings also mount by twos – from the 'small craft warning' at force 6, to 'gale warning' at force 8, 'storm warning' at force 10, and 'hurricane warning' at force 12.[88]

The Beaufort scale is simple, elegant and easy to remember, which gave it a head start. But the main reason that it continues to survive, even in the face of modern meteorological techniques, is that it is, by accident or design, also astonishingly profound.

Beaufort assessed each point on his scale as an increment, as something additional that happened to, or had to be done to, his man-of-war. The move from force 1 to force 2, meant a change from barely having steerage way, to seeing the sails fill. An increase to force 3 brought the frigate, close-hauled with all twenty-two sails set, up to a speed of four knots. A progression from force 3 to 4, meant a change from a ship that merely 'careened' or leaned to one side, to one that really heeled over with a good 'list', moving at five or six knots, still under full sail. But the next increment brought a qualitative change.

At force 5, fabric on a frigate 'in chase' – being pushed for speed in pursuit of an enemy – began to strain, and a captain had to think about shortening sail. 'Reefing' on a man-of-war in the early nineteenth century meant one of four things. Each of the square sails on all three masts was fitted with four rows of light line or reefs, sewn to the sail so that it could be gathered in and tied to the cross bar or 'yard'. So there were four intermediate positions – single reef, double reef, treble reef and close reef – between full sail and furled sail. Which gave Beaufort the six logical steps he needed – exactly half the range of the scale – to bring his ship from full sail in a fresh breeze at force 5, down to no sail at force 10 in a whole gale.

The beauty of Beaufort's system is that it approaches the whole question of wind energy in a very modern and comprehensive way, but without resort to instrumentation. Even the most sophisticated anemometer is a limited device, doing nothing more than measuring how fast air is moving at a particular point in space and time. While what really matters to a ship or an organism exposed to such wind, is the wind's total effect on a system.

Beaufort's numbers are not simplistic measurements of wind speed, though these have been added to the scale since his time to satisfy more gadget-happy sailors and scientists.[256] His numbers are complex assessments of the power of the wind, of the product of the mass and acceleration of a body of air on a particular system – in his case on the sail area of an English frigate in 1805. But the glory and genius of the Beaufort scale is that it is not limited to antique warships.

Force 7 is as definite a physical quantity as seven tons or seven kilograms per square metre. If there were still an eighteenth-century frigate in good condition, and a captain skilled enough to handle her, the ship could be used to verify a Beaufort estimate in the same way as a standard weight or measure can be used to check the accuracy of a pound or a metre. It is not, however, necessary to do so. The numbers come out the same when applied to any vessel, creature, plant or structure – because the canny Admiral hit on a progression which makes universal sense.

Wind pressure increases in proportion to the square of wind speed, so although the air speed at force 5 is only double that at force 3 (increasing from about sixteen to thirty-two kilometres per hour), the pressure of that air against an object will be four times as great.

An adult human, for example, exposes roughly one square metre of skin (half the body surface) to the wind at any time. So, using one of us as the chosen object, wind acting against a person at the various Beaufort force numbers will produce these pressures, in kilograms per square metre:

Force:	1	2	3	4	5	6	7	8	9	10	11	12
Pressure:	0.05	0.4	1.3	3.2	6.3	11	18	26	36	50	68	85

Simple arithmetic shows that the pressure increases in these proportions:

Ratio:	1	8	27	64	125	216	343	512	720	1000	1331	1728

215

Which means that a wind at force 10, though it blows no more than 25 times as fast, exerts 1000 times as much pressure as a wind at force 1.

But the most interesting thing about these figures is that there is a curious relationship between the pressure proportions and the Beaufort numbers. The pressures are precisely the cube of each force number. For instance, force 6 exerts 216 times the pressure of force 1, and $6 \times 6 \times 6 = 216$.

This is more than just a mathematical curiosity. No matter how you calculate wind pressure, and no matter what system it may be acting on, there is always a direct relation between the actual force and the Beaufort number. In other words, if you plot the one against the other on a graph, they produce a beautiful straight line.

The relevance of this to humans is that we, and apparently everything else larger and heavier than a dandelion seed, react to wind pressure in much the same way as sailing ships. At force 2, we first become aware of the movement of air across the skin as it, in effect, begins to fill our sails. At forces 3 and 4, gentle and moderate breezes tilt things a little, keeping twigs and leaves in motion, raising dust and consciousness of the presence of wind about us. At force 5, simple tasks such as opening an umbrella, become quite complex. And suddenly, at the upper limit of force 6, we seem to run headlong into a real threshold.

When wind speeds reach about forty-five kilometres an hour, we and porcupines and pedigree cows all begin to run into the same pressure problems as sailing vessels, and may find it necessary to reef in a little, perhaps even lie down.

The original Beaufort scale of 1805 included specifications for use at sea. It was amended later for use on land with the addition of signs that at the time were common and useful enough, such as the removal of chimney pots from the roofs of buildings.

The following table is an attempt to make the measurements more relevant to all living things.

It is, in effect, the first Biological Beaufort Scale.

No actual wind speeds are included, because these tend to be measurements of air movement at particular points, and are of little relevance to habitat as a whole. Prevailing wind conditions are better represented by the Beaufort force numbers and quite adequately described by the terms 'calm', 'breeze' and 'gale'. The threshold between forces 6 and 7 is very real, marking the point at which humans begin to struggle, whole trees start to sway, loose

216

FORCE	DESCRIPTION	HUMAN ACTIVITY	WHOLE PLANTS	SEED & LEAF	BIRDS	INVERTEBRATES
0	CALM	Smoke rises vertically	Still	Plumed seeds fall in less than 10 secs.	All active	Gossamer seen in air
1	LIGHT AIR	Smoke drifts	Still	Light plumed seeds airborne	Thermals with many soarers	Aphids fly Spiders take off
2	LIGHT BREEZE	Wind felt on face	Leaves rustle	Willow seeds in clouds	Few thermals	All species active
3	GENTLE BREEZE	Dust raised	Twigs move	Winged seeds glide	No thermals	Hoppers, aphids spiders grounded
4	MODERATE BREEZE	Hair disturbed Clothing flaps	Small branches move	All plumed seed airborne	Peak activity for sea soarers	Beetles grounded Mosquitoes, gnats stop biting
5	FRESH BREEZE	Eye discomfort from airborne matter	Tumbleweed roll Small leafy tree sways	Leaves airborne	Nocturnal migrations stop	Flies grounded, except horse & deer flies
6	STRONG BREEZE	Arms blown out from sides	Large branches move	All leaves & seeds airborne	Few small perching birds in flight	Moths & bees grounded
			BIOLOGICAL WIND THRESHOLD			
7	MODERATE GALE	Walking becomes difficult	Whole trees move	Loose matter airborne	Small perching birds grounded	Butterflies & deerfly grounded
8	FRESH GALE	General progress impeded	Twigs break	Remains so	Swifts, ducks, swallows, few raptors flying	Only dragonflies still airborne
9	STRONG GALE	Children blown over	Branches break	—	Only swifts airborne	All insects grounded
10	WHOLE GALE	Adults blown over	Trees uprooted	—	All birds grounded	—

plant material becomes airborne, and small perching birds and most insects get grounded.

The Biological Wind Threshold lies at the point where the Weather Bureau finds it necessary to issue its first 'small craft warning'. This is not an arbitrary point. It corresponds to a rise in turbulence which is not a linear function of wind speed, but shows a marked stepwise increase in intensity. This occurs at a wind speed of about ten metres per second (thirty-six kilometres per hour), which is the velocity at which a wind begins to exert a pressure equivalent to force 6. The result is that any force greater than 6 becomes universally relevant, and the 'small craft warning' gets broadcast even on our own internal nervous network. The 'gale warning' not only keeps shipping in harbour, but effectively keeps all birds and insects out of the air and all prudent humans indoors. But by the time 'storm warnings' become appropriate, at wind forces greater than 11, the scale loses biological relevance. Such conditions are rare on land and, when they do occur, lead to general organic and inorganic devastation.

For each species and every wind force there are of course differences appropriate to that species' needs. Amongst humans, there seem even to be sex differences.

Confronted with a wind above the threshold between force 6 and 7, men tend to face into it, leaning against the pressure, head tilted up, often squinting and grimacing, seeming almost to take a wry pleasure in accepting the challenge as a test of strength. Women, even when similarly dressed in jeans and sweaters, behave differently. Most tend to turn away from the wind, head down, with arms folded protectively across the breast.

This may be nothing but a cultural twitch, a response conditioned by long and painful experience with wayward skirts and flimsy blouses, but the reaction seems to be cross-cultural and there could be biological grounds which make such behavioural differences almost instinctive. Women's breasts and nipples are very much more sensitive to cold than the slabs of muscle on a masculine chest and, given the fact that chill can shock a nursing mother to the point where her flow of milk ceases altogether, the face-about response to strong winds could have high survival value.[370]

Both sexes respond equally to winds above the critical threshold, by moving more quickly. During a study of pedestrians in a small American town, 100 adults a day each day for eight months – selected so as to exclude the elderly, the crippled or those carrying packages – were timed crossing an open space between the main street and the local post office. The results were highly significant. Between forces 2 and 5 there was considerable variety, but starting suddenly at the upper limits of force 6, there was a general and precipitous increase in walking speed. People not only moved much more quickly, but did so with obvious increase in their output of physical energy, some literally running for cover.[68]

Part of the reason for this response, is almost certainly due to the close relationship between wind and temperature.

Every species of plant or animal has a characteristic climatic range, a set of conditions under which it can function effectively. It cannot go very far beyond this range without undergoing some kind of evolutionary change, which in its turn sets new limits. Perhaps the most limiting of all environmental influences is temperature, and in the constant struggle for advantage, a lot of the gains in territory and adaptability have gone to those creatures which have become warm-blooded and carry their own steady microclimates around with them.

Birds and mammals are now the most cosmopolitan of creatures and, amongst them, humans enjoy the widest geographical range of all. We have become unusually abundant for an animal of our size, and the reason for our runaway success, is that we have learned how to exceed physiological limits without physical change. We can adjust to every climate on Earth, and to several beyond it, by changes in clothing and shelter drawn from that vicarious and cumulative experience we call culture.

The limits, however, remain. As a species, we have an inner body temperature of 37° Centigrade, which cannot be allowed to fluctuate by more than a few degrees. Normal rhythms produce a slight rise in the afternoon and an equivalent fall a few hours after midnight and, superimposed on this cycle, there may be monthly or menstrual variations produced by hormonal adjustments. Natural upper limits for our species are set by the fact that our bodies produce heat continuously, enough each hour to bring a litre of ice cold water to the boil, and to maintain a constant temperature, we must lose some of it. Even at rest, the temperature of a completely insulated human body rises 2° Centigrade in every sixty minutes, killing even the least heat sensitive of us in a few hours.

Lower limits are imposed by our inability to produce enough heat in some circumstances, no matter how hard the body works. Hands are always some 8° Centigrade and feet 10° Centigrade cooler than the body core, and when either of these extremities reach zero, the tissues freeze and blacken and die. Unconsciousness occurs when the whole body temperature falls to 30° Centigrade, and the heart stops altogether at 26° Centigrade. A naked body dropped into freezing water reaches this point in less than five minutes.[505]

Thermal neutrality, the point at which we neither sweat nor shiver, seems to rest at about 30° Centigrade in still air, but comfort levels change dramatically as air begins to move.

Our bodies lose most of their excess heat by direct radiation or evaporation through sweating, but about 15 per cent is carried off by convection. A thin layer of air lies in direct contact with the skin and, as this warms and rises, it is replaced by cooler drier air. The process stops when surrounding air gets close to body temperature, but it increases rapidly whenever the wind blows. Our air blanket is up to eight millimetres thick in calm weather, but thins to just a millimetre deep at force 2 and, by the time a wind reaches force 6, only a third of a millimetre remains, and this is continually detached by passing gusts.

Windy days feel far cooler than calm days of the same ambient temperature. As long as there is no wind, it is possible to sit comfortably in the sun with air temperatures below freezing, but even a light breeze disturbs the thin insulating layer around us and produces a pronounced *wind chill*.[85]

The combined effects of wind and air temperature are calculated by a 'wind chill index', which charts the sensible effects of different wind speeds. If, for instance, the temperature is 25° Centigrade and a gentle breeze of fifteen kilometres per hour starts to blow, it will feel as though the air temperature has fallen by 20° Centigrade. And if air is already at 20° Centigrade and a full gale blows up, the net effect will be a chill equivalent to a fall in temperature of a further 25° Centigrade. Which means that a naked body in such a gale, loses as much heat each minute as it would if air were still and the temperature 5° Centigrade, below freezing.[448]

When the wind chill index is superimposed on the Beaufort scale – and air temperature is assumed to be 10° Centigrade, which is the annual mean in London, Paris and New York – it produces some interesting results:

A calm creates conditions which most people would experience as 'pleasant'. At force 1 or 2, a young healthy adult would describe the sensation as 'cool'; at force 3, as 'very cool'; at force 4, 'cold'; at force 5, 'very cold'; at force 6, 'bitterly cold'; and beyond the biological wind threshold at force 7, exposed flesh begins to freeze.

Frostbite naturally takes place at lower wind speeds, when the air temperature is lower to begin with. At 0° Centigrade it need only be force 5, at −10° Centigrade force 4, at −20° Centigrade force 2, while at −40° Centigrade force 1 is sufficient. Sir William Parry, in his search for the Northwest Passage, recorded temperatures of −44° Centigrade, of which he said,

> Not the slightest inconvenience was suffered from exposure to the open air, by a person well clothed, so long as the weather was perfectly calm; but in walking against a very light air of wind, a smarting sensation was experienced all over the face, accompanied by a pain in the middle of the forehead, which soon became quite severe.[375]

Our skins contain a network of fine nerve endings which are sensitive to temperature changes as small as one hundredth of a degree Celsius. Cooling increases the frequency with which they send messages to the temperature-regulating centre in the hypothalamus of the brain. This mastermind can restore equilibrium by instructing the skeletal muscles to produce local heat by shivering, or by cutting down on heat loss by reducing blood flow to the extremities. Or it can concentrate blood and reroute it to the internal organs so that someone who feels cold actually shrinks in size. The exact response depends to a great extent on experience, on the degree and nature of adaptation to cold.

Charles Darwin was astonished to find Yaghan people in the Beagle Channel of Tierra del Fuego wearing no more than an otter skin 'about as large as a pocket handkerchief', which they tied around their chests and shifted from side to side, 'according as the wind blows'. He described a woman 'suckling a recently-born child, who came one day alongside the vessel, and remained there out of mere curiosity, whilst the sleet fell and thawed on her naked bosom, and on the skin of her naked baby!'[95]

A recent physiological study amongst canoe people of the channels around Cape Horn, discovered that their basal metabolism was 160 per cent higher than that of a European, and fell during sleep in the open to about the same level as a white person might reach when speeded up to combat a night of cold exposure. And although the average skin temperature of the two was about the same on the following morning, the Indian's naked feet were 2° to

3° Centigrade warmer than those of an unacclimatised white control.[192]

A similar study was made with central Australian aborigines who are exposed to extremes of both heat and cold. These people sleep naked between two small fires that do no more than temper the freezing cold of a clear desert night. It was found that their skin temperatures fell steadily throughout the night, but that they nevertheless slept soundly, without the bouts of shivering which increased metabolism in white control subjects by 90 per cent, and kept them awake most of the night.

The high temperatures during the desert days make it impossible for native Australians to combat cold with a higher general metabolic rate, so they have evolved a different solution. They adapt to occasional cold by insulation, by slowing down the rate at which heat passes outward from the body core. Measurements made on sleeping aborigines show that conduction between the core and the skin surface is 30 per cent lower than in white controls under the same conditions.[426]

Humans are essentially tropical animals, whose naked skin belies a surprising natural ability to cope with environmental extremes. Arguments still rage about the reasons for our loss of body hair. These range from a need to keep cool during the rigours of the hunt, to a need to keep dry during an aquatic phase of development. But whatever the stimulus may have been, it is likely that our nakedness will prove to be a very brief interlude in human evolution. It was not long before we began to borrow other creatures' coats and took to wearing fur again. In fact, we seem even to have been pre-adapted to do so.

One of the basic responses to exposure to cold is a reduction in general circulation produced by a decrease in systemic blood pressure. A naked body in a cold wind very soon has a heart that beats more slowly, in order to minimise heat loss from surface layers. But a recent experiment shows that it is not necessary for the whole body to experience that triggering wind. It is enough to have the air play only on the forehead of someone otherwise warmly wrapped from head to toe in protective clothing. Within thirty seconds of such exposure, heart rates fall by up to eight beats a minute. And it does not matter whether the air temperature is 20° Centigrade or 0° Centigrade to begin with; but it is important that the wind be blowing across the critical threshold represented by the upper limits of force 6.[279]

Evolution, as it were, saw clothes coming. And took care to

concentrate the vital wind sensors, with their small craft warning system, on the relatively hairless face, which is the only area of skin to be more or less constantly exposed.

Fur normally does an excellent job of keeping the thin laminar layer of warm air next to the skin intact, free from disturbance by the wind. Experiments show that a wind of force 5 penetrates no further into the coat of a horse than a wind of force 3. But there is a qualitative difference, with a marked increase in penetration, at the magic number 6 – which seems to be just as critical for those creatures that grow their own coats.[488]

The heat we produce is confined, in still air, to a boundary layer that is thin around our ankles, but grows in thickness with height, extending into a plume half a metre long over our heads, turning each of us into a flickering human candle.[284]

A normal healthy adult human, awake but at rest, produces about fifty kilogram calories per hour per square metre of body surface. For convenience, this quantity of heat is called 'one metabolic unit' and is described as basal metabolism. To be comfortable, even in still air, in a room at 21° Centigrade with a humidity of 50 per cent, such a person would need to be insulated. These particular conditions were selected in 1941, because it was discovered that they were the ones at which a man wearing the standard three-piece business suit of the day, could maintain a constant skin temperature. The degree of warmth represented by that old-fashioned western suit and its underwear has now become the standard unit of insulation in the international clothing industry and is known everywhere as one *Clo*.

For comparison, a bikini represents a Clo value of 0.04, a pair of shorts 0.10, jeans and T-shirt 0.33, a warm skirt and blouse over a slip total 0.45, and a light modern suit about 0.7. Add about 0.30 Clo for knee boots, 0.37 for a V-neck sweater, and a further 0.70 Clo for a complete set of thermal underwear. It takes four layers of clothing, including a greatcoat, gloves, hat and earmuffs, to bring the Clo value up to 3.0, and a down-lined polar suit to reach 3.5, and yet both still fall short of the 4.1 Clo routinely worn by any husky dog. Which begins to make sense of the old trapper's rating of winter weather into 'one, two or three-dog nights'. The average polar bear, incidentally, wouldn't be seen alive or dead in a coat of less than 8.0 Clo.[360]

There is, of course, a direct relationship between Beaufort and Clo numbers. The United States Army Quartermaster Corps describes a temperature of − 10° Centigrade with a wind of force 2, as a 'two-Clo day'. And gives the same rating to a day of $+5°$ Centigrade with a wind at force 3.[408] The higher the wind, the lower the temperature and the greater the need for insulation. In the days before central heating and climate control, this relationship was clearly evident in fashion.

Between three and five thousand years ago, the Aegean was dominated by a maritime power based on rich harbour cities and palaces in Crete. Starting at about 2600 BC, the Minoans built an empire that extended from Assyria and middle-kingdom Egypt all the way to the European mainland. It was the end of a warm period in climatic history, one of the best since the great Ice Age, and the high temperatures were reflected in the style of dress. Statuettes and paintings from Knossos show men wearing a sort of kilt, knee length with a tight belt; and women clad only in a brief apron, or bare-breasted above a bell-shaped skirt drawn in sharply at the waist.[4]

The climate, however, was starting to change. By 1500 BC, it had become far more erratic and stormy. Pine and beech trees took over from fair weather forests of oak and elm, and nomadic barbarians spilled out of the increasingly inhospitable interior of Europe and Asia to live on the kinder coastal plains. These people, who called themselves the Achaeans, arrived at about the same time that Crete was shattered by a series of devastating earthquakes, and it was not long before they displaced Minoan rule, adopting many of the old empire's customs, but not their style of dress, because it had by now become too cold and windy for such skimpy attire.[443]

The new people, we know them as the Greeks, dressed more warmly. Everyone wore a woollen robe or pleated shirt called a *chiton*. Over this, women put the *peplos*, a piece of woven wool about three metres wide draped across the back like a shawl, then brought through under the left arm and fastened to the right shoulder. In addition, there was also a heavy cloak called the *himation*, which fell all the way to the ankles. This was sometimes worn by older men, but young ones covered their shirts with an elaborately decorated short cloak or *chlamys*. After Alexander's conquests in the fourth century BC, some Greeks adopted the comfortable Persian custom of wearing long tight sleeves on windy winter days.[259]

Wind proved equally critical of the free and easy fashions which prevailed in western Europe after the French Revolution. A long series of warm winters ended in the first two decades of the nineteenth century with bitterly cold and windy weather. The flowing, classical lines which dared 'expose the person', were suddenly cut more decorously, underlain with woolly 'bosom friends', and covered by a short jacket that became known as a 'spencer'.[272]

In a letter dated 1814, which was the coldest winter in England since the seventeenth century, Jane Austen said, 'I am amused by the present style of female dress. It seems to me to be a more marked change than one has seen – long sleeves appear universal, and as far as I have been able to judge, the bosom is covered.'[511] A clergyman of the time confirmed that 'it took the north wind to enforce a return to modesty in women's dress'.[32]

When prevailing temperatures are high, winds can have the opposite effect of encouraging fashions which are even less modest.

We each have an estimated two million sweat glands, capable of producing up to two litres of liquid an hour. Complete evaporation of this fluid, would correspond to a heat loss of nearly 700 kilocalories per square metre per hour, which is fifteen times as high as the basal metabolic rate. But when air temperature is higher than skin temperature, the wind simply augments heat gain and makes everything more difficult and less comfortable.

It should be possible to compile a 'wind bake index', showing how, above a certain temperature, body heat increases with the wind, probably varying with the square root of wind speed, and eventually reaches lethal levels. It is certainly no accident that many hot desert people have taken to wearing voluminous robes that both inhibit water loss and the heat gains produced by wind.

Clothing of this kind is both insulation and windbreak and, surprisingly, performs both functions better if it is slightly permeable to wind. Solid windbreaks are not very efficient. They produce violent eddies that break up the airflow, but these are almost entirely eliminated, producing smooth laminar flow, if the barrier is about 30 per cent permeable.

The same principle applies to windscreens and shelterbelts erected to protect farmlands or urban areas. The best ones are narrow, have an irregular profile on top, and are about as permeable as a hedge. They can be fences, palisades, natural woodlands, or shrubs and trees planted in strategic combinations. A good windbreak set at right angles to the prevailing wind, works on a leeward area whose width is up to ten times the height of the break. Close to the barrier, the air is almost calm, and wind speeds may be reduced by as much as half along a band another ten times the break's height – on the windward side.[242]

Provençal farms in France are protected from the direct effects of the *mistral* by a dense row of cypress trees, and those on the windy plateau of Caux by stands of beech and elm. It is interesting that all these traditional, slightly permeable barriers have proved to be a great deal more effective than the solid cement walls which have, in some areas, now come to replace them.

An isolated tree is a windbreak of a different kind. It creates a sheltered area directly behind the foliage, but has the effect of increasing airflow both over and under the crown. Which is why it is so pleasant on a hot day to sit in the air-conditioned shade beneath such a mature tree. An even faster flow can be created by planting shrubs on the windward side of the tree, narrowing the space between the foliage and the ground. Umbrellas, by comparison, perform very poorly. They may provide convenient protection from direct sunlight, but have none of a tree's ability to stimulate a cooling air flow.

Deliberate inducement of airflow is the secret of all successful ventilation. When wind blows against a building, the front wall is subject to positive pressure, while the sides and back experience a negative pressure – they come under suction. The roof is always subject to suction. The Pantheon in Rome is an object lesson on how best to take advantage of this situation. At the crown of its superb dome is a round open eye, placed precisely where wind flowing round the roof creates maximal negative pressure. The result is a strong suction effect, which pulls air through the front door and up out of the building regardless of wind direction. A similar principle is employed in the venting towers placed on roofs of buildings in Iran. Vertical slots on all sides of these towers allow wind to flow through the structure from any direction, producing negative pressure which sucks air up shafts from the rooms and spaces below.[330]

At Hyderabad in the Sind district of Pakistan, where summer temperatures are regularly around 50° Centigrade, the roof of every building bristles with angular windscoops that turn the skyline into a strange geometric mushroom field. The afternoon breezes always blow from the same direction, so these *badgir* all face that way, funnelling air down into the rooms below and out through doors and windows, keeping temperature inside the houses down to a liveable 35° Centigrade. The origin of these windchutes is unknown, but they have been serving the people in

Sind, both as air-conditioners and intramural telephones, for at least 500 years.[412]

For anyone tired of breathing canned air, or interested in harnessing the wind to defray power bills, it is worth looking at the design of other ancient buildings that have grown organically, that welcome rather than try to conquer the vagaries of nature. Remembering, for a start, that the word 'ventilation' comes from the Latin *ventus*, a 'wind', and that 'window' derives from Middle English *windoge* and Old Norse *vindauge*, both meaning 'the eye of the wind'. The eyes of a house are not just for looking through, and were never meant to be sealed shut.

These are the time-honoured rules:

One wind eye or aperture is not enough, but if there is only one, it ought to face directly into the prevailing wind.

Two small apertures are far more efficient than one large one and, if placed on the same wall, should be as far apart as possible.

It actually works best if this wall is built obliquely, at an angle of about sixty degrees to the most frequent wind.

Flow through the interior space is trebled if baffles or wind guides are built out at right angles to the leeward side of each aperture.

The most efficient arrangement is to have apertures on opposite walls, but not directly opposite each other.

Hot air near the ceiling is moved most readily if the inlet aperture is high and the outlet low.[159]

Architects have learned from the destruction of the Tacoma Narrows bridge to make proper allowances for aerodynamic wind forces, but as recently as 1968, the John Hancock Tower in Boston popped half of the 10,344 huge plate-glass windows out of its façade. The building had to be reinforced with energy dissipators, and while this was being done, the temporary replacement of glass by hardboard earned it the lasting nickname of the 'Plywood Palace'.

In 1973, the slender steel structure of the Gulf and Western Building in New York swayed so alarmingly that it had to be evacuated during a winter gale. Two of the world's tallest buildings, the twin World Trade Centres in lower Manhattan, each rest on 10,000 giant neoprene shock absorbers, and have not yet given any cause for alarm.

It is not of course enough now for an architect to consider only the building under construction. It is vital also to test, preferably in a wind tunnel, its effect on spaces and structures around it.

The infamous John Hancock building in Boston may have learned to live with its panes, but its namesake in Chicago has raised another problem by creating a tremendous suction force of almost 250 kilograms per square metre on the Water Tower Palace, the world's tallest reinforced concrete building, which stands 70 metres away in its wake.

People in the turbulence that develops at the bottom of wind canyons between tall buildings are also at risk.

The 'Monroe Phenomenon', named for Marilyn's problem with her skirts in *The Seven Year Itch*, occurs when wind slides down buildings with smooth surfaces and bounces back off the pavement.

The 'Venturi Effect' is a blast that develops in the gap between two buildings, which can involve wind forces three times greater than those in the open, and many times higher than those in sheltered streets nearby.

It is always unpleasant, can be dangerous, and seems best avoided by the old-fashioned solution of building around piazzas or squares that enjoy the wind shadows created by their own 'Cell Effect'.

But no matter how well designed they may be, cities are going to create problems of their own by forming 'heat islands', and by brewing up strange and dreadful chemistries.

Chemical Effects

The presence of almost five thousand million people, milling around, going about their daily business, amounts to the existence on the surface of Earth, of a perpetually active human volcano.

Windblown soil from farmland, smoke from slash-and-burn cultivation, dust from barren land ruined by bad farming practices, pollution from factory chimneys, and from car and airplane exhausts, all add their burden to the atmosphere. At any one time, there are approximately fifteen million tons of human debris suspended in the air, being distributed through the atmosphere by winds, contributing to the curtain of dust that keeps out some of the sunlight, cooling the globe.

Where people are gathered together in large numbers, however, they tend to form their own climatic domes which are warmer than the rest of the environment, producing isolated 'heat islands'. Outside these islands, in the words of an early weather-watcher,

> the blue sky laughs over the landscape, while in the city all is covered with gray and the sun shines only with a weak yellowish-red light. Outside, it is possible to see church towers several kilometres away; inside, the houses on long streets soon disappear in impenetrable gray. The larger the city, the denser, heavier and more resistant is its haze hood.[153]

In a city such as New York, people consume about 500 million kilowatt hours of energy per square kilometre every year. This is almost half as high as the total amount of solar energy falling on such an area in the year – so New Yorkers, in effect, live under a sun-and-a-half. In addition, paved areas, walls and the roofs of buildings absorb and re-radiate more heat than soil or plant-covered ground. Water runs off these surfaces more rapidly, allowing little cooling by evaporation. And heat is produced by the bodies of the people themselves and leaks out of every building. By rights, New York should long since have been burned to a cinder. Most of the rest of the United States believes it deserves to. But what saves the city year after year from cooking itself to death, is the wind. All it takes is a force 6 once in a while to eliminate the heat island effect altogether.

London needs a similar strong breeze to keep it cool. Montreal manages with a fresh force 5, Bremen in Germany is restored by a moderate breeze of force 4, and all it takes to carry the waste heat away from Palo Alto in California is an occasional breeze at a gentle force 3.[368]

There is, nevertheless, always a difference in temperature between the surrounding country and inner-city sites. In New York this averages 1.1° Centigrade higher in the city, in London 1.5° Centigrade and in Berlin 1.7° Centigrade. On fine summer evenings or clear winter nights, the disparity in all three cities may rise as high as 5° Centigrade, and the same seems to be true of all communities with more than 100,000 inhabitants. With the result that each produces more convection and enjoys roughly 10 per cent more rain than the countryside around it.[271]

To a very real extent, a city behaves like an organism in its own microclimate. The roughness of the skyline reduces the speed of winds passing over it, but increases their turbulence, channelling eddies and flows, sometimes at high speeds, through the concrete canyons. When winds are light, the irregularity of heat flow inside the city island creates its own wind field. It has been suggested that if the surface air motions in a city were to be recorded from above with time-lapse photography, the patterns would resemble the motions of 'an amorphous, slowly-pulsating jellyfish'.[350]

The creature is, however, fed from the outside and turns on itself when isolated from the whole environment by completely still air or by an inversion.

The warmest air is normally closest to the ground, but when it lies above a layer of cool air, an inversion is said to exist. This may come about as a result of radiation on a clear still night, when Earth cools rapidly. Or when evaporation from a summer shower or a pool of water creates an 'oasis effect'. Or when air at higher levels is warmed by compression as it falls. Or when a warm front rolls across a cold land surface. In every case, the result is a cool body of air trapped, for days or even weeks, beneath a warmer air blanket. And when such a pool forms around a city, it rapidly becomes stagnant.

The main product of fuel combustion is the gas sulphur dioxide (SO_2). In air this tends to take on more oxygen and become sulphur trioxide (SO_3), which reacts chemically with water vapour to form a mist of dilute sulphuric acid (H_2SO_4). The acid then combines with other substances to form particles which accumulate in the air as a *sulphurous smog*. Which is precisely what happened in December 1952, when fog collected beneath an inversion that settled over London, and trapped the fumes of millions of coal fires in what became a lethal crucible. Over 4000 people died in just four days.[522]

Inversions which sometimes make it hard to breathe in Los Angeles or São Paolo, are different. These brew up a complex *photochemical smog*.

The sequence begins with nitrogen oxides (NO_2 and NO), which are encouraged by ultraviolet light from the sun to combine with hydrocarbons released from vehicle exhausts. Together these form

230

an atmospheric factory, churning out a wide range of toxins that include ozone, ethylene, ketones and even formaldehyde. The result is a brownish haze with a distinctive smell that poisons plants, irritates the nose and throat, and can kill susceptible people.[366]

It has been said that as long as you can still smell the pollution, you do not need to worry about it. There is some truth in this, because when the chemicals reach higher and more lethal concentrations, our sensitive olfactory receptors are overwhelmed and simply stop sending their messages to the brain. There is, however, a case to be made for early detection of potentially harmful pollutants by something like the canary in a cage that miners once took down pits to warn them of the explosive presence of odourless methane.[331]

Amongst the best living indicators are the lichens. These strange assemblages of algae and fungi are pioneer plant communities, moving into hard new habitats on mountain tops and desert floors ahead of all other species. But they are highly sensitive to sulphur dioxide, and become more and more scarce as concentrations increase. The complete absence of these normally robust organisms from most city centres should already be cause for concern about long-term effects on our own health.

Those of us who grow up in cities develop some tolerance to prevailing poisons, but we ought to recognise that the presence of such substances on the wind is acting as a strong selective pressure, and changing the nature of our species in the process. To demonstrate the mechanism involved, it is worth looking at the recent history of bent grass *Agrostis tenuis* in parts of Wales.

Mining for heavy metals reached its peak there as early as 1780, which means that some older sites have been covered with toxic spoils for over two centuries. Concentrated copper and zinc are lethal to plants and until quite recently these waste dumps were barren, but some have been covered once again by a thin growth of spiky grass. This is still *Agrostis tenuis*, but unlike other populations of bent grass nearby, these stands have become metal tolerant. Evolution has taken place along a very narrow front, separating individual grass plants that in some cases grow just a metre apart.[50]

The copper mine at Drws-y-Coed is a small one, in a steep-sided valley which runs from west to east. Grass is wind-pollinated and the prevailing wind blows up the valley from the west. Genes flow with it, bringing non-tolerance into the mine area, and carrying genes for copper tolerance out and up the valley. Tolerant grass disappears just one metre west of the mine dump, but grows in normal soil as much as 180 metres downwind to the east, where its newfound copper resistance is wasted.

Natural selection dislikes wastage of this kind, which reduces the reproductive fitness of a population, and has in an astonishingly short time (evolution normally works in thousands or millions of years), come up with a solution. Bent grass growing in the mine area now flowers eight days earlier than grass in non-contaminated pastures nearby. Which means that copper tolerant plants cross only with each other, ensuring the concentration, selection and survival of the new trait.[311]

We have been lulled into a complacent belief that evolution is something that only takes place in textbooks or becomes apparent by analysis of long fossil records. This is patently untrue. Major changes in the genetics of populations can take place in just a few generations, and seem to have the capacity to screen themselves from dilution in the gene pool of the species as a whole.

Pollution provides powerful selection pressures in favour of change, even in unlikely and otherwise unnatural directions. Who knows what it could be doing to human populations in inner city areas, who are exposed to constant chemical pressure, and who may already be reproductively isolated by living in, and breeding only with, members of their own group?

Even the worst big city smogs are eventually dispersed by wind, bringing an end to, or at least providing a respite from, their unpleasant pressures. But the winds which break up inversions, keep on going with their cargoes of pollutants to produce more widespread, and in the end possibly even more serious, long-term effects.

Most studies of dispersal have concentrated on what happens to a plume of smoke from a single tall chimney stack. On fine summer days without much wind, the smoke 'loops', rising and falling in a sinuous track that touches the ground in small areas and bounces back into the air, so that the plume soon breaks up. When wind is stronger, the plume 'cones', grow gradually larger as it spreads further from the stack. When there is a marked and normal temperature gradient, the plume 'lofts', spread quickly upwards above cold stable air near the ground. And when there is an inversion and very little wind, smoke 'fans' out into a thin flat layer just beneath the temperature ceiling. But if the air under the inversion is disturbed by hills, or eroded by the re-heating of the ground after sunrise, the smoke falls widely and evenly and 'fumigates' the whole downwind zone.

Beyond fifty kilometres from the source of pollution, all such simple plumes lose their identity and join the planetary boundary layer. This tends to thicken in the early morning and evening, and to be generally more dense in winter, except when it is fractured by a passing front. On a global scale, cities form giant plumes that behave in much the same way, but can be detected at least 250 kilometres downwind.

In densely-populated temperate areas of the northern hemisphere, where urban sprawl may bring one city close to another, the plume never completely disperses. A westerly airflow starting in the foothills of the Rockies at about 40° north, passes directly over Kansas City, St. Louis, Indianapolis, Cincinnati, Pittsburgh and New York, growing in foulness as it goes, producing a megalopolitan plume that stretches halfway across the Atlantic. Hardworking crews of at least one of America's Cup yacht competing off Newport in 1983, were forced at times to resort to clean canned air.

International trade in pollutants by wind is becoming a matter of global concern, particularly in the northern zone of prevailing westerly winds. These pass directly across the world's largest concentrations of heavy industry, where they turn into airborne streams of dilute nitric and sulphuric acid. Later, somewhere a long way downwind, this burden is shed in devastating 'acid rain'.

In eastern Canada, in Scotland and Scandinavia, the fish are dying. Their death sentences are sealed in winter snows which concentrate the acids, releasing them in a sudden spate during a spring thaw that saturates the soil, dissolving out toxic aluminium and washing it straight into the lakes and rivers. Aluminium clogs the gills of fish and they soon die, frothing as they gasp for breath. Salmon started to disappear everywhere after the First World War. Brown trout became scarce in the 1920s and have now disappeared altogether from 2000 Norwegian lakes. All species of fish have gone from over 4000 Swedish lakes, now so acidic they have been declared officially dead.

In Bavaria, 1500 hectares of forest have died in the last five years, and 80,000 hectares more are seriously damaged. The same situation exists in East Germany, Poland and Czechoslovakia. In all these nations, sulphur from power stations hundreds of kilometres away is raining down at a rate of 100 kilograms or more on every hectare each year, transforming forest floors from pools of nutrients into cocktails of toxic metals. Here too, aluminium is the major culprit, damaging the endodermis of the roots, destroying their ability to extract moisture from the soil. Death comes slowly, by starvation. In Germany alone, two million hectares of forest may die in this way in the next decade.

The situation seems to be most serious in eastern Europe, but the democracy of world winds ensure that it is by no means confined to the major industrial countries. Aluminium exists in vast quantities in soils all over the world, where it is normally fixed as a harmless organic compound, but everywhere that acid rains fall, it becomes a killer. Huge new conurbations in the Middle East and India put the farmlands of China and the paddy fields of southeast Asia equally at risk. The Taj Mahal is already beginning to show signs of the same acid corrosion that has all but destroyed the Parthenon.

The quantity of acid being washed out of the wind is astonishing. We measure acidity on the pH (for hydrogen ion concentration) scale, which ranges from battery acid at 1 to the highest degree of alkalinity at 14. Pure distilled water is a neutral 7. 'Clean' rain usually comes in at 5.6, already slightly acidified by carbon dioxide in the air. Orange juice is 4.5 and lemonade about 4.0, but one thunderstorm recorded at Pitlochry in the southern highlands of Scotland recently, produced a shower with a pH of just 2.4. Which means that the rain that fell that day was more acid than vinegar. You could have made pickles in a puddle of it.[377]

The hint of carbonic acid in 'clean' rain has never been a problem, but it could become one if the level of carbon dioxide in the atmosphere were higher. And it is getting higher. Carbon dioxide is not normally considered an atmospheric pollutant. You cannot see it or smell it, and it does nothing unpleasant to the mucous membranes or the colour of the sky. But it is beginning to be seen as one of the most important of all environmental concerns.

The present concentration of carbon dioxide in the air is about 0.035 per cent, but it has been rising ever since we began to use large quantities of fossil fuel. Coal and oil are plant products that contain old carbon dioxide trapped by ferns and palms over periods of millions of years. And we are, by burning them, turning loose these stores in just a few decades. The forests and oceans of today simply cannot cope, and each year an excess accumulates in the atmosphere.[34]

The buildup began during the industrial revolution in Europe at the end of the eighteenth century, and now puts something like 20,000 million tons of carbon dioxide into the atmosphere every year. About half of it gets used again, but the rest stays there, mixed in well by the winds. To measure its progress round the world, a monitoring station has been set up by the Scripps Institution of Oceanography on a mountain in Hawaii – about as far from major sources of pollution as it is possible to get. Measurements made there during the last twenty-five years, show an annual rhythm, with peaks in the summer months produced by the active breathing of northern forests, but also reveal a steady long-term rise in overall carbon dioxide concentration, from 315 parts per million in 1957 to 335 parts per million in 1980. Carbon dioxide still represents only a small proportion of the total atmosphere, but the possibility exists that it could be contributing directly to a major climatic change.[181]

Roughly half of the solar radiation that comes our way, reaches Earth's surface. This is eventually returned to space, but outgoing radiation is slowed by the atmosphere, keeping us warmer than we would be without an air blanket. Most of the delaying tactics are practised by water vapour and by carbon dioxide, so the more of these there are in the atmosphere, the warmer we are likely to be. This, in essence, is the *greenhouse effect*. And the reason there is concern about it, is that some scientists believe it could make the world warm enough in our lifetimes to produce dramatic changes in the size and shape of the circumpolar vortex, the westerly airflow in each hemisphere that shapes the climate of the world's temperate zones. This would certainly alter patterns of rainfall, and therefore of agriculture, and might even melt part of the polar ice caps, flooding places like Florida, Holland and Singapore.

Such concerns may be exaggerated, but the 10 per cent rise in carbon dioxide during the last quarter century is real enough, and has already produced an increase of 0.3° Centigrade in the global temperature. There is every reason to believe that this warming influence will continue, and perhaps even be accentuated by other atmospheric forces working in the same direction.

Methane, for example, is also on the increase and is another excellent insulator. Most of it is produced by swamps, rice fields and coal mines, but a growing quantity is being farted about by ruminant animals. Cows create methane as a by-product of digestion by bacteria in their rumens, and there are more and more of them doing it. The world cattle population has doubled in the last forty years to over 1250 million. The present concentration of methane is just 1.6 parts per million, but this only has to double to raise world temperature by another 0.2° Centigrade.

Dust raised by human and natural volcanoes may be cooling Earth's brow, but everything else seems to be working toward a warmer world, one destined to be at least 2° Centigrade hotter before another century has passed. If so, it will be warmer than the Early Medieval epoch between AD 1150 and 1300, which was one of the best in post-glacial times and seemed, with the lifting of storm winds and bitter winters, to give the whole world a rush of blood to the head.

The Vikings established colonies in Greenland and were able to bury their dead there, deep in soil that has since then been permanently frozen. Oats and barley were grown in Iceland, and vineyards in England flourished so well that the French trade tried to have them abolished. Cathedrals, churches and abbeys sprang up all over Europe, and the Sahara flowered. A five-arched bridge had to be built to span a broad river at Palermo in Sicily where there is barely a trickle today, and Holland's Zuyder Zee was formed. People grew bigger and lived longer. The average age at death rose from thirty-two in classical times to forty-eight in England in 1276, and then fell back to thirty-eight by the fifteenth century when the weather cooled once again.

By the middle of the twenty-first century, we may have gone well beyond this 'little climatic optimum' of a thousand years ago, and be living on an Earth warmer than it has been at any time in the past 125,000 years.[271]

This is bound to affect economic and political stability and to change our coastlines and our lives, but it could also be the making of a new world – one worth getting excited about all over again.

Magnetic Effects

The banquet of early medieval warmth was followed, like a swift account from the caterer, by a Little Ice Age.

Starting in the late thirteenth century, westerly winds grew in both strength and frequency. Arctic ice spread into the north Atlantic, cutting the sailing routes to Greenland and isolating Iceland. Tree lines were lowered on all the mountains of Europe, grapes stopped growing in England, villages and farms were abandoned, and there were outbreaks of famine and plague.

This bitter onslaught, which spread right across the globe, and continued until the eighteenth century, seems to have begun in eastern Asia. Chinese weather data go back over three thousand years to the Chou dynasty, and show that the most severe conditions, the coldest winters on record, occurred there during the twelfth century. The same is true of Japan, where the best old records are floral ones.

Because spring festivals and temple rites are timed each year to coincide with the first appearance of cherry blossom, the date on which trees in the garden of the Imperial Palace in Kyoto burst into bloom, is carefully noted. This has been done since AD 812, and is still an auspicious moment for many Japanese, with television weathermen following the northward progress of the bloom up through Kyushu and southern Honshu as closely as their counterparts in Florida track an approaching hurricane. The average date so far this century has been April 14th, but blossoms in the twelfth century seldom appeared before April 28th, and were delayed following the severe winter of 1323 until the fourth day of May.[7]

The same glacial grip seized Russia around 1350, crushed Poland and Bohemia during the fifteenth century and only reached out towards England during the reign of Elizabeth I. The river Thames froze on several occasions in Tudor times, but did not do so on a regular basis until after 1683, well into the Stuart Restoration. This Eurasian sequence shows a slow westward roll, against the windflow, covering five or six degrees of longitude in each decade, at which rate it would have taken over 600 years to complete a full circuit of Earth.[271]

Wind and weather fall naturally into cycles – monthly ones associated with the Moon, and annual ones produced by our rhythmic orbit about the Sun. There is good evidence for longer cycles with periods of eleven and ninety years in connection with sunspots and solar flares. But, perhaps because they take so long to detect, there are very few known influences with a period longer than six centuries. The only likely candidate seems to be our planet's own mobile magnetic field.

Earth has a field similar to, but somewhat more complex than, that surrounding a bar magnet. It moves and, on occasion, even reverses polarity altogether. The northern magnetic pole at the moment is actually a south pole and is adrift somewhere amongst Canada's Queen Elizabeth Islands, over 1500 kilometres from the geographic pole. It is moving steadily westward at about the same rate as the Little Ice Age once spread across Asia, a coincidence which prompted a British meteorologist to take a close look at the connection between weather and magnetism.

Joe King of the Appleton Laboratory noticed that the contour lines in Earth's magnetic field bear a striking resemblance to isobars on a map of atmospheric pressure drawn around the true north pole. He suggests that the two are actually linked, and that this is why the circumpolar flow of westerly wind, which dominates temperate weather, is sensitive to solar activity. Whenever there are sunspots, there is magnetic disturbance in our atmosphere, a strengthening of Earth's field, and a tendency for severe wind storms to break out over the sea. And the winds do seem to drop, producing calmer warmer weather everywhere, when the Sun is quiet and our field is correspondingly weak.[253]

The precise nature of this connection between Earth's magnetic field and the wind remains a mystery, but an answer could lie in the oceans. Sea water is a good conductor and the huge rivers of salt water carried along by ocean currents cut across our field, inducing a weak electric current that charges the entire planet, turning the world into a gigantic dynamo.

Evidence in support of this theory has been provided in recent years by Goesta Wollin, a Swedish–American meteorologist who, acting on a hunch, compared the changes in magnetic field strength over the years with seasonal sea surface temperatures in the North Pacific, and found to his delight that the graphs matched perfectly.[181] The ocean temperatures rise and fall in perfect harmony with the fluctuations in Earth's field, but they lag precisely three years behind. It takes this long for a strengthening of the field to be translated into an increase in speed of the Pacific gyre, that huge clockwise swirl which results in better mixing of heat picked up in the tropics, and a higher overall sea temperature. And conversely, a weakening of the field lets the ocean currents slow until eventually sea temperatures fall proportionately.

A strong magnetic field produces fast ocean currents which spread heat in the ocean sink more evenly, cutting down the disparity between hot and cold zones of water and air, and eliminating some of the winds which normally rush to restore equilibrium. But the process of course goes on. Calm warm weather results in a slowing down of ocean currents, producing a weaker field, uneven distribution of heat, more frequent winds and therefore cooler weather.

The result is an elegant feedback system in which temperature, rainfall, ocean currents and electromagnetic fields are all held in delicate equilibrium on the responsive see-saw of world wind.

8. THE PERCEPTION OF WIND

Wind is invisible.

Which immediately puts it into a category of things like love, hate and politics that we find difficult to explain, and impossible to ignore. We experience each of them directly as elemental forces that shape our lives, but we know them only indirectly, by their effects on us and on the world.

Our panoply of senses is normally dominated by just one sense, by sight. This is the only one we really trust. Under the dictatorship of the eye, all information is translated into a visual code. 'Show me,' we say. We expect truth to conform to observed experience. Anything that cannot be clearly seen, has not been sensed. It is nonsense.

And yet the effects of wind are undeniable. It moves us. Being moved, we are impressed, we look for meaning and take measurements. We know the weight and chemical composition of air. We begin to understand the roles of pressure and humidity. Our models even make it possible to predict roughly what will happen if . . .

But sooner or later the models break down. Forecasts miss the mark and theories are confounded. We are left with an enigma. Something so strong it shapes continents and seas, yet so insubstantial that it continues to elude us, and probably always will.

Which does not stop us from thinking, wondering and worrying about wind – and from picturing it in our mind's eye.

Each culture does this a little differently, endowing the wind with characteristics that have a special and local significance. But the astonishing thing about perceptions everywhere is that they have so much in common.

All myths are deeply rooted in biology and geography, which provide common ground, and come in time to grow together. In the case of wind, they converge in an artistic consensus that is global – and loaded with meaning.

Wind Words

Meteorology is the study of meteors. Or at least that is what Aristotle intended when he first used the word in 330 BC.

The Greek *meteoros* literally means 'something raised up' from below. It is only in this century that its meaning has become restricted to fragments raining down from above.

Aristotle believed that the four basic elements of earth, water, fire and air were in a state of constant transformation, and that heat from the Sun drew up two sorts of evaporations. One kind came from water. These he called 'moist vapours', which produced the 'meteors' of cloud, rain, hail, snow, mist, dew and frost. The other kind were the 'dry exhalations' of earth. These included the 'meteors' of thunder, lightning, comets, earthquakes – and the winds.

Wind in his cosmology therefore was simply 'a body of dry exhalation moving about the earth'. Assuming, as he intended, that wind is dry only in comparison to the moistness of rain and mist – that is a definition we can still accept. With our present knowledge of the contribution made to the atmosphere by the 'degassing' of the planet, it even makes good sense.

It was not until four centuries later that this relative simplicity was disturbed by the older Pliny, whose indiscriminate curiosity added myth to meteorology with the suggestion that 'there be certain caves and holes which breed winds continually and without end'.

To be fair to him, he was only following the accepted Greek and

Roman practice of explaining difficult things in mythological rather than purely logical terms. So lightning, if you did not feel like thinking too hard about it, became the wrath of Zeus or Jupiter; the rainbow was a multicoloured robe worn by the goddess Iris; and each of the winds was either a bearded old man 'of wrathful countenance', or a beautiful young man, according to their disposition.

The Latin poet Ovid in his *Metamorphoses*, skilfully rearranged the myths of both Greece and Rome, retelling them in a new and narrative way which had a tremendous influence on all later interpretation. Most representation of the individual winds in Medieval and Renaissance art, in Elizabethan and Romantic literature, and in modern thinking about classical concepts, follows Ovid's conventions in which the attributes of each wind are very precisely defined.

So, the north wind *Boreas*, personification of winter, became portrayed as an old man with flowing grey locks, blowing through a conch-shell trumpet. He was usually described as strong in body and harsh in disposition. Shakespeare called him 'that ruffian Boreas'. Sometimes, because of his association with the snake god Ophion, he was shown – as in a Rubens painting of *Neptune Calming the Waves* – with serpent tails instead of feet. He loved Oreithyia, daughter of a king of Athens, and carried her off naked and against her will. But he is said also to have mated with the mares of Erichthonius, producing foals so light of foot they ran across fields of standing corn without bruising an ear of grain. Boreas is identified with the element Earth, the humour of melancholy, and with autumn and middle age. Amongst the Apache Indians his colour was black, though to the Irish it was merely dark grey. In Egypt, his symbol was the ram.

The west wind *Zephyrus* or Favonius, was very different, gentle, almost effeminate. It is said that he was once a savage and baleful wind like his brother Boreas, but that he mellowed under the influence of Chloris or Flora, the goddess of spring. Which is why he is usually shown gliding sweetly by with his mantle filled with flowers. Zephyrus is identified with water and phlegm, and surprisingly, with winter and old age. His colour amongst the Apache, the Irish and the Chinese, was white. In China he was a tiger, but in Egypt was more often shown as a serpent.

The south wind *Notus* or Auster, brought rain to Greece and Rome and is shown as a young man bearing an inverted jar, whose contents he spills over the land. He normally flew wrapped in cloud, in Ovid's words 'all darkish was his head, with water streaming down his hair'. Notus was an air element and was identified with blood, spring and youth. His Apache colour was red, and his Egyptian symbol was a lion.

Eurus or Solanus, the east wind and last of the four cardinal airs, was seen as a dark-complexioned old man, sometimes crowned with a shining sun. He was generally described as gloomy, with a fierce countenance. His element was fire, his attributes summer and infancy, and his humour bile – which is accurately reflected in the Apache's choice of yellow as his colour. In Egypt he was seen as a falcon, the sign of the sun god Horus.

The ancillary winds seldom appear in tradition, but when they did, *Skiron* of the northwest was seen as old and intemperate, bearing a fire-pot of burning coals. *Lips* in the southwest was shown as a serious young man bearing an *aplustre*, the ornament that was commonly fixed to the stern of Greek boats. *Apeliotes* of the southeast was represented as an eager young man, bringing gentle rain and ripe fruit. And *Kaikias* in the northeast, who brought cold and rain, was portrayed as a stern older man carrying a shield covered in hailstones or ripe autumn olives.

Classic tradition played very little part in the medieval world, except where it touched Christian doctrine. They built windmills, and even towards the end learned to tilt at a few of them, but with their other gothic concerns, medieval scholars were not much interested in wind for its own sake.

> The medieval mind saw nature as hostile, dwelt on its horrors rather than its beauties, sought safety rather than the wilderness. The great forests were inimical to human life, dangerous and wolf-infested, an obstacle to be hacked and cleared and burned down before man could survive. The population was tiny, the waste expanses vast, and each little settlement was a precarious endeavour. Walled cities shut out marauders, both animal and human: nature could be perceived as beautiful only when men had the leisure and the skill to build walled gardens.[108]

This process of transformation began with the Italian humanists of the fourteenth century, who revived Greek scholarship and an interest in nature. No man before Petrarch, the first Renaissance man, is known to have climbed a mountain for the sake of the view. And the fact that this awakening coincided with the invention of printing, meant that many of the first words to be published anywhere, were concerned with things like the rediscovery of 'meteors'.

In 1563, William Fulke in England produced his influential book – *A Goodly Gallery with a Most Pleasant Prospect, into the Garden of Naturall Contemplation, to Beholde the Naturall Causes of All Kind of Meteors*. He revived Aristotle by defining wind as 'an exhalation whote and drie, drawne up into the aire by the power of the sunne, and by reason of the wayght thereof being drive down, is laterally or sidelongs caried about the earth'. And paraphrased Pliny with the contention that, 'within the globe of the earth be wonderful great holes, caves, or dongeons in which ayer abondeth . . . that cannot abide to be pinned in, findeth a little hole in or about those countries, as it weare a mouth to break out of'.

Thus armed, the Elizabethan writers made much of meteorology.

It was Edmund Spenser who started the revolution in the late sixteenth century, proving that the English language need not be confined to the earthy realism of Chaucer. *The Faerie Queene*, in particular, abounds with new and subtle sound effects. He represents, for instance, knights in battle as a conflict of the cardinal winds, that come:

> . . . breaking forth with rude unruliment,
> From all foure parts of heaven doe rage full sore,
> And tosse the deepes, and teare the firmament.
> And all the world confound with wide uprore,
> As if in stead thereof they chaos would restore.

Christopher Marlowe soared still higher, breaking the stiff wooden form of blank verse and creating instead his 'mighty line'. He drew on meteorology, as Spenser did, but transformed the images with his own rich vitality. For him it was no longer enough that fame should be likened to a fleeting meteor. In *Hero and Leander*, he conjured up a wind whose wings were still dripping with Aristotle's 'exhalations':

> Swifter than the wind whose tardie plumes
> Are reeking water and dull earthlie fumes.

Of all the Elizabethans, it was William Shakespeare who left the most enduring mark on both literature and the weather. All his work bristles with meteorological references that show his familiarity not only with mythology and weather theory, but also with nature lore. As King Henry IV waits for the battle that ends his first play, the audience is prepared for the impending violence by portents of a storm as:

> . . . the southern wind
> Doth play the trumpet to his purposes,
> And by his hollow whistling in the leaves
> Foretells a tempest and a blustering day.

Shakespeare uses storms at sea as effective dramatic devices in *The Tempest*, *Othello* and *Pericles*. He introduces a waterspout in *Troilus and Cressida*, and turns loose the famous gale that cracks its cheeks for *King Lear*. He uses winds as images of speed – with Oberon urging 'go swifter than the wind'; of freedom – with Prospero promising 'thou shalt be as free as mountain winds'; and even of abandon – with Titania remembering 'we have laughed to see the sails conceive, and grow big-bellied with the wanton wind.' He is amongst the first, in *Henry VI Part Three*, to record the well-worn proverb 'Ill blows the wind that profits nobody.' And he anticipates biometeorology by having Hamlet confess, 'I am but mad north-northwest; when the wind is southerly, I know a hawk from a handsaw.'

'Of all his contemporaries, Shakespeare used the widest range of meteorological information, he used it most frequently, and he used it most effectively. Like Marlowe, he could fill the heavens with awe-inspiring splendour.'[201]

John Milton was next on the wind scene with lofty words which, for all his virtues, seem pale and puritanical beside the Elizabethans:

> The windes with wonder whist,
> Smoothly the waters kist,
> Whispering new joyes to the milde Ocean,
> Who now hath quite forgot to raise,
> While Birds of Calm sit brooding on the charméd wave.[338]

It was wispy flights like that that led, toward the end of the seventeenth century, to a neoclassic revival of greater dignity and restraint. Poetic technique was the thing under the refined literary dictatorship of Alexander Pope, who, in his *Essay on Man*, railed

against those 'whose untutored mind sees God in clouds, or hears
him in the wind'. To Pope, balance, symmetry and finish were at
least as important as content:

> 'The sound must seem an echo to the sense.
> Soft is the strain when zephyr gently blows.
> And the smooth stream in smoother numbers flows.'[386]

The early eighteenth century on the whole preferred to view nature
through the drawing-room window, but in time the pendulum
swung again and the romantics re-emerged, led by Thomas Gray
and William Wordsworth on behalf of all those who 'hear a voice in
every wind, and snatch a fearful joy'[171] – even in the midst of
industrial and political revolution.

Most of the early nineteenth century was dominated by the
romantic revival, emphasising imagination over reason and the
intellect, flouting the tyranny of machines. There are no stormy winds
of violent social change in romantic poetry, only Lord Byron's
gentle airs that 'come lightly whispering from the west, kissing not
ruffling the blue deep's serene'.[65] Or Sir Walter Scott's soft breeze
that 'just kiss'd the lake, just stir'd the trees'.[429] 'Nought', said
John Keats, 'but a lovely sighing of the wind along the reedy
stream; a half-heard strain, full of sweet desolation – balmy
pain.'[248] Producing an echo across the Atlantic where Henry
Longfellow could 'hear the wind among the trees, playing celestial
harmonies'.[291]

It was impossible for the visionary pantheist Percy Shelley to be
quite so pliant:

> O wind West Wind, thou breath of Autumn's being,
> Thou, from whose unseen presence the leaves dead
> Are driven, like ghost from an enchanter fleeing.[440]

Or for Emily Brontë's passion to see the wind as anything less than:

> An universal influence,
> From thine own influence free;
> A principle of life – intense,
> Lost to mortality.[54]

And by mid nineteenth century even the church, or at least Charles Kingsley's muscular Christianity, was in headlong rebellion against romantic gentility:

> Welcome, wild North-Easter!
> Shame it is to see,
> Odes to every zephyr,
> Ne'er a verse to thee.[254]

The Pre-Raphaelites succeeded briefly in urging a return to sensuality and nature, persuading Christina Rossetti that it was worth asking, 'Who has seen the wind?' Even if the rather wet response was:

> Neither you nor I:
> But when the trees bow down their heads
> The wind is passing by.[409]

But with Queen Victoria on the throne, there was bound to be some more worthy literature and robust, character-building winds to go with the call of Empire. The Scots, as always, were amongst the first to answer. 'I feel you push, I heard you call!' said Robert Louis Stevenson:

> Whenever the moon and stars are set,
> Whenever the wind is high,
> All night long in the dark and wet,
> A man goes riding by.[461]

Others followed. Charles Sorley, convinced that, 'The wind visited us and made us strong', pleaded:

> Wind that has caught us, cleansed us, made us grand,
> Wind that is we,
> We that are men – make men in this land.[455]

'Tis the hard grey weather', agreed Charles Kingsley, 'Breeds hard English men.'

In balance to this chauvinism, however, there was always Thomas Hardy, perhaps the greatest writer on rural life and landscape in English. He could, it is said, 'hear the wind whispering in the dried harebells, and tell each tree from the distant rustling of its leaves'.[108]

The turn of this century was effortless, bridged by John Masefield with his combination of realism at sea, 'where the wind's like a whetted knife', and romance on land:

> It's a warm wind, the west wind, full of birds' cries;
> I never hear the west wind but tears are in my eyes.
> For it comes from the west lands, the old brown hills,
> And April's in the west wind, and daffodils.[323]

But we are still too close to the twentieth century itself to see the patterns.

There are unrepentant romantics like Alfred Noyes, for whom the wind remains a 'torrent of darkness among the gusty trees';[364] William Merwin, who waits while, 'All through the dark, the wind looks for the grief it belongs to';[333] and Duncan Scott, celebrating an, 'Indolent wind on the prairie, wind that loiters and passes.'[428]

We have our moody poets like Archibald MacLeish who can still tell how:

> The west wind blew
> Day after day as the winds on the plain blow
> Burning the grass, turning the leaves brown, filling
> Now with the bronze of cicadas.[310]

There are realists, such as Edwin Robinson in New Zealand, 'Where the wind is always north-northeast, and children learn to walk on frozen toes.'[401] Or John Burroughs recounting 'a tremendous wind from the north; day like a raving maniac bent on demolishing the world. It almost blows the hair off a dog.'[63]

There are classicists such as Robert Bridges, who sees the south wind:

> Sucked by the sun from midmost calms of the main,
> From groves of coral islands secretly drawn,
> O'er half the round of earth to be driven,
> Now to fall on my face.
> In silky skeins spun from the mists of heaven.[52]

And there are the rest of us, voyeurs all, who like Federico Lorca lurk and watch as:

The girl of beautiful face goes gathering olives.
The wind, that suitor of towers, grasps her round the waist.[292]

In the midst of this general growth of awareness during the last two centuries, a process that began with Wordsworth's romantic vision of the natural world as a vital formative influence on us, there arose a sort of subculture. A very select group of enthusiasts who responded to the effect of wind over water, to the call of the sea.

Just as 'landscape' can be seen to have been a sixteenth-century invention, so 'scenery' seems to have appeared suddenly in the eighteenth century. The European vogue for Alpine holidays would have been impossible any earlier, when travellers regarded mountains as dangerous obstacles, which were invariably cold and wet and always aesthetically unpleasing. An appreciation of the grandeur and beauty of the ocean was a similarly romantic innovation. Before the eighteenth century, houses were built safely inland, often with their backs to the sea. There was no such thing as a 'sea view' or 'waterfront property'. There were no coastal resorts and no theories about the curative properties of sea air. Then, suddenly, Byron, Tennyson, Swinburne and Matthew Arnold were all creating 'seascapes'. And it was in this atmosphere that two young writers, both electing to serve as merchant seamen, wrote the greatest of all sea stories. These are, in essence, the purest of perceptions of wind, from a world dominated by it in every possible way.

Richard Henry Dana was the first. He served his two years before the mast and published, while still only twenty-five years old, an account which includes a great battle with the notorious winds of Cape Horn:

We lay hove to under a close-reefed topsail, drifting bodily off to leeward before the fiercest storm that we had yet felt, blowing dead ahead, from the eastward. It seemed as though the genius of the place had been roused at finding that we had nearly slipped through his fingers, and had come down upon us with tenfold fury. The sailors said that every blast, as it shook the shrouds, and whistled through the rigging, said to the old ship, 'No, you don't! – No, you don't!' . . .

Toward the middle of the week, the wind hauled to southward, which brought us upon a taut bowline, made the ship meet, nearly head-on, the heavy swell which rolled from that quarter; and there was something not at all encouraging in the manner in which she met it . . . I stood on the forecastle, looking at the seas, which were rolling high as far as the eye could reach, their tops white with foam and the body of them a deep indigo blue, reflecting the bright rays of the sun. Our ship rose slowly over a few of the largest of them, until one immense fellow came rolling on, threatening to cover her, and which I was sailor enough to know, by the 'feeling of her' under my feet, she would not rise over. I sprang upon the knight-head, and, seizing hold of the forestay, drew myself up upon it. My feet were just off the stanchion when the bow struck fairly into the middle of the sea, and it washed the ship fore and aft, burying her in the water. As soon as she rose out of it, I looked aft, and everything forward of the mainmast, except the longboat, which was gripped and double-lashed down to the ringbolts, was swept off clear. The galley, the pigsty, the hen-coop, and a large sheep-pen which had been built upon the fore-hatch, were all gone in a twinkling of an eye – leaving the deck as clean as a chin new reaped, and not a stick left to show where anything had stood.[90]

Herman Melville came next and his personal experience at sea shows in equally rich descriptive detail, but it is at times so heavy in allegory that it is difficult to see the wind for the symbols.

The master, the author of the greatest lyric descriptions of wind at sea, was unquestionably Joseph Conrad.[49]

His early work is filled with the enchantment of *Youth*, with an animal vigour that almost welcomes storms as a challenge. The famous October gale 'blew day after day: it blew with spite, without interval, without mercy, without rest. The world was nothing but an immensity of great foaming waves rushing at us, under a sky low enough to touch with the hand and dirty like a smoked ceiling.'[79]

Later, through the eyes of those on the *Narcissus*, he runs the gamut from: 'Gentle sighs wandering here and there like forlorn souls, made the sails flutter as in sudden fear, and the ripple of a beshrouded ocean whisper its compassion afar – in a voice mournal, immense, and faint . . .'

To a fierce squall that: 'Seemed to burst asunder the quick mass of sooty vapours; and above the wrack of torn clouds glimpses could be caught of the high moon rushing backwards with frightful speed over the sky, right into the wind's eye.'[80]

But it is on the *Nan Shan* that he reaches his peak of maturity, recognising along with stoic old Captain Mac Whirr 'the dirty weather knocking about the world' that cannot be avoided:

> This is the disintegrating power of a great wind: it isolates one from one's kind. An earthquake, a landslip, an avalanche, overtake a man incidentally, as it were – without passion. A furious gale attacks him like a personal enemy, tries to grasp his limbs, fastens upon his mind, seeks to rout his very spirit out of him.[81]

> But a mere trader ought not to grumble at the tolls levied by a mighty king. His mightiness was sometimes very overwhelming; but even when you had to defy him openly, as on the banks of the Agulhas homeward bound from the East Indies, or on the outward passage round the Horn, he struck at you fairly his stinging blows (full in the face, too), and it was your business not to get too much staggered. And, after all, if you showed anything of a countenance, the good natured barbarian would let you fight your way past the very steps of his throne. It was only now and then that the sword descended and a head fell; but if you fell you were sure of impressive obsequies and of a roomy, generous grave.[82]

There have been others since Conrad who have used great winds as narrative vehicles. George Stewart in 1941 followed the growth of a single tropical storm across the Pacific and its impact on the coast of California – making nicely the point that air is so much a part of our lives that we take it for granted, until it does something unexpected.[462] Desmond Bagley in 1966 reworked the 'Junior Meteorologist' character into the hero of a Caribbean thriller who uses his inside knowledge of an approaching hurricane as an instrument of revolution.[14] And John D. MacDonald, easily the most accomplished writer of the three, in 1956,[298] and on a grander scale in 1977, made his storm the nemesis of unscrupulous land developers in Florida:

> The oily waves lift high and come racing in, and they turn, tumble, thud against reef and rock and sand like a great slow drum. The fiddler crabs move inland in small brown torrents, one large claw held on high. The seabirds circle nervously, crying out, getting ready to head away from the oncoming drop in pressure. Fish turn ravenous, storing food against the tumbled

251

days ahead. Primitive man looks at the streamers in the sky, hears the slow boom of the surf and feels an uneasy dread.[299]

Languages themselves become windblown.

Something as elusive as the wind is obviously fair game for those tired of the usual similes. Shakespeare made his winds swift, free and wanton. Others have called them fast, fickle and wild. Wind is almost anything you want it to be. There are few things as steady or as changeable, as fierce or as gentle, as unstable or as undeviating, as light or as bold, as wroth, as balmy, or as protean as wind.

Someone who is *windy*, is either frightened, lacking in substance, or talks too much. A *windbag* in German is a *Windbeutel*.

Casual talk is *shooting the breeze* – in French *causer en l'air* and in Italian *parole al vento*. An easy task is *a breeze*. *To speak to the wind* is futile, whether you *parler en l'air* in French, *predicare al vento* in Italian, or *in den Wind reden* in German.

An unexpected gift of money is a *windfall*; to *raise the wind* is to find financial support; and to be *between wind and water* is to cut it a little fine; but to *go down the wind* is to be bankrupt. 'I am unwilling to mix my fortune with him that is going down the wind', said Samuel Pepys. '*Gone With The Wind*', said Margaret Mitchell.

An opportunist is someone who *blows with the wind. Einer der den Mantel nach der Wind hängt* in German is 'a person who hangs his coat according to the wind' – a turncoat. If you are canny, you already know *where the wind sits*, what is *blowing in the wind* or *which way the wind blows*. A shrewd German *er weiss aus welchem Loch der Wind bläst*, knows 'from which hole the wind comes'. But in Seneca's opinion, 'When a man does not know what harbour he is making for, no wind is the right wind.'

You breathe through a *windpipe*, chase a *windswift* or hare, and find out which way to land from a *windsock*.

You are a little drunk if you find yourself *three sheets in the wind*. Very drunk if you indulge in a *whirlwind* courtship. But the chances are that after either you will have time to repent at leisure, once *the wind is taken out of your sails*.

'Puff not against the wind', they warn. 'Blow not against the hurricane.' 'He that spits against the wind, spits in his own face.' And, because such warnings are seldom heeded, they usually get the chance to add, 'You have sown the wind, and you shall reap the whirlwind.' But the outcome may have been inevitable anyway for, according to the Chinese proverb, 'There is no wall through which a wind cannot pass.'

The word *wind* in English has an impressive pedigree and relatives everywhere.

It is unchanged in Old English, Old Frisian, Old Saxon, Dutch and German; and alters by only one letter to *wint* in Medieval Dutch, Middle and Old High German; and to *vind* in Danish and Norwegian. In Icelandic it is *windr*, and in Old Norse *vindr*. So in all the Germanic languages, it is virtually identical and descended from a common root, which seems to be the Old Aryan *wa*, which means 'to blow'.

The Celtic winds – *gwynt* in Welsh, and *gwins* in Cornish, certainly have a similar parentage.

The Italic tongues took a slightly different direction. Wind in Italian and Portuguese is *vento*, in Spanish *viento*, and in French simply *vent* – hence the English 'ventilate'. These all derive from the Medieval Latin *ventus* or *uentus*, which in turn seems to go back to the same Indo–European root *we* that produced the Aryan *wa*, and the Sanskrit *va*. *Nirva*, incidentally, means 'to stop blowing', and *nirvana* 'to be extinguished altogether'.

The Baltic and Slavic languages went from *we* to *wed*, which seems to be a more liquid than an airy root, meaning 'to flow' or 'rush along', rather than 'blow'. This produced the Polish *wiatr*, the Czech *vitr*, the Serbian *vetar* – and the infamous English 'weather' and 'water' with which it has become synonymous.

Of all the languages in the Indo–European family, only Greek went its own way, with *anemos* – which provides English 'anemones' and 'anemometers'. It is from the Greek *atmos* for 'vapour', through Old High German *atum*, Old English *aethm*, and Middle English *braeth*, that we draw 'breath'. And from Greek *aer*, for 'haze', through Latin *aer* and Old French *air* that we get both the 'air' we breathe and our 'aura'.

The word for wind in most Dravidian languages is based on *kat* or *gal*. In Malay it is *angin*, which seems to lead directly to the *matangi* or *matagi* of most Oceanic languages. In Tonga, wind is *agi*, and in Hawaiian *ani*. In Tagalog of the Philippines, it becomes *hangin*.

And in all languages there is the same blurring of boundaries, roots and meaning between the words for wind, spirit, breath and soul – as though each felt they clustered about the same essential mystery, and were loth to get too close for fear of startling or trampling on it.

'You cannot catch the wind in a net', says the proverb.

Nor keep it in a cage of words.

Wind Pictures

There are no photographs of the wind.

Even the classic shots of tornadoes bridging the gap between the dark cloud base and a flat midwestern landscape, have a paste-up feel to them that fails to capture the sensation of being in a wind. Photographs of hurricane-lashed seas and palms with fronds pulled back like tightly-braided hair, convey something of the scope of wind effects, but nothing of the full grandeur of the storm itself. And satellite pictures of the spiral immensity of a whole weather system, are just pretty patterns.

Moving pictures, with appropriate sound effects, are more successful – but even these seem to suffer from a lack of perspective. The studio winds have no fetch. You are painfully aware of the machine that made them, standing just out of shot. It takes an artist, most often a great painter, to do justice to the invisible wind. To capture its spirit in space and time in a way that tells us what it felt like to be touched by that wind, in that place, on that day – and in doing so, to relate it to all wind everywhere.

The representation of winds in medieval art as bodiless heads with puffed cheeks, or as gods with butterfly wings scattering flowers across sylvan glades, has little or nothing to do with the perception of wind itself. There are wind messages in the conventions and attributes connected with these myths, but little wind substance in the endless procession of playful *putti* and chubby little *amoretti* blowing on horns.

True Wind in art manifests itself only in its proper place, in the landscape.

Landscape painting is almost a modern invention. Prehistoric rock art is largely confined to the representation of animals in action, with occasional human figures, usually in hot pursuit. There are geometric and abstract forms in rock engraving that may, if we can ever decipher them, turn out to be spirit and wind symbols – the swastika seems to be one of these – but in Europe at least, the landscape painting as an exercise in realism, started with the Renaissance.

Painters of the Venetian school in the sixteenth century, particularly Giovanni Cima and Jacopo Tintoretto, gradually displaced the incident in their pictures deeper into the canvas, allowing it to disperse and be dominated by its surroundings. But the first landscape painter, who set up a scene as an object for separate aesthetic contemplation in its own right, was probably Joachim Patinir in Belgium. His active country scenes with jagged mountains, very blue in the distance, were the direct precursors of the more naturalistic work of the first master of the genre – Pieter Brueghel, the Elder.

It is significant that the art of landscape should have been invented by a lowland people who had no real landscape of their own. It is essentially a romantic art, born of need and transported by the prevailing wind to give the artist a god's-eye view, an aerial perspective on life. Something very similar can be seen as an influence this century on the work of the American artist Grant Wood, who grew up on the flat cornlands of Iowa.

The hard, direct effect of the climate on Brueghel is clearly evident in paintings like his famous *Hunters in the Snow*, which was painted in 1565 – the first of the brutal winters that marked the onset of 150 years of deteriorating weather which climatologists now know as the Little Ice Age. In 1563, he painted a nativity scene in which the family are shown surrounded by the warmth of a group of homely shepherds, soldiers and wise men. In 1567, after the freeze set in, he painted another version in which the family are barely visible, huddled in a rude shelter in one corner of a bleak windswept snowscape through which distant peasants fight their way, leaning hard against the weather.

Climatologist Hubert Lamb wondered how faithful such an artist's work might be as a mirror of the climate of his time, and analysed 200 paintings done by Brueghel in the sixteenth century; and by two other Flemish landscape painters, Jacob van Ruisdael and Meindert Hobbema, in the seventeenth century. He estimated the cloud cover in Brueghel's work as 40 per cent, while in the work of the two painters of the following century, when summers were even wetter and cooler, it soared to higher than 70 per cent.[269]

The meteorologist Hans Neuberger, took this survey a stage further by analysing 12,000 paintings completed between 1400 and 1950. He rated the blueness of the sky, the clarity of the air, cloudiness, and the degree of shelter from the wind provided by any structures shown. The results show predictable regional differences, with Mediterranean artists depicting bluer skies and greater visibility than the British school. There was not one British painting from any period with a clear sky.

Average cloudiness ranged from 33 per cent in the fifteenth century, to 82 per cent in the seventeenth century, with a marked jump from 50 to 77 per cent when the Little Ice Age set in around 1550. Cloud cover fell again to 67 per cent in the nineteenth century, when the present warming trend began. He concludes that 'In general, these results leave little doubt that the artist's collective experience of cloudiness and transparency of the air in a given region during a given period can be deduced from their paintings', which make them 'conscious and unconscious chroniclers of their environment'.[358]

These general tendencies of course conceal individual preferences. It is said of Ruisdael that 'he never painted a hot day'. All his Dutch panoramas show the sun breaking fitfully through the clouds and lighting up isolated patches of duneland. Something similar is true of the romantic landscapers Gaspard Poussin and Gellée Claude, but it was in Britain in the following century that an artist, for the first time, met and dealt with the wind on its own terms.

John Constable said that he wanted to paint what he saw with his own eyes, to capture the 'light, dews, breezes, bloom and freshness' of the English countryside.

He did, and more. He painted the invisible, doing for the perception of nature in paint what Wordsworth did for it in verse. He discarded the weight of myth and allegory and substituted a natural vision. He caught nature in the act of change, which is the essence of windiness. He was enthralled by the shift of shadows, vapour and cloud in air, and determined he said, 'to arrest the more abrupt and transient appearances . . . to give to one brief moment, caught from fleeting time, a lasting and sober existence'.[414]

Constable became a wind-watcher, the most dedicated cloud sketcher in history.[483] His interest was fuelled by the growing awareness of meteorology in his day. Luke Howard had just published his classification of cloud types,[217] and Thomas Forster's *Researches About Atmospheric Phaenomena* was a best-seller. 'I have done a great deal of skying', said Constable. 'It will be difficult to name a class of landscape in which the sky is not the key note, the standard of scale, and the chief organ of sentiment . . . It is the source of light in nature, and governs everything.'[281]

In 1820 and 1821, he produced a mammoth series of cloud studies in oil and on the back of each recorded the place, date, time and wind direction. This scientific approach may have focused his attention, giving him a formal framework, but it in no way impedes his vision. Using meteorology, drawing on his own childhood experience of working in his father's windmill, but above all by watching what actually happened when the wind blew, Constable captured the essence of moving air. His sketches single out small moments of wind time, no two the same, but each one filled with a compelling, almost uncanny quality, that makes them more airy and more real than any photograph could ever be.[13]

John Ruskin, the influential nineteenth-century art critic, reviewed landscape painting in his time with the comment that.

> The first thing that will strike us . . . is their cloudiness. Out of the perfect light and motionless air of medieval art, we find ourselves on a sudden brought under sombre skies, and into drifting wind . . . we are reduced to track the changes of the shadows on the grass, or watch the rents of twilight through angry cloud. And we find that whereas all the pleasure of the medieval was in stability, definiteness and luminousness, we are expected to rejoice in darkness, and triumph in mutability; to lay the foundation of happiness in things which momentarily change or fade; and to expect the utmost satisfaction and instruction from what it is impossible to arrest, and difficult to comprehend . . . If a general and characteristic name were needed for modern landscape art, none better could be invented than 'the service of clouds'.[413]

Constable was first in this service, but the romantic revival reached its peak with his contemporary Joseph Turner, whose whole art was dedicated to catching the most fleeting effects of light and air, to painting the unpaintable. Turner, said Ruskin, 'is an exception to all rules . . . he rushes through the aethereal dominions of the world of his own mind – a place inhabited by the spirit of things – seizing the soul and the essence of beauty'. Doing this, very often, by capturing the quality of changing light in wind and air with what Constable called 'tinted steam'.[394]

Turner was obsessed with atmosphere. He painted clouds, waves and mist as billows of form, dramatically lit. He reduced all the elements to windblown abstractions and was, as a result, not easily understood. Some contemporary critics dismissed his landscapes as 'pictures of nothing, and very like'. But he is now credited with being 'the greatest natural colourist the world has known' and with the creation of a vast body of work that represents 'perhaps the greatest revelation ever made of the power and majesty of nature'.[392]

Turner fathered the modern movement in art and, by his influence on both Edouard Manet and Claude Monet, was ultimately responsible for the development of the Impressionists, though even the best of them was static by comparison. By stepping out of the studio and into the wind and rain, Constable and Turner between them created a new experience of space, extending our environment into infinity, giving us an aerial perspective which colours the way in which we see the world today. We are still involved in their search for an understanding of Earth as a whole, through the wind, in time, and from the sky. A quest for what Wordsworth called 'the sentiment of being spread o'er all that moves'.

This sensitivity extended from painting to the more environmental arts of sculpture and architecture. As the Little Ice Age drew to a close at the beginning of the eighteenth century, there were brighter years and a lifting of spirits and horizons. Older settlements tended to cling to the safety of sheltered sites in the valleys, but around 1750 people began to want to see further and to build their homes on hills. The bright and airy terraces of Bath in England date from this period. But it was not until after Turner's time, perhaps even in the early years of this century, that the long look became fashionable and penthouses were retrieved from the servants.

It is from a height that the pattern in things becomes apparent. Random activity, as of people in a park, is ordered and abstracted by distance, so that it becomes a meaningful ebb and flow. Isolated elements of behaviour merge into biological rhythms and cycles. Stray breezes become part of an integrated weather system. This idea of transformation in time is basic to an understanding of wind and nature, but foreign to sculpture until 1920 when Naum Gabo, at the height of unrest and civil war in Russia, produced his *Standing Wave* – the first piece of deliberately *kinetic* art.

There were early Oriental 'wind chimes' of porcelain, glass and bamboo; and there had been mechanical figures, such as those in medieval clock towers that moved to strike the hours; but Gabo started something new. It was just a simple metal rod that produced a sinuous harmonic vibration when activated by an electromagnet, but it denied the limits of its material and set sculpture free, giving it a rhythm and a cadence that were almost musical.

In the United States, Alexander Calder took up the challenge and after a few early motorised models, realised that the choreography was best left to nature, preferably to the wind. Calder created the *mobile*, a system of fluid shapes reminiscent of leaves, flowers, planets and tropical fish hung in delicate balance that was easily disturbed and set in motion by a passing breeze, by rising air or simply by the movement of someone walking by. Some of his larger pieces are anchored to the ground and tend to be more stately, moving only within prescribed orbits, but all provide intricate ecologies of movement, transforming themselves in time like living organisms. Each becomes, in Jean-Paul Sartre's words, 'a strange kind of top, which only exists in motion'.

George Rickey has brought the skill of a precision tool-maker to a series of kinetic sculptures, such as *Windflower*, which are even more spirited. Jean Tinguely, with his self-destroying sculptures, contributes a vital and lively sense of humour. Lynn Chadwick in England produced architectural sculptures of heavy metal so beautifully balanced that, like his *Fisheater*, they become animated, almost frothy, in the wind. And Arnaldo Pomodoro in Italy is turning the kinetics of windpower into an artistic expression. Through all these attempts to communicate the eloquence of change runs an awareness, finely expressed by T. S. Eliot, that:

> Only by the form, the pattern,
> Can words or music reach
> The stillness . . .

Without wind, without its constant reminder of the changes being made by time, we would find it difficult to see nature clearly as a process of flow.

We would be more like astronomers, condemned much of the time to looking at the static old light of dead or distant stars.

Wind Sounds

Without air, there would be no sound.

All sounds are caused by vibration which compresses molecules of air. And it is waves of such compression, spreading rhythmically out through the air in all directions, that carry news of the vibration in the form of sound signals.

Our ears are tuned to vibrations between 16 and 20,000 cycles a second. Anything slower than that is 'infrasonic', anything faster 'ultrasonic'. All frequencies travel at the same speed, about 332 metres per second, at 0° Centigrade at sea level. The speed is higher in warmer air, which is a better conductor, and lower at altitude where there are fewer molecules to do the conducting. In water, which is denser than air, sound travels roughly four times as fast.

The distance covered depends partly on the frequency. High-pitched sounds travel only a fraction as far as low-pitched sounds of the same power.

And wind, by producing rival disturbance in the air, can alter the speed and the frequency and the reach of any sound, imposing new patterns and rhythms on it. Most wind sounds are borrowed and orchestrated in this way, but recent research suggests that the wind may actually have a voice of its own.

Artificial sounds in the very long-wave, low-frequency range are limited to sonic booms and large chemical or nuclear explosions. As part of arms-control agreements, an international system of listening posts has been set up to monitor such sounds, and has in the last decade begun to pick up mysterious echoes from sounds of unknown origin.

These infrasounds range in frequency from 0.001 to 0.1 cycles per second and are very hard to pin down. They seem to come from areas in which severe storms are active, and are believed to be produced by violent turbulence in the air, intense 'shear' forces produced as air flows over an obstacle in its path, or perhaps even from the windblown surface of the sea. Though their intensity is low, they travel vast distances, rebounding off the tropopause, touching everything in their path. We cannot hear such infra-sounds in the normal way, but it is possible that we, and other living things, are sensitive to them and are conscious at some level of the sporadic deep-throated rumble of distant winds that might be heading our way.

Sailors in Brittany say that the sea is a living thing, feels the impending weather in its water, and tells by its sound what is likely to happen as much as a week in advance. Fishermen on the Moray Firth in Scotland call this sound the 'song of the sea' and, when they hear it, know that a southeast wind will soon start to blow. A long song from the bar of Banff means a gale within twenty-four hours straight out of the heart of the North Sea.[391]

In the Marshall Islands, they say that the great sky god *Lówa* made a magic wind sound over the ocean – something like 'Mmmmmmmm . . .' – and then the islands were created.[96] In the Kalahari, the San 'Bushmen' say that the call of the air is that which sounds like a person, humming to windward.[37] And in Africa and the Pacific, they make a similar sound to attract the attention of the supreme being. They use a 'bull-roarer'.

When a thin board, usually of carved wood, is whirled around on the end of a cord, it produces a throbbing sound. The smaller the board and the faster it whirls, the higher its pitch. The effect is not so much like the lowing of a bull, as it is like the wind in all its moods, from a breeze to a howling gale. To the aborigines of Australia it was the voice of *Daramulun*, the 'master of thunder'.[220] In West Africa it was the voice of *Oro*.[276] In the Amazon, the board was made of the frontal bone of a human skull and was used to summon the wind.[250] It was a vital part of the rainmaking ceremony amongst Apache, Navaho, Zuni, Ute and Kwakiutl people.[334] A practice which may well have been inherited from Stone-Age ancestors – similar oval pendants of bone, perforated at one end, have been found in Palaeolithic deposits at Laugerie in the Dordogne.[240]

In the Szechwan and Yunnan provinces of China, they used to call the wind for winnowing by making the sounds '*Ja Da Da, Ja Da Da*', which meant 'wind come home'.[169] Elsewhere, almost everywhere amongst coastal or seafaring people, the accepted practice for those in need of wind seems to have been just to whistle, softly for a breeze, loudly for a gale. Casual whistling, of course, was frowned on.[391]

The wind frequently whistles back. 'The wind stood up, and gave a shout; And whistled on his fingers . . .'[460] Part of this effect is produced simply by fluting through gaps and cracks of the right size and shape in doors and windows, but often there seems to be more to it than that. W. H. Hudson gives a vibrant account of an assault by the *pampero* on his childhood home in Argentina:

> The wind beats unceasingly on the exposed roof with a succession of blasts or waves which vary in length and violence, causing all the loose parts to vibrate into sound. And the sounds are hissing, whispering, whistling, muttering and murmuring, whining, wailing, howling, shrieking – all the inarticulate sounds uttered by man and beast in states of intense excitement, grief, terror, rage, and what not. And as they sink and swell and are prolonged or shattered into convulsive sobs and moans, and overlap and interweave, acute and shrill and piercing, and deep and low, all together forming a sort of harmony, it seems to express the whole ancient dreadful tragedy of man on earth.[224]

Out of doors it seldom sounds quite so bad. But there are places where the wind rubs up against natural sounding boards to produce terrifying chords. On a number of beaches and dunes all over the world – wherever there are smooth, well-rounded, close-packed, grains of quartz – a squeaking sound with a frequency between 500 and 2500 cycles per second occurs when the sands are disturbed. This is best heard while dragging your feet through a dune.

In a smaller number of places – in the deserts of China, Afghanistan, Sinai, Sahara, Kalahari, Namib, Chile, Baja California and Kauai Island in Hawaii – there are more formidable 'booming dunes'. Humming, buzzing and moaning sounds fill the air when these sands are played on by a passing wind. And when someone walks or rides through them, they respond with a trumpeting, rumbling, roaring, thundering boom that is enough to fray the steadiest nerves. These sounds lie between twenty and fifty cycles per second and seem to need slightly smaller grains of highly polished, perfectly spherical quartz or carbonate, that are absolutely dry. Such conditions may be relatively common on Mars, which could give that planet's formidable winds a voice throaty enough to be heard throughout its thin atmosphere.[287]

Trees provide the wind with another natural instrument. The

diameter of each leaf and twig determines the specific pitch. Guy Murchie likens the apple tree to a cello, an old oak to a bass viol. The cypress is a harp, the willow a flute and the young pine a muted violin. 'Put your ear', he suggests, 'close to the whispering branch and you may catch what it is saying: a brittle twitter of dry oak leaves in winter, the faint breathing of the junipers, the whirring of hickory twigs, the thrumming of slender birch clumps, the sibilant souffle of the cedars, the mild murmuring of the sugar maple, and behind them all the trafficky thunder of whole bare trees torn in a headlong tide of air.'[351]

The conifers are probably the most responsive and quite different in tone from the purr of the wind in the willows. Redwoods and cypress tend to build up a dense background of white sound; casuarinas exaggerate shamelessly, amplifying the easiest sea breeze into an incipient gale; but the virtuoso performers are undoubtedly the broad-needled pines. Minute eddies form on the lee side of each needle, causing ripples which set the air vibrating in a pure soft tone each time wind, 'that grand old harper, smote his thunder harp of pines'. On summer afternoons, the Japanese make pilgrimages to the woods to hear what they call *matsukaze*, the wind that knows the song of the pines.

The wind produces some of its most memorable effects once it reaches force 6, the beginning of a gale, and strums telephone and power lines like the strings on a giant guitar. When the evoked tone coincides with the natural harmonic of the wire, then the volume increases and holds its pitch in an eerie concert. If this moves into a minor key, it has the ability to produce physiological changes, lowering the blood pressure of anyone nearby. Wires strung between poles whose interval is appropriate for major key tones, have the opposite effect.

This is the principle behind one of the oldest of stringed instruments, the only one actually designed for the wind to play – the *aeolian harp*. Rabbinic records tell that King David deliberately hung his harp above his bed at night so that the wind could soothe him with its own harmonies. There are Indian poems which suggest that the *vina*, a form of stick zither, was sometimes used in the same way. Kite-makers went for a more lofty rendition, setting up whole orchestras of soaring string sounds that still resound over Malay villages during the monsoon. Saint Dunstan, the Archbishop of Canterbury, is said to have provoked suspicion of sorcery in the tenth century by experimenting with windblown strings in his monasteries.

The first detailed plans for a harp designed for the exclusive use of Aeolus, are given in *Misurgia Universalis*, a seventeenth-century treatise on music by Father Athanasius Kircher. They show a set of twelve cat-gut strings of different thickness strung across a rectangular box with a perforated sounding board at the back. At the front are a pair of hinged doors that can be opened at the best angles to funnel the wind directly across the simple harp. The ghostly sound was a perfect accompaniment to the romantic period and such instruments were much in favour in parks, on roofs and in the ruins of castles and follies during the late eighteenth and early nineteenth centuries:

> A certain music, never known before,
> Here sooth'd the pensive melancholy mind.[481]

Wind instruments which depend on human breath or air pumped from a bellows, are of two basic types: 'free aerophones' produce pulsations directly in open air, usually by the vibration of a reed, as in an accordion or harmonium. Early organs like the *regal*, worked in this way. All other wind instruments depend on the vibration of a column of captive air in a pipe or tube; the shorter the tube, the higher its pitch.

Straight bamboo or reed pipes must have been the earliest forms, but the spiral of modern horns is already anticipated in the natural shape of conch shells that are still blown all over the world. It is interesting that those who live in such communities, do not hear the sea in a shell when it is placed to the ear – they all say it sounds like the wind.

The device which sets air vibrating in a wind instrument may be compressed lips of the player, as in a trumpet, or the passage of air over a sharp edge, as in a flute. But the invention of the reed was an early one. Boys everywhere know how to stretch a blade of grass between their thumbs and make it squeak by blowing on it. The brighter ones roll a long broad blade up spirally to form a funnel with the thin end free to vibrate across its upper orifice. This is a direct precursor of the clarinet, or with the reed split into two, of the oboe. Pan-pipes are simply sets of flutes tied together in a raft or bundle.

Given the natural origins of most orchestral instruments in shells, reeds, leaves and membranes, it is hardly surprising that composers should at times return to nature for inspiration – and in particular to the sounds produced by winds.

264

This is difficult to hear in music composed only for the human voice, but becomes apparent with the development of instrumental music, and with the addition of drama to music, in the late sixteenth century.

Amongst the first to experiment with new descriptive forms was the Venetian priest Antonio Vivaldi, whose 500 odd concertos include *The Four Seasons* – in each of which he manages to evoke the mood of movement in air characteristic of a particular time of the year. In *Summer*, the violas whisper of a gentle west wind, only to be buffeted by squally scales from the whole ensemble as the stormy north wind hints of approaching autumn. And in *Winter*, vigorous chords mark the stamp of frozen feet as rival, equally icy, winds break into open conflict.

Bach was far too involved in exploring the permutations of polyphony to write purely descriptive music, but Beethoven sometimes used his equally powerful melodic gifts in unashamedly romantic ways. His *Pastoral Symphony* includes what is probably the best of all descriptions of the calm before a storm. The fourth movement opens with an atmosphere of marvellous suspense. The peasant dancers in the tavern stop, and one can almost feel the early flurries of wind and the first big drops of rain as the ominous calm is broken and thunder begins to roar.

The early nineteenth-century Romantic movement in literature, had its counterpart in music. Form became less important than content, and composers in search of new freedom of expression, turned with enthusiasm to literary and natural themes. Franz Liszt, except perhaps in *The Four Elements*, remained closest to the classical mold. Felix Mendelssohn introduced a more poetic sensibility – particularly in *Fingal's Cave*, where he has his oboes howling up a Hebridean gale, as gulls clamour and waves of strings crash against the rocks.

But the giant amongst romantics was unquestionably Hector Berlioz, the perfect painter in sound. His uncanny mastery of orchestration is best illustrated in *The Trojans*, which includes perhaps the finest and most sustained piece of nature painting in all music. From the first distant sound of thunder, usually provided by two kettledrums placed deliberately offstage, through the full fury of the storm, to the carolling woodwinds that restore calm again once it passes, the imagery is complete, theatrical and wonderfully evocative.

Modest Moussorgsky in *A Night on Bald Mountain*; Peter Tchaikovsky in *Francesca da Rimini*; and Nicolai Rimsky-

Korsakov in *Sheherezade*; all conjure up great storms, decorating their descriptions with melodic, often sentimental, Russian folk forms. But it was another national movement – that of the French musical impressionists at the end of the nineteenth century – that succeeded best in capturing natural atmosphere.

In his three symphonic sketches that are collectively called *La Mer*, Claude Debussy records the echoes of creation in the passage of wind over water. *The Dialogue of the Wind and the Sea* uses all the wind instruments to superb effect, gathering aerial forces into a climactic brassy clash of new waves in an old storm.

The last great romantic was a German – Richard Strauss. Often vulgar, sometimes inspired, his powers of natural description are nowhere better displayed than in the *Alpine Symphony* – where his climbers ascend through fog and cloud and experience both the eerie premonition and the harsh reality of a passing thunderstorm.

Opera, with its dramatic potential, provided the perfect setting for violent winds and storms as metaphors of emotional disturbance.

Mozart just touched on the meteorological potential by having his lovers in *Cosi fan Tutte* escape together during a storm that is miraculously stilled. But the master of drama – Richard Wagner – never missed a chance to have the elements accentuate his ceremony with their stormy comment.

The Flying Dutchman is filled with airy open fifths that whistle up a wind in wild, wavelike figures about the unhappy captain – 'doomed to all eternity to hunt the watery waste for treasure that gives him no satisfaction'. And *The Valkyrie* actually opens in storm, building tension right from the start as wind sweeps through the forest of the world, gathering strength for one last furious onslaught on mankind.

Italian opera was more concerned with human passion in its own right, but Vincenzo Bellini wrecks his *Pirate* in a storm that does its work entirely in the overture. Giuseppe Verdi gets more mileage from his wind by having a storm rage constantly around all those who meet to plot the death of the Duke in *Rigoletto*.

All these musical and dramatic allegories play upon an old theme. On the fact that music, like smell, seems to go more directly to the heart of memory than either sight or touch can do. 'When people hear good music, it makes them homesick for something they never had, and never will have.'

It is a way we have of coping with the intangibles – of coming to terms with mysteries like the wind.

In Polynesia, traditional history is divided into two periods. The most recent of these is called 'The Hearing of the Ears' and covers all information contained in folklore and oral memory. There is also an earlier period known as 'The Wind Clouds', which consists of misty intimations of an ancestral home somewhere over the sea. Information about this place and time is picked up in moments of natural knowing that come to some people, usually when they are alone and exposed to the wind.

The people believe that winds have voices and carry songs. They speak of 'tale-bearing winds that gossip afar', and try to listen to the songs and learn the messages they carry.[296]

The Swedish–American poet Carl Sandburg came as close as anyone in our industrial time to understanding:

> Long ago I learned how to listen to the singing wind,
> and how to forget and how to hear the deep whine . . .
> Who can ever forget listening to the wind go by,
> counting its money and throwing it away?[420]

Perhaps if he had lived in Polynesia, it would have seemed more relevant, less painful.

Perhaps we, with a wider knowledge of the wind in all its guises, can learn to bridge that gap.

PART FIVE
Wind and Mind

We have trouble with space.

In western thought, space is empty and has to be occupied with matter. Time is empty and must be filled with activity. 'Don't just sit there, *do* something!' We like our lives to be carpeted wall to wall.

Silence makes us uneasy.

Radio silence is described as 'dead air', something to be avoided at all costs. So we plug it with words or music, often words and music, finding comfort in the clutter, however meaningless it might be. We have lost the meaningful pause. Silence in music is invariably destroyed by applause from some idiot who thinks the concert is over.

Our philosophy, our culture, assert the exclusive integrity of things. Between them there is no-thing, nothing worth being concerned about. We negate and nullify the interval, shattering its integrity with a half-time show. Even our sciences are misled by comforting misapprehensions, half-truths such as 'Nature abhors a vacuum.'

Other cultures are more fortunate.

The Japanese describe empty space as *kukan*, which comes close to our understanding of it; but they also recognise *ma*, which is something very different.

There is little in any of the precise Indo–European languages, certainly nothing in English, to compare with *ma*. In that syllable is a guarantee of equal rights for the interval, an assurance that empty space and free time have their own value and integrity. *Ma* is an ability to see and feel space, which is perceived as an area of change, of hue and brightness. It recognises the huge burden of meaning which is carried by small silences, and the fact that the interval calls for more, rather than less participation, simply because it is incomplete.

It is this concept of nothingness-as-somethingness that made it possible for early Indian philosophers to perceive the integrity of non-being, to name free space and give us the *zero*. And it is in this understanding that we can find a way of coming to terms with the spirit of the wind.

It never occurred to the Babylonians, Egyptians, Hebrews, Greeks or Romans to give 'nothing' a symbol. And, as a result, they all ended up with cumbersome number systems which were very difficult to manipulate and made it impossible for mathema-

271

tics to flower. The Indians came to the problem from a different direction, with the advantage of a background in the *vedas* – a Sanskrit word that simply means 'knowledge' and refers specifically to a collection of ancient Aryan hymns composed sometime in the third millennium BC.

Central to all the vedas is a belief in *brahma*, the universal spirit, to which access is gained by meditation. Brahma is described as 'everything and nothing, the source of all power and strength, the minutest of the minute and the greatest of the great'. And in the *Atharva-Veda*, as transcribed in Kashmiri around 500 BC, brahma is represented by two different symbols – a small dot to indicate absence or emptiness, and a circle to show presence or fullness. The same symbols also appear in astronomical texts from that period, where they are used to represent the planets, whose circular motions and endless journeys linked them with the notion of infinity.[348]

Babylonian mathematicians in 1400 BC were using cuneiform symbols for twelve different number names and simply leaving blank spaces to indicate the relative place of different digits. The number twelve was written 12. And the number one hundred and two as 1 2. The Mayans in Central America had a complex symbol that was used to express the time elapsing between dates, but the genius of the early Indian mathematicians lay in their recognition of the need for a separate and simple symbol to represent the concept of 'nothing in this column'. And it came very naturally to them, once this need was apparent, to choose the physical symbol that in their culture already stood for the dual spiritual concept of absence and presence all at the same time.

So the concept of 'nothing' came to be represented as a little dot, the closest approach to nothing that was still visible, and as a circle or hollow dot, which stood for 'nothing-as-something'.

Alexander and the stream of Greek merchants who followed him to India may have brought the notion back to Europe with them – there are places in the works of Archimedes and Ptolemy where a circle is used to indicate a marginal gap – but it was the Arabs who realised the full implications.

They called the symbol *sifr*, meaning 'empty', a direct reference to the empty column in an abacus, and used it in their numeric system. The *sifr* or 'cipher' soon became our 'zero' and immediately made it possible to distinguish at a glance between 12, 102 and 120. It was this positional notation that made Arabic numerals superior to any previous system, and was directly responsible for the development of *al-jabr* or algebra. It turned long-division from a subject for experts only, into something any child could do, and it gave people with real ideas the time and space to think.

The Japanese *ma* and our zero have a common ancestor in a philosophy born on the wide Asian steppes which makes something of nothing. Which, in Shakespeare's words, 'gives to airy nothing a local habitation and a name'. There is enough of this understanding still in the eastern roots of western religion, which speaks of 'having nothing, and yet possessing all things'. But such concepts become increasingly alien to us and would mean nothing at all if it were not for our own continuous exposure to, and sensitivity to, the wind.

It is no accident that those western artists that are best able to understand space and interval (and are very popular in Japan), are products of windswept landscapes in which 'airy nothing' becomes an elemental force. Joseph Turner's ability to create intense psychological volumes; Hans Holbein's talent for linking blank spaces of paper together in apparently casual tensions between mark and void; and Piet Mondrian's dogmatic austerity – are all wind-inspired. They are demonstrations on canvas of an awareness of the power and importance of space, of our continuing need to 'enter into the small silences between'.[341]

The wind, as moving air, touches us directly, sometimes brutally. But there is another more subtle and pervading influence which moves us.

Our relationship to it is like that of a pawn on a chessboard facing a bishop confined to the black squares. On the dark diagonals we are directly vulnerable, but even on the white squares the force has influence through the mediation of other pieces. We are never free from its pressure, which controls our mental weather and has shaped much of our minds and lives.

9. THE PSYCHOLOGY OF WIND

When air is completely still and cool, we can enjoy bright starlit nights and the respite from manic combustion. But when calms are warm, or go on too long beyond the dawn, they become awkward and disturbing. They tend, we have learned, to precede the storm.

So wind, when it comes, is usually welcome. But not always and not for everyone.

There is an old English proverb that says:

> When the wind is in the east,
> 'tis good for neither man nor beast.

Voltaire, in exile in London, wrote.

This east wind, is responsible for numerous cases of suicide . . . black melancholy spreads over the whole nation. Even the animals suffer from it and have a dejected air. Men who are strong enough to preserve their health in this accursed wind at least lose their good humour. Everyone wears a grim expression and is inclined to make desperate decisions. It was literally in an east wind that Charles I was beheaded and James II deposed.[100]

Hippocrates was convinced that west winds were worse, and that people exposed to them became pale and sickly with digestive organs that were 'frequently deranged from the phlegm that runs down into them from the head'. Theophrastus noted that it was in southerly winds that 'men find themselves more weary and incapable' due to thinning of the lubricant in their joints. While the north wind was described by Spenser as 'bitter, black and blustering', by Shakespeare as 'wrathful and tyrannous', and held responsible for 'gout, the falling evil, itch and the ague'.[202]

The direction, it seems, is unimportant. But there is general agreement amongst Byron, Columbus, Dante, Darwin, Humboldt, Luther, Michelangelo, Milton, Mozart, Napoleon, Nietzsche, Rousseau, Schiller and Wagner that wind does make a difference, turning both body and mind. Goethe thought it 'a pity that just the excellent personalities suffer most'. But the winds are nothing if not democratic. Roughly 30 per cent of all people everywhere are sensitive in some way to their touch.[467]

Ill Winds

There is something about wind, quite apart from its cooling influence, that directly affects our well-being.

One study of performance in physical fitness tests at a variety of temperatures, found an efficiency peak with wind blowing at twenty-five kilometres per hour (force 4), with energy falling off at both lower and higher wind speeds. Observation of the behaviour of children in the playground of an American school, revealed that the average number of fights per day doubled when wind speeds crossed the biological threshold above force 6.[68]

As a species, we seem to have a high awareness and a surprisingly low tolerance of wind. There are individual differences of course, some of which are apparently sex linked.

Most women, very sensibly, seek shelter from the wind. But there is something about an approaching gale that makes men very restless. Almost as though the sight of swiftly-driven cloud or the sound of air rushing through the trees were stimuli that triggered some deep-seated response. A fisherman on the Dalmatian coast of Yugoslavia described *hukovi*, the shrill scream that warns of an approaching *bora*, as 'a desperate sound that causes a man's heart to tremble'.[509]

There is no doubt that days with a lot of wind were once dangerous ones, destroying shelters, dispersing warning scents, and masking the sound of an approaching predator. And it may well be that, even in our modern microclimates, men in particular are still excited and disturbed by the old signals.

The physiology seems to involve the classic alarm reaction of an increased production of adrenalin. Metabolism speeds up, blood vessels of the heart and muscles dilate, skin vessels contract, the pupils widen and the hair shows a disturbing tendency to stand on end, producing prickles of apprehension. This is a fine and useful response to an emergency, a good prelude to instant action; but when it is provoked by an alarm that goes on ringing, by a wind that blows for hours or even days on end, it puts a lot of strain on the system.

The feelings to begin with, may amount almost to euphoria. The Yale geographer Ellsworth Huntington recalled seeing a small boy, 'who was usually very quiet, climb to the top of a tall tree when a violent wind came up, and swing in the branches, singing at the top of his voice'.[227] It has been suggested that the controversial Bishop James Pike, who walked to his death in the desert near the Dead Sea in 1969, was over-reacting in this way to the stimulus of a strong warm wind known locally as the *sharav*.[467]

Sailors and fishermen, people who live constantly in the wind, become habituated. Their bodies learn to deal with wind stress. But many city dwellers lose the ability to adapt and find that the wind seeks them out even in the comfort of their homes, where some of them die of myocardial infarction, extra-systolic contractions and angina pectoris.

All of these 'heart attacks' and 'strokes' are the result of over stimulation. They are far more common in men than women, and they frequently take place on windy days. In one study of general blood vessel disorders, it was found that 50 per cent of all myocardial infarctions and strokes occurred when wind was blowing at force 4 or 5.[241] Higher wind speeds, strangely, seem to put less strain on the heart. Perhaps adrenal fatigue eventually sets in, or fear cancels out excitement and leads to some sort of withdrawal.

All winds seem to be unsettling to some degree, but a few have a positively evil reputation.

In the middle of the eighteenth century, with their first Empire at its height and the founding of the British Museum, tourists from

England began to visit classic sites. One of these early travellers, in a 'Series of Letters to William Beckford Esquire of Somerly in Suffolk', sent back a wind description so vivid it is still worth quoting at length.

The most disagreeable part of the Neapolitan climate is the sirocc or southeast wind, which is very common at this season. It has now blown for these six days without intermission; and has indeed blown away all our gaiety and spirits; and if it continues much longer, I do not know what may be the consequence. It gives a degree of lassitude, both to the body and the mind, that renders them absolutely incapable of performing their usual functions.

It is perhaps not surprising, that it should produce these effects on a phlegmatic English constitution; but we have just now an instance, that all the mercury of France must sink under the load of this horrid, leaden atmosphere. A smart Parisian marquis came here about ten days ago: he was so full of animal spirits that the people thought him mad. He never remained a moment in the same place; but, at their grave conversations, used to skip from room to room with such amazing elasticity, that the Italians swore he had got springs in his shoes. I met him this morning, walking with the step of a philosopher; a smelling bottle in his hands, and all his vivacity extinguished. I asked him what was the matter? '*Ah, mon ami*', said he, 'I am near to death. I, who never knew the meaning of the word *ennui. Mais cet exécrable vent*, if it lasts even two days more, I will hang myself.'

The natives themselves do not suffer less than strangers; and all nature seems to languish during this abominable wind. A Neapolitan lover avoids his mistress with the utmost care in the time of the sirocc, and the indolence it inspires, is almost sufficient to extinguish every passion. All works of genius are laid aside, during its continuance; and when anything very flat or insipid is produced, the strongest phrase of disapprobation they can bestow, is that it was written in the time of the sirocc.

I have not observed that the sirocc makes any remarkable change in the barometer, and it is certainly not the warmth of this wind, that renders it so oppressive to the spirits; it is rather the want of that genial quality, which is so enlivening. The spring and elasticity of the air seems to be lost; and that active principle which animates all nature, appears to be dead. This principle we have sometimes supposed to be nothing else than the subtle electric fluid that the air usually contains; and indeed we have found, that during this wind, it appears to be almost annihilated.[59]

For someone writing two years before Franklin flew his famous kite and twenty years before Priestley wrote his book on electricity, let alone someone putting pen to paper in the time of a 'sirocc', this electric insight is astounding.

Today we know a great deal more about our physical environment, but it has only been in the last few years that we have begun to realise just how closely living things are tied to subtle variations in Earth's electromagnetic field, many of which are directly connected to particular movements of air.

Sirocco, in Arabic, means 'easterly'. In meteorological terms it is a wind that appears most often in the spring, bringing warm air in from the Sahara and the Arabian Peninsula as a series of low-pressure systems moves eastward across the Mediterranean. It has many local names – *levante* in Spain, *leveche* in Morocco, *chergui* in Algeria, *chili* in Tunisia, *ghibli* in Libya, *khamsin* in Egypt, *sharav* in Israel, *sharkiye* in Jordan, and *shamal* in Iraq. In every guise it is warm, with a temperature more than 10° Centigrade higher than the seasonal average; and dry, with a relative humidity that is always less than 30 per cent, falling sometimes to zero. And the combination of these two characteristics seems to do something fundamental to the wind's electrical properties, and in turn to us.

At the Hebrew University in Jerusalem, pharmacologist Felix Sulman has been studying physiological responses to Israel's version of the *sirocco*. He finds that almost one third of the population experience some kind of adverse reaction to the *sharav*. And of these, 43 per cent show an unusually high concentration of serotonin in their urine. This is a powerful and versatile hormone which causes the constriction of peripheral blood vessels, including those in the brain, controls sleep, and is responsible for the development of mood. It is a natural tranquilliser, but too much of it produces clinical symptoms which include migraines, allergic reactions, flushes, palpitation, irritability, sleeplessness and nausea.[456]

Another 44 per cent of the wind casualties, most of them women, have little or no adrenalin in their urine, and complain of fatigue, apathy and depression. Together, these two groups amount to a quarter of the total population, which is a considerable

279

number of people to be feeling out-of-sorts all at the same time.[468]

Sulman concludes that wind sensitivity is due to 'a neurohormonal reaction based on the regulatory functions of the hypothalamus, the pituitary gland, the thyroid and the adrenal gland', and that women, who tend anyway to produce less adrenalin, are particularly susceptible.[467]

> Maidens will waver,
> Falter and quaver,
> With air in mind.[478]

A Victorian lady on a visit to Algiers in 1868 decided that 'This awful wind is a moral as well as a physical poison.' But her evident distress did nothing to curb her formidable powers of description.

> The leaves of trees fold before the eye. After minutes of a heavy and suffocating calm, succeeded squalls of stinging wind. The clouds of flying sand soon eclipsed the obscured disc of the sun; and the different shades of yellow, orange, saffron and lemon melted into a mass of copperish colour impossible to describe. The covers of my books were shrivelled as if they had been lying a whole day before a fire. When in company with some officers, I happened to touch the sword of one of them, my hand was seared as if by a hot iron.[118]

Part, at least, of the ill effects of various *sirocco* winds, must be due to the dust they carry. An army officer exposed to the *ghibli* in Libya complained that, 'the eyes become red, swelled and inflamed, the lips parched and chapped; while severe pain in the chest is generally felt, in consequence of the qualities of sand unavoidably inhaled. Nothing, indeed, is able to resist the unwholesome effects of this wind.'[56] His symptoms have now become recognised as those of a seasonal disorder known as 'ghiblitis'.

There is little respite from such pain even when the winds move out to sea. Philippe Cousteau, while working in the Red Sea, produced a gritty description of his experiences with the *haboob* that hurls sand out of the Egyptian desert:

> At about two o'clock every afternoon, the sky above the western horizon would turn a reddish-gold and the sea would cease to live, the surface becoming absolutely motionless, seeming almost solid. The already stifling temperature became intoler-

able; our bodies ran with sweat and every movement was torture, aggravated by the rash of prickly heat with which we were all afflicted. Then the storm was on us and the howling wind raised little sprouts of water that mingled with the sand and covered everything with a coating of yellowish, destructive mud . . . Our eyes red and swollen, we moved about like automatons in a sandy, unbearable universe.[86]

When the *sirocco* moves north from Africa and crosses the Mediterranean, its character changes a little. It sops up moisture over the ocean and reaches southern Italy and Sicily not only hot, but now also insufferably humid. Which apparently does nothing at all to mellow its inimical nature. Neapolitan crime rates soar above their already impressive levels and even missionaries are moved. Ellsworth Huntington remembered one, 'a man of unusual strength of character, who during a particularly oppressive *sirocco* locked himself up in his study for fear that he might say something disagreeable to his colleagues'.[227]

The naturalist Norman Douglas in his novel *South Wind*, used the *sirocco* as a dramatic force, like Nemesis in a Greek tragedy, manipulating his cast of decaying English exotics on Capri:

'This south wind! This African pest! Is there no other wind hereabouts? Does the *sirocco* always blow?'

'So far as I have observed it blows constantly during the spring and summer. Hardly less constantly in autumn. And in winter often for weeks on end.'

'Sounds promising. And it has no influence on the character?'

'The native is accustomed, or resigned. Foreigners, sometimes are tempted to strange actions under its influence.'[107]

The ageing writer in Thomas Mann's *Death in Venice* found that

An offensive sultriness lay over the streets. The air was so heavy that the smells pouring out of homes, stores, and eating houses became mixed with oil, vapours, clouds of perfumes, and still other odours – and these would not blow away, but hung in layers. Cigarette smoke remained suspended, disappearing very slowly. The crush of people along the narrow streets irritated rather than entertained the walker. The farther he went, the more he was depressed by the repulsive condition resulting from the combination of sea air and *sirocco*, which was at the same time both stimulating and enervating. He broke into an uncomfortable sweat. His eyes failed him, his chest became tight, he had a fever, the blood was pounding in his head. He fled from the crowded business streets across a bridge into the walks of the poor. On a quiet square, one of those forgotten and enchanted places which lie in the interior of Venice, he rested at the brink of a well, dried his forehead, and realised that he would have to leave here.[316]

The *sirocco* continues, despite changes in humidity and temperature as it travels, to be an ill wind, without 'spring or elasticity'. Something fundamental, some quality that may well be connected with its 'subtle electric fluid', prevails, tainting the air it carries, turning it into *simoom* – one of the 'poison winds'.

There are others, amongst which the most infamous perhaps is the *föhn*.

When a low pressure frontal system approaches the Alps from the west across France, it sucks in air from the Mediterranean, pulling it up the southern slopes, cooling it and condensing its moisture into a bank of stratocumulus cloud that settles on the peaks. This is the 'föhn wall', which drops dehydrated air down the northern slopes where it is warmed by compression, gaining heat at 1° Centigrade for every 100 metres it falls, pouring through traditional 'föhn windows' into the valleys of Switzerland, Bavaria and the Austrian Tyrol as a hot and very dry wind.

A winter *föhn* feeds on snow, clearing drifts in minutes, but it is otherwise an exceeding ill wind that blows nobody much good.

When it blows strongly, it can cause terrible damage, and, as soon as they feel its approach, the mountain dwellers hasten to gather in their flocks and herds. They extinguish all fires and the Alpine village closes in on itself as long as the storm lasts. Crossing the tops of the mountain, this down-driving wind beats into the valleys on the northern flank of the Alps and then sweeps over the whole of the Swiss plateau, tearing the roofs off chalets, flattening crops, uprooting trees and devastating the forests . . . Brunnen, on the Lake of Lucerne, has a special port for use on days when it is blowing. Sometimes it is so strong that even steamers have to stop to take shelter there.[100]

It was once thought that the *föhn* was 'a fiery child of the Sahara', come to free the Alps of their snow, but it in fact generates its own heat and has no connection at all with the *sirocco*. It has however come, perhaps by a similar mechanism, to share the same properties and produce similar unpleasant effects. And these are certainly not due to dust or pollution. On *föhn* days, visibility is uncommonly good, the air is clear, and the Alps seem crisp and close, with hard steel-blue lines. But plants wilt, cattle go off their feed and people become disgruntled and surly.

Along with a rise in temperature of more than 10° Centigrade, and a fall in relative humidity to less than 20 per cent, there are abrupt pressure and electrical changes. These seem, once again in approximately one in every three or four people, to produce abnormally high concentrations of serotonin in blood and urine, or to lead to stress and in the end, adrenal exhaustion. The symptoms include anxiety, insomnia, irritability, tension, migraine, colic and apoplexy. And the results, even amongst the stoic Swiss, can be devastating.[467]

Human reaction time is affected so that, according to the Touring Club Suisse, traffic accidents in Geneva in 1972 rose by over 50 per cent when *föhn* conditions prevailed.[456] In 1976, the medical department of the West German Weather Station in Freiburg published the results of a four-year study proving that industrial accidents during a *föhn* required surgery 16 per cent more often, and other medical treatment 20 per cent more frequently, than at any other time.[196] At Bad Tölz in Bavaria, internists report increases in hypotension, coronary crises, migraine and psychic disturbances both during and on the day preceding a *föhn*.[133] The incidence of postoperative deaths due to both heavy bleeding and thrombosis during a *föhn*, has become so high that in some hospitals in Switzerland and Bavaria, major surgery is now postponed wherever possible, until the wind has passed. But suicides and suicide attempts still soar to epidemic proportions all through Switzerland and into Austria, wherever the 'witch's wind' touches ground.[490]

The whole northern network of valleys that drain the Swiss plateau are scoured by *föhn*, but föhn effects are not confined to the Alps.

Where air from the Mediterranean is drawn through the gap in the Corbières Mountains at Carcassonne, it drops on to the plain of Toulouse as the dry *autan*. The seventeenth century author of a book on sexual disorders, blamed most of his patients' problems on this wind, explaining that 'This hot, burdensome and oppressive gale benumbs and prostrates both men and animals. It renders the brain torpid, robs a person of his appetite and seems to bloat up the body.'[12] Numbness, prostration, torpor and bloat are enough to turn anyone off.

In New Zealand, the Southern Alps have a *föhn* of their own, breeding a *northwester* that devours winter snow on the Canterbury Plains. In Australia, the *brickfielder* brings the hot breath of the interior to New South Wales on the skirts of a trough of low pressure. Trade winds falling over the Javanese highlands give rise to the warm *kubang* and *gending*. In Sumatra, they call their föhn *bohorok*. Falling winds off the Great Karroo and the Drakensberg produce the enervating *berg* winds of South Africa. And the Andes generate the *zonda* that periodically stifles southern Chile and Argentina.

In North America, the Rocky Mountains deliver the snow-guzzling *chinook*, while the Sierra Nevada in California have an even more notorious product in their lusty *Santa Ana*, named according to your preference, after the mountains through which it passes; after General Santa Ana, whose Mexican cavalry stirred up similar clouds of dust; or from *santanta*, an American Indian name which means 'devil wind'.

Between five and ten times every year, when pressures are high over Utah and Nevada, air spills off the Mojave desert, rushes down the valleys, through the narrow mountain passes of Santa Ana, and out on to the coast around Los Angeles. Within minutes, the sea off Long Beach is whipped into white-caps and gales of up to 100 kilometres an hour buffet the coast between Santa Monica and Oxnard. In a single swat, one such blast flattened 252 oil derricks, while another rained fifteen million tons of dust on downtown Burbank. Sailboats are capsized, sailplanes sent soaring up to 14,000 metres, and helicopters slammed into the ground.[431] Devastating fires sweep through the hills and homes. Moods change. Skin turns taut. Aches come back to old scars in the night. And, in the words of Raymond Chandler, 'On nights like that, every booze party ends in a fight. Meek little wives feel the edge of the carving knife and study their husband's necks. Anything can happen.'[69]

Murder often does. In 1968, Willis Miller of California Western University, collected statistics for homicides in Los Angeles county and compared them with weather records. There were fifty-three days during 1964 and 1965 when the *Santa Ana* blew and humidity, which is normally around 43 per cent, fell below 15 per cent. On thirty-four of those fifty-three windy days, there were more deaths than normal. And during the longest sustained *Santa Ana*, which blew from October 20th to 26th in 1965, the total was 47 per cent higher than in any other windless week.[337]

The FBI recognises a 'long, hot summer' phenomenon, expecting more murder, aggravated assault and rape between June and September than at any other time. But the *Santa Ana* seems to supersede this annual cycle, producing short-term local effects whenever it blows, winter or summer, turning any time at all into a season of discontent. In California's early, and to this extent more enlightened days, defendants in crimes of passion were able to plead for leniency, citing the wind as an extenuating circumstance.

The *Santa Ana*, like other ill winds, is perfectly capable of violence without human help. When it blows, there is the usual increase in ulcer perforation, embolism, thrombosis, haemorrhage, myocardial infarction and migraine – not to mention theatrical failures, lower industrial production and loss of milk in cows. All these disorders are the result of tension, the product of an atmosphere heavy with menace and imminent catastrophe. Of dry air so filled with static electricity that even a handshake becomes shocking.

It is in such air that unnatural charges accumulate – and it seems to be these that have the most negative effect.

The Body Electric

The Greek philosopher-physicians explained human variety and disposition by resorting to 'humours' – fluids with mystical properties that flowed through the body. If these were in balance, you were 'sanguine'. But if one or another was predominant, that left you 'melancholic', 'phlegmatic', or downright 'choleric'.

285

We smile at such quaint conceits and speak instead of emotion being controlled by the rival activity of enzymes or hormones.

We now know that the hormone adrenalin is amino-hydroxy phenyl-proprionic acid, and can even synthesise it. We can describe the nature of the nerve impulse, and are beginning to unravel the anatomical complexities of the nervous system which lead to integrated 'higher' functions, such as memory and mood. We have every reason to feel smug, and yet despite all the pride and excitement, there are still some things which obstinately elude understanding.

The biggest mystery of all is cell differentiation, the force which turns one cell in an embryo into an eye, while another, apparently identical and derived from the same fertilised egg, becomes a tooth or a toenail. There is nothing in physics or chemistry that can account for this manifest destiny, this awareness of purpose and design.

The blueprint is in the genes, but there are so many other rival plans filed there, that some power still needs to be exercised to turn to the right page, to select the relevant part of our elaborate inheritance. And as of now, we do not have a single physical or chemical candidate capable of carrying out such a complex piece of organisation.

It seems certain that life is more than just an assembly of bits and pieces, that an organism is more than the sum of its parts and has properties and talents which do not exist in, and cannot be predicted from, its components. There are things about life which are apparently not subject to the normal physical and chemical laws, and may never be explained on a physicochemical basis alone. They may, however, be amenable to an explanation that takes account of the ability of air-living things to produce, and respond to, electric potentials.

In 1943, the Nobel Prize-winning biochemist Szent-Györgyi announced that biological knowledge was considerably less complete than advertised by the establishment. He said, 'It looks as if some basic fact about life is still missing, without which any real understanding is impossible.' He thought it might be electricity, and made the startling suggestion that proteins in living organisms could act as semi-conductors, forming crystal lattices, carrying or amplifying electric currents in precisely the same way as transistors.

This introduced solid-state electronics to biology for the first time and began to provide answers to problems, particularly in neurophysiology, that were poorly explained by biochemistry. But even now, forty years later, the electrical potential of living things is still largely ignored.

The facts, however, speak for themselves.

A sure sign of the actual flow of an electric current, is the magnetic field it creates. And it has now been shown that a beating human heart produces such a field.[22] Its intensity is so low that it was necessary to measure it in a rural environment far from other electrical interference, but it is definitely there, pulsing along. And there is a similar field around every human head, carrying a wave pattern that is directly influenced by brain activity.[75]

The conclusion is clear. We produce subtle electric currents that are carried through the entire body by the central nervous system. These create a total body field that is gentle, but complex enough to control and organise functions such as growth and development. It is a field that changes constantly with our internal state and pervades the space around us, setting up an individual biomagnetic net that carries news of us as swiftly and surely as a press release.

'I sing', said Walt Whitman, 'the body electric, The armies of those I love engirth me and I engirth them, They will not let me off till I go with them, respond to them, And disrupt them, and charge them full with the charge of the soul.'[520]

The existence of such natural electricity may account for the symptoms of irritability, fatigue, dizziness, headache, and changes in heart function and blood pressure which seem to plague all those (like workers in factories making permanent magnets) who are exposed for any length of time to conflicting sources of electromagnetic energy. And, within this response, may lie the answer to and the explanation for our strange, and often pathological, reaction to certain bodies of air.

All air, like everything else, consists of molecules. Each molecule has a core of positively charged protons surrounded by a field of negatively charged electrons. In a stable molecule, these two forces are in equilibrium, they cancel each other out. But the lighter electrons are inherently unstable and easily displaced from their orbits. They have a habit of slipping off like wayward planets and joining other systems. This produces new molecules with an excess of electrons and a negative bias, and it leaves behind the original donor molecules with too few electrons, and a positive charge. All these unbalanced 'maverick' molecules, both positive and negative, are called *ions*, and it is these that produce the active electricity in air.

There are very few of them. In normal clean dry air over land, there may be between 1000 and 1200 ions in amongst twenty-seven million million million other gas molecules in every cubic centimetre, but they exercise an influence out of all proportion to their concentration. For a start, they form rafts, gathering up to twelve other uncharged molecules around each ion in a large slow-moving cluster. And, unlike ordinary molecules, they move and react in an electric field. This makes each ion, despite its size and rarity, as conspicuous as a flash of bright light on an otherwise dark night, and brings it more readily, and more credibly, to the attention of an organism.

It is worth bearing in mind that sensitivity of this order is not unusual amongst living things. A male moth responds vigorously to traces of female pheromone in the air even when concentrations are as low as 200 molecules per cubic centimetre – five times weaker than most ionised air.[261]

Natural ionisation occurs in a number of ways.

Most ions in the upper atmosphere were liberated by cosmic radiation. Those in the lower levels of air are created by radioactivity, by lightning, or by friction. Contact between two air masses, or between airflows at different temperature and density, between air and the ground, or between moving air and the particles of dust, water vapour and other debris it contains – all knock off loose electrons. These are fast-moving and have a tendency to attach themselves not just to other air molecules, but to all foreign particles in the air, as a result of which they lose their charges altogether.

Earth as a whole is negatively charged, tending to repel loose negative ions, driving them off up into the ionosphere; and attracting positive ions, concentrating them in the lower levels of the troposphere. So the net result is that most of the natural air we breathe has a slight preponderance of ions with a positive charge. Of 1000 ions present in each cubic centimetre of a standard sample of clean air, about 550 or 55 per cent are normally positive.

Life has evolved in air of this quality and seems to be very sensitive to changes in either concentration or charge.[260]

Plants grow best in air with a high concentration of ions of either charge. The United States Department of Agriculture has produced cucumbers fifty centimetres longer than normal in ion-enriched air.[456] Seedlings of barley, oats and lettuce kept in air with 10,000 ions per cubic centimetre, increased in both length and weight, growing 50 per cent more quickly than those in ordinary air.[265] While other seedlings in air with just sixty ions per cubic centimetre, grew more slowly and all became flaccid, with none of their normal rigidity.[266]

A long series of experiments on *Puccinia striiformes*, the fungus that produces stripe rust in wheat, found that germination on fields in Montana was reduced, and in some cases stopped altogether, when the concentration of natural air ions was low.[436] The reason for this response may be that normal cell division requires polarity and the control of something like a life field. We know that the surfaces of healthy cells carry a negative charge, but this seems to decrease and even disappear when such cells become diseased.[78] Micro-organisms react rather differently. *Micrococcus pyogenes*, one of the bacteria that produce skin abscesses, is effectively removed from the air by a high concentration of ions.[264] Soviet workers claim that a concentration of just 10,000 ions per cubic centimetre is sufficient to remove all trace of most bacteria from the air inside buildings.[48]

Part of this antibiotic effect is a purely physical one, in which ions effectively 'wash' bacteria out of the air. Negative ions in particular are 'sticky' and readily attach themselves to airborne bacteria, building up coagulations of cells and debris that become too heavy to remain airborne and simply fall to the ground.

The air-cleansing effect of ions is not confined to bacteria. Dust, pollen and other pollutants are removed in the same way, making ionised air apparently easier to breathe for those who suffer from asthma, hay fever and other allergies. But ions are effectively limited by their concentration, and soon get used up by highly polluted air. One researcher who set up an ion counter in San Francisco, found such peculiar readings that he thought at first his equipment was faulty. 'There was a funny fluctuation in the morning and early evening.' So he repeated the tests and found that during rush-hour traffic on the Bay Bridge approach road which lay 400 metres from his laboratory, the already small ion count totally disappeared. 'There simply weren't any to be counted.'[456]

This artificial depletion in our cities may have a great deal to do with urban ills. A number of studies show that people become more uncomfortable in rooms with low ion counts, even though temperature, humidity and carbon dioxide concentration are effectively controlled by air conditioning.

The ill effects on occasion go well beyond discomfort. One study on mice in ion-depleted air, found that they caught influenza more easily, and died from it more often, than mice in normally-ionised air.[262]

Alfred Krueger, the bacteriologist responsible for these and a number of other pioneering studies on ions, concludes that humans face the same risks as his experimental mice.

> If they live in a typical city, travel to and from work in polluted air, work in an ordinary office or toil in a factory where pollutants are generated freely, and spend their leisure hours in homes providing essentially the same ionic environment, people are going to breathe ion-depleted air for a very substantial portion of their lives. This is clearly unnatural and probably unhealthy.[456]

Just how unhealthy, is still a moot point. A great deal seems to depend on the relative abundance of positive and negative charges.

Volunteers in a double-blind test who breathed air with a concentration of 32,000 positive ions per cubic centimetre for twenty minutes, all complained of dry throat, husky voice and itchy or obstructed nasal passages. While those who received an equivalent concentration of negative ions, experienced no ill effects.[529] The tracheal tubes which lead to our lungs are lined with tiny whiplike hairs or cilia which carry dust and other debris trapped in the mucous lining up and back to the throat at a rate of about two centimetres a minute. Positive ions slow down the cilia, cut down on blood supply to the trachea, and increase the rate of respiration, while negative ions speed up the cilia and restore blood supply and breathing.[263]

In an elaborate Soviet experiment, forty athletes were put through a series of physical tests before and after spending a month breathing normal, negative, or positively-charged air. The reaction time, endurance and sense of balance of the group living with a preponderance of negative ions, were always better than those under the other two regimes, sometimes by as much as 400 per cent. And they are said also to have had better appetites, slept more soundly, and been generally more cheerful and energetic. The report concludes somewhat ominously that 'negatively ionised air in doses employed in medical practice at a number of clinical and polyclinical institutions in the USSR, can be used for increasing the physical work capacity of the people'.[477]

The literature on air ions and the ion effect is now vast and largely partisan. The misgivings of those who opposed it from the start, have been fuelled by the existence of a commercial lobby with a vested interest in selling a variety of negative ion generators, all guaranteed to improve health, prolong life, enhance sex, and even ward off evil.

A lot of the early work on ions was slipshod anyway, inviting criticism. The ions themselves were crudely made and counted, and there was little or no control over other factors which might have influenced the test subjects. But there is something about the blanket opposition to all subtle electrical effects in biology, which ought to arouse the suspicion of anyone coming to the field for the first time. It has a desperate, last-ditch flavour which is reminiscent of Bishop Wilberforce, or the Astronomer Royal who insisted, just a few years before the first *Sputnik* was launched, that space travel was impossible.

This much is beyond dispute:

There are air ions. They do carry electric charges – and the nature of these is determined by a number of environmental factors. There is a life field. It is subject to outside influences – and the effect of these depends on the organism and on its initial state.

It is difficult to avoid the conclusion that Earth and life are linked electromagnetically. We seem to pick up very quickly on the parent pulse. But we still need to discover exactly how the connection is made.

And this may be where wind comes in.

The ill winds are mostly dry winds. Dry air is a poor conductor of electricity, so the longer it blows about, the more charged it becomes. If it also happens to be a gritty wind, filled with sharp fragments of desert dust, then it will spawn unstable molecules, losing the 'sticky' negative ones and accumulating a preponderance of positive ions.

Nathan Robinson, at the Israel Institute of Technology in Haifa, finds that there is a noticeable increase in the number of positive ions in air, starting twelve hours before a *sharav* arrives, and reaching concentrations of over 5000 ions per cubic centimetre when the wind is at its peak.[402]

Felix Sulman in Jerusalem sees this as the missing link between winds and their ills. The positive ions are mostly molecules of oxygen which, he suggests, are inhaled and bind even more readily than uncharged oxygen to red blood corpuscles in the lungs. This lowers partial oxygen pressure in the alveoli and allows carbon dioxide pressure to increase. People breathing positive ions certainly do begin to pant and struggle for breath and this, concludes Sulman, is what releases the stress hormone serotonin, with all its characteristic symptoms.[91]

Support for his theory comes from a psychiatric study on the opposite effects of negative air ions. Twenty subjects were involved in a double-blind test in which half were allowed to breathe air with a strong concentration of negative ions. Electroencephalographic tests on the whole group showed that the frequency of alpha brain waves fell to seven or eight cycles per second in all those who, without knowing it, had breathed the charged air for ten minutes or more, and stayed at that relaxed frequency as long as the ionisation continued.

A slow alpha rhythm is non-specific, but it is generally regarded as a good sign of lack of stress, and this diagnosis was confirmed by blood tests that showed little or no serotonin in those breathing negative ions – as long as they continued to do so.

The study concludes with this note: 'The ioniser is now in operation in our laboratory during the whole working day, and people in the negatively ionised atmosphere find much relief and improved working capacity on days when they are harassed by incoming desert winds.'[11]

The same laboratory is responsible for examining candidates who apply for jobs in Israeli government offices and technological institutes, and reports that those tested on *sharav* days have 'higher scores for neuroticism and extroversion, and score significantly lower than controls in intelligence tests and tests of mechanical comprehension'.[482]

An excess of the hormone serotonin clearly has something to do with wind stress. The symptoms are the same. And, while negative ions go some way towards alleviating the tension that accompanies stress, they are not in themselves enough to counteract an ill wind that works on a broad front, producing mechanical and physical effects in addition to its electrical sorcery.

For patients particularly sensitive to 'Serotonin Irritation Syndrome', Sulman also prescribes a drug, usually one of the ergotamine derivatives, which is antagonistic to the hormone. These are similar in action to the antihistamines that have been found to be effective as 'föhn pills', but there are major differences between the *sharav* and the *föhn*.[467] The *föhn* is equally hot and dry, but lacks the dust burden and the long desert fetch of the *sharav*. It is therefore less positively charged, though an advancing *föhn* always produces a 50 per cent increase in the total number of ions, and there are local readings in Switzerland that record as many as 4000 positive ions per cubic centimetre.

The most active ingredient in a *föhn*, and in all other falling winds, may however be another peculiar electrical phenomenon known as *sferics*.

'Sferics' is short for atmospherics, which are defined simply as 'electromagnetic waves of atmospheric origin'. These are long waves, with a frequency between 4000 and 50,000 cycles a second, and a wavelength of ten to a hundred kilometres. They pass invisibly and effortlessly directly through brick walls and human bodies, accompanied only by 'static' of the sort that sometimes interferes with radio reception. On a clear quiet day there are less than 10,000 such pulses passing by, but when a storm is imminent, the long waves travel in bombardments 650,000 strong.[482]

Sferics are usually associated with thunderstorms, but they are attached to a variety of weather conditions, most noticeably and dramatically to bodies of cold unstable air that move in polar fronts or loom over the Alps as the first sign of an impending *föhn*.

Reinhold Reiter is director of a bioclimatic laboratory at Farchant in the foothills of the Bavarian Alps, which has been keeping daily records of sferics for more than twenty years. In addition to the usual meteorological instruments in his laboratory, Reiter also employs a highly sensitive network of human amputees who feel a variety of aches and pains in their stumps prior to a change in the weather. They warn him of showers, thunderstorms and the passage of cold fronts with reasonable accuracy, but his switchboard really lights up when there are unusually high emissions of sferics. In one sequence of 20,000 daily reports from his network, the correlation between amputation pain and atmospheric electrics is so high, that there is not more than one chance in 1000 of the connection between the two being purely coincidental.[393]

Reiter has been able to replicate weather reactions by subjecting amputees in his laboratory to artificial electric fields, and finds that sensations begin after a latent period of around ten minutes, as though the whole body has to go through a period of electrical warming like an old-fashioned radio valve. He has applied sferics locally to selected parts of live human guineapigs and found that there were marked changes in the acidity of the tissue being irradiated in this way, but that no alteration of pH value could be induced in tissue taken from a dead animal, no matter how recently it was killed or how strongly it was irradiated. The implication is that the response is not just chemical, but depends on an electrical interaction between the atmospheric field and the body field of a responsive living organism.[393]

The likelihood that *föhn* sickness is produced by a sensitivity to electromagnetic radiation, is enhanced by studies of a similar condition that has become known as 'shortwave hangover'. People who are exposed for long periods to radio transmitters that are insufficiently screened, complain of headaches, nausea, insomnia and depression. And there may be worse to come. One study of 598 suicides in the English Midlands, showed that an abnormally large number of them, 40 per cent more than a control group of random addresses, lived in houses that lay inside a field of high strength produced by nearby powerlines.[319]

In an experiment which involved exposing golden hamsters *Mesocricetus auratus* to far weaker localised electrical fields, all of the animals dismantled and moved their nests beyond the limits of the field within seventy-two hours. Those hamsters which had litters of young, moved in less than twenty-four hours.[393]

These artificial sources involve frequencies higher than those in any natural situation, but there is good evidence that all approaching weather fronts generate electric fields strong enough to affect living organisms.

During a ten-year study of mortality in New York, deaths on days during which a front touched the city were always higher than on any other less turbulent day in that month.[267]

In Bavaria, Reiter found that suicides in his area occurred most often on days with high natural electrical activity. He also found that, compared to fair weather days with low sferics; reaction time was slower, traffic accidents up by 70 per cent, industrial and mining injuries by 32 per cent; there was an increase in general illness of 100 per cent, and a rise in the incidence of deaths other than by suicide, by a total of 20 per cent.[393]

Slowly and steadily a general picture begins to emerge.

Gaia is subject to electromagnetic moods involving fluctuations in sferics, ionisation and atmospheric fields that accompany the more obvious and easily-measured weather changes.

And, despite anything we can do, we in our sparky bodies are touched by these electrics, not equally, but locally and casually on the breath of a handful of irritable winds.

Witch's winds – crones, scolds and shrews, like the *föhn*, the *sirocco* and the *sharav*.

Well Winds

Not all the news is bad. There are warm dry winds with pleasant personalities that turn some deserts into places people go to for their health.

> There was something clean about the sun in Las Vegas. Even in February there was a searing, blinding white light that made you feel as if you were being sterilised, even cauterised, so there wasn't a germ that could stick to you. Everything extraneous would be burned off your skin, desiccated and sucked dry, its empty husk blown clattering away in the hot wind out of the desert. Even the air itself felt like that – a breeze that carried with it tiny abrasive particles of ground-up quartz and topaz too small to see. You could feel them buffing and polishing away at you.[380]

The rationale may be fanciful, but the well-being is undeniable and must have something to do with the fact that such winds have not travelled far enough to be ionised, or fallen from high enough to be hung about with sferics. Sometimes, however, it is difficult to understand how they could avoid such pitfalls.

In the days of Empire, Sierra Leone and the Guinea Coast of West Africa were known as 'The White Man's Grave'. One of its victims said,

> Nothing can be compared with the feeling of utter prostration that overcomes a European. Though he sits motionless in an armchair, he perspires as after violent toil; his fatigue is not like what is felt after work, but rather a weakness in the limbs, and especially in the bones – an indescribable discomfort which precludes all movement, all bodily or mental work, but yet forbids sleep.[56]

Into this torment each November comes an arid, dusty wind known as the *harmattan*, red hot from the sands of the Sahara. Its desiccating breath scalds the skin, cracks the lips, produces nose bleeds, kills plants, curls up the corners of books, covers everything with a fine layer of red dust, and 'sometimes causes glasses to crack and fall to pieces as they stand on the table'. And yet, so great is their relief at being spared the humidity that normally turns salt into liquid, devours iron with rust, and fills the air with the heavy odour of putrefaction, that the colonists called this suffocating winter wind 'The Doctor'.

The *harmattan* is a northeast trade wind, one of those that made the Sahara in the first place, and counts as a 'well wind' only by comparison with the conditions it replaces. But in Ashanti and Dahomey, which have less oppressive climates, there are Africans who share the opinion of white settlers further west. 'The Doctor' has never had a breath test, but it would be very interesting to know what its ion count is, and whether there are factors which render this wind less poisonous than its hot-headed relatives north of the Sahara.

There have been a number of attempts to devise a scale on which weather could be rated in terms of human comfort. The 'wind chill index' is one of these and the 'temperature humidity index' another. The United States Weather Bureau uses a 'discomfort index' based on effective temperature – which is an arbitrary figure that combines the effect of temperature, humidity and air movement on a human body, and calculates the temperature of still air that would induce an identical sensation. Ten per cent of all Americans are found to be uncomfortable at an index figure of 70, half feel uncomfortable at 75, almost all find 80 unpleasant, and at a discomfort index figure of 86, government regulations require that employees be dismissed for the day.[479]

The problem with all these systems is that they are devised on the basis of physical responses such as perspiration and breathing rate, and take little account of psychological factors which play just as large a part in a subjective assessment of comfort. What little evidence we do have, suggests that physical and mental well-being in any place is very much a product of 'something in the air'.

That something may be a radioactive gas called *radon*. This is a product of the disintegration of the element radium which is found naturally in Earth's crust. Chemically, it is described as an inert gas, but it has quite definite biological effects.

Radon diffuses slowly through the soil, wherever the surface is not frozen or composed of solid impermeable rock.

Its concentration in air is always higher in mountains than on the plains, probably because the folding of the ground exposes a greater surface of soil per unit area, and it falls to almost nothing over the ocean. But it is highest of all in areas with thermal springs, many of which have now become successful commercial spas.

In these fashionable resorts, 'therapeutic agents' offer courses of radioactive baths, drinks and aerosols, which are reputed to correct circulatory and endocrinal problems and ease the pain of arthritis.

It is very difficult without long and carefully controlled tests to prove or disprove any of these claims, but physiological tests leave little doubt that radon from the springs is absorbed by the body. About 40 per cent of the radioactive particles in an aerosol of spring water get into the lungs, and half an hour after such treatment begins, the radon concentration in a patient's blood is 30 per cent as high as the loaded air being offered in this way by the clinic.

The result is a feeling of well-being which is difficult to quantify or trace physiologically, but seems to be based on a general improvement in cell functioning brought about by increased efficiency of the coenzyme ATP (adenosine triphosphate) which releases energy for use by the body. It is recognised by those who have experienced it simply as the 'spa response'.[393]

This reaction takes place within days in most of those who pay for the treatment. Paying in itself seems to play quite a large part in the success of such therapies, but there are benefits also for those who only came along for the ride. After a week or two, all visitors to places like Badgastein in Bavaria seem to share the response to the spa, simply by being there and breathing the air.

Since 1958, Reinhold Reiter has been making daily measurements of radon in the air and comparing its concentration with weather patterns. His monitoring stations are at Farchant and Wankpeak in Bavaria, neither of which is a spa town, but there are days on which radon concentration in the air at Farchant is as high as it is at Badgastein, which is 2.3 times higher than the average for continental mountain air.

The highest concentrations occur in winter and spring when there are fewer convection currents and more inversions to keep air captive near the ground. But the most interesting finding is that the radon content soars when winds blow toward the test areas from regions with exposed rocks, or soils formed from rocks, that are composed of granite or rhyolite and contain high proportions of silica and free quartz. There is something as yet undetermined about this chemistry that favours the release of radon and spices local winds so that they spread well-being for hundreds of kilometres downstream.[393]

The alpine location of many spas and other places where people feel good, may not be coincidental. The decrease in partial oxygen pressure at altitude, forces us to breathe faster in order to keep up, and within two hours of climbing any decent-sized mountain, our body starts to compensate by producing more red blood cells and circulating more haemoglobin. Even so, the maximum amount of oxygen that any body can consume, declines by 10 per cent with every kilometre climbed. So everyone, no matter how well they may be acclimatised, tires more quickly at altitude. In theory, mountain tops ought to be places of severe stress for us, but at least since the days of the psalmist, people have been lifting their eyes, and dragging their bodies, up into the hills for help and inspiration.

Apart from adaptations designed to increase oxygen intake, our bodies also respond to altitude by producing extra hormones and a higher concentration of the immunoglobulins which tackle infection. Our general resistance may be lowered in the mountains, but the concentration of bacteria and other pathogens is also greatly reduced. Microorganisms are rare at 500 metres, and only the hardiest survive above 5000, largely because radiation in the thin air is increased and there are many more negative ions about. And because the air is cleaner, the ions stay there rather than being filtered out by pollution. Ion counts at Davos in Switzerland and at a series of Russian mountain spas, all show a preponderance of negative ions, with a concentration in one case as high as 100,000 per cubic centimetre.

The combination of high negative ion concentration and low dust and pollen content, makes mountain air very attractive to those suffering from respiratory problems. Galen, the Greek physician who settled in Rome in AD 164, sent many of his asthmatic patients to convalesce on the slopes of Mount Vesuvius. And before antibiotics and vaccination, the only effective therapy for tuberculosis was time spent at altitude.

The only other natural source of negative ions is moving water, which traps positive ions and tosses out the negatively-charged particles in fine spray. There are concentrations of negative ions along the seashore, near breaking surf, and around the base of waterfalls, which may account for the popularity of beaches and the success of Niagara Falls as a honeymoon resort.

The best combination would be a high mountain waterfall at the base of a granite cliff. Any wind that blew from such a source would be as auspicious a breeze as it is possible to get. Aficionados claim that one of this kind exists at Yosemite Valley in California's Sierra Nevada, and that it blows across a line of places sacred to the Miwok Penutian Indians.

The well winds of the world have never had the sort of press that ill winds receive. No catalogue exists of their individual virtues, but there are clues of a kind in the names they still go by:

There is no mistaking the gentle qualities of *feh*, the breeze that cools summer evenings in Shanghai; *nasim*, whispering across the desert at dusk in Saudi Arabia; a Scottish *waff* coming through the rye, or an Irish *creithleag* just moving the heads of ripe barley; a *cat's paw* rippling the surface of a summer millpond in the United States; *sz*, the first faint touch of autumn in Peking; or *i tien tien fung*, literally a 'sigh in the sky' of China.

There is a caress of kinder ions in the various sea breezes known in Italy as *libeccio*; in North Africa as *imbat*; or Gibraltar as *datoo*; in Majorca as the *skysweeper*; on the coast of Angola as *cacimbo*; in China as *kai*; and in Hawaii as *kapalilua* or, if it carries a little mist, as wispy *waimea*.

There is a fond familiarity in the Algonquin description of a southerly puff as *shawondasee*, the 'lazy wind'; in Moroccan recognition of the winnowing wind as *laawan*, the 'helper'; and in the relief with which people in Provence greet the soft *aspre* after the rigours of a *mistral*.

But the best and most welcome sounds to ears tuned to classical tones, must always be *zephyr*, the flowery west wind of a Roman solstice; and *aura*, that delicious first breath of air at any Greek dawn.

10. THE PHILOSOPHY OF WIND

Of all natural forces, wind is the most enigmatic.

Heat and light have visible sources, the difference between sun and shade is obvious, even without any conscious knowledge of radiation. Water and earth are material things, with predictable properties, which do not depend on an intellectual awareness of gravity. Sunset, full moon, lightning, a forest fire are all cause for wonder and alarm, but they have a natural focus, they excite curiosity and demand explanation. Hot springs, waterfalls, geysers, waves and tides have rhythm and reality, and they occur within recognisable natural confines. You can take them or leave them alone.

But wind is different.

For a start, it is invisible. It can creep up out of nowhere and tickle the back of your neck, or throw you flat on your face. It has no shape, size, smell, taste or sound of its own. All its properties are borrowed, all our experience of it comes at second hand. We have nothing to work with but indirect effects. It is big enough and strong enough to tear the largest living things on Earth out by their roots, and yet it can seep through a hairline crack. Wind is elusive, shifty, fugitive, difficult to define – and impossible to ignore.

This combination of something and nothing, of a force that cannot be apprehended by the senses and yet has an undeniable existence, was our first experience of the spiritual. Once human beings became complex enough to look for cause and effect, to ask questions about the world, they ran headlong into this dilemma.

'If we trust to language', said Freud, 'it was the movement of the air that provided the image of spirituality, since the spirit borrows its name from the breath of the wind.' In Latin, *spiritus* described an intake of breath by a god, literally an inspiration. The Greeks used the word *pneuma* for both wind and spirit. And in Hebrew and Arabic, *ruh* means both breath and wind.

'The idea of the soul was thus born as the spiritual principle in the individual . . . Now the realm of spirits had opened for man, and he was ready to endow everything in nature with the soul he had discovered in himself.'[148]

The wind, as always, gave us just the nudge we needed.

Wind Myth

In the beginning, there was no such thing as 'the wind'.

There were gentle breezes off the sea that made hot summer afternoons more pleasant, there were icy blasts that turned toes and fingers blue, there were muscular gales that punched their way through the thickest undergrowth, and whirling columns of dust that seemed to have a substance and an anger of their own. There was no reason to lump all these things together, no way of knowing that they shared the qualities and properties of moving air.

It was only in the seventeenth century, after the invention of the barometer and the thermometer, that anything like a unified theory of wind began to emerge. But right from the start, there was something about all winds that gave the first questioning minds pause for thought, and reason to look for common cause.

An astonishing number of the earliest wind myths, for instance, have them living in, or coming from, a cave or a hole in the wall or ground.

The Bakitara of Uganda identified their sacred hill of Kahola as the home of the winds, and pointed out that it had four holes from which the breezes blew. All the local rainmaker had to do when

301

plying his trade, was to wait for a likely-looking cloud to arrive overhead, and then cover the holes with red barkcloth, carefully weighted down with stones to stop the winds from blowing the cloud away.[407]

Inuit of the Lower Yukon, tell of a childless couple who made a wooden doll that came to life and travelled 'to the edge of the day, where the sky comes down to earth'. There it found a hole in the sky wall covered with a skin and cut it, releasing the east wind and a herd of live caribou. A similar skin-covered hole in the south produced a warm wind with rain, one in the west turned loose trees and bushes in flower, and another in the north a freezing gale with snow. Gathering all the four winds together, the doll instructed them to 'sometimes blow hard, sometimes light, and sometimes not at all'.[357]

The Iroquois believed that Gaoh, the spirit of all the winds, held his captives in a mountain cave called 'The Home of the Winds'. The Algonquin shared the same belief and recognised the prisoners as *Wabun* from the east, the 'morning bringer'; *Kabeyun* from the west, the 'father of winds'; *Kibibonokko (Kabibnokka)* from the north, the 'fierce one'; the *Shawondasee*, the 'lazy wind from the south'.[494]

The Batek Negrito people of Malaysia held that the winds were kept in a cave on the peak of Batu Balok, and guided out from there along ropes arranged by Gobar, the supreme being.[127] The Dusuus of Sarawak had a hero-blacksmith called Kinorohingan who forged and welded the seven parts of the soul that go to make up every individual, and when these became worn, he hammered them out into winds, which he stored in a mountain cave.[129] The Maori of New Zealand have a host of stories about the hero Maui, the clever one of many brothers, who captured all the winds and confined them in caves – all except for the persistent west wind, which continues to this day to elude him.[518] On the nearby Chatham Islands, the hero was Tawhaki, who collected his winds in a basket.[435] And in Samoa, they were held captive in a coconut.[458]

In the Southern Cook Islands, the sky was represented as a calabash with holes drilled in it and stopped with plugs. 'Should the wind be unfavourable for a grand expedition, the chief priest began his incantation by withdrawing the plug from the aperture through which an unpropitious wind was supposed to blow. Rebuking this wind, he stopped up the hole, and advanced through all the intermediate apertures, moving plug by plug until the desired windhole was reached. This one was left open.'[157] On Hawaii, the 'gourd of constant winds' was in the keeping of an hereditary *kahuna* who knew all the names and secrets of the hundreds of winds that could be summoned from it.[26]

Elsewhere in Polynesia, it was held that there are a series of holes at the edge of the horizon itself, some large and some small, through which Raka, the god of winds, and his children loved to blow. And people there still use the phrase *ruamatangi* or 'wind-hole', when speaking of the source of the weather.

They do the same thing in the Swiss Alps, where rain-bearing winds are said to come through recognised *Wetterloch* or weather holes.[376] In England there are a number of narrow gorges or funnels in the hills that seem to accelerate a wind and bring it, with rain, down on a particular community. In Shrewsbury they say, with good reason, 'There'll be some rain, for the wind has got to Habberley Hole', which lies due southwest of the town right on the line of the wettest westerlies. Bodjham Hole near Ashford Vale in Kent, and Flammer's Hole in the Chilterns above Dunstable, have similar soggy reputations – and deserve them.[321]

All the holes, the caves, the chambers, the gods and the heroes seem by the first millennium BC to have slid together into a more precise mythology, one that is familiar to us from ancient Greece. Aeolus became the warden of the winds. He was the son of Poseidon the sea god and lived on the bronze-walled island of Aeolia, where he kept the 'wrestling winds and soaring tempests' chained in a cave. Their constant chafing and moaning in this underground prison seem to have begun to interfere with his eternal feasting, for when Odysseus and his crew arrived, Aeolus was only too happy to pack the winds up in an oxhide sack and give them to the hero. But when Odysseus left and fell asleep in sight of home, his avaricious crew opened the sack in search of treasure and their ship was swept away in the resulting tempest. With the exception of Odysseus himself, none of them ever saw Ithaca again.

The story is one that any Maori or Mayan would recognise. Only the names have been changed. The hero, the warden, the winds and the place of captivity are common to cultures that are widely separated. Even the sack has been borrowed. In ancient Chinese lore, the god of the wind is Feng Po, an old man with a white beard and a blue cap. He holds a yellow sack called the 'Mother of the Winds' and points its mouth in any direction he pleases.[515] And in Japan he becomes Fu Jin, also dressed in blue, who carries a large bag from which he pours the wind in any required strength, according to how wide he opens the neck.[5]

There is more than coincidence or cultural connection in the similarity of these myths. That people everywhere should come to have and hold such analogous beliefs about something as intangible as wind, is fascinating. The stories are not explanations of wind. Nor can they be dismissed as the delusions and fallacies of a few early philosophers who were led astray by a lack of information and scientific know-how. The accounts have to be seen, like all myth, on several levels.

Myth has two main functions. The first is to answer awkward questions in a graphic and positive way. And the second is to justify an existing social system, accounting for traditional rites and customs in a way that gives authority to the local priesthood, to the scientists of the day. These stories do that, creating allegories that are at least good enough to keep the children quiet, and leaving room for those with special knowledge to exercise their skills as wind wizards or weathermen. But there is a third function, another more fundamental level on which the beliefs have to be scrutinised for meaning.

When human minds are faced with things beyond the range of understanding, they tend to take refuge in symbols. They seize on forms, patterns, colours, shapes and numbers that carry more than their normal load of meaning. There is an ancient and common code in which things that cannot be precisely defined or fully explained, are given an image which conveys at least part of this mystery to others. It opens doors, leading to ideas that lie beyond the grasp of reason.

Wind myths are full of such implications.

Caves, sacks, bags, baskets, calabashes, gourds and coconuts are only thinly veiled symbols for the womb in which the winds wait for release. And the situation of so many of these maternal caverns in mountains with voluptuous, pregnant shapes emphasises the identity of winds as the direct offspring of Mother Earth. They are, however, no ordinary children, but wayward, somewhat rowdy, progeny over which there is little or no possibility of control. Old man Aeolus and his cronies, including all the great wind-gathering heroes of history, come out of such insights rather badly. They look less like wind gods and more like midwives, hanging around the entrance to the cave, looking a little ambivalent and shifty, with very little control over what happens next. Meteorological doormen, rolling rocks to and fro as the residents decide where and when to wander out.

The emphasis in all the early mythological rationale is on the source, on the origin of wind in nature, flowing from the Great Mother. The winds are not seen as gods, nor as having been made by gods, though they may come under the nominal stewardship of a number of familiar gods and local heroes. The winds themselves seem to be elemental. There is something robust about the images used for them, a hint of raw energy that needs, for its own good and ours, to be chained and restrained. A potent and creative force, a very seminal thing, but incomplete and lacking in social graces.

The winds, in other words, are unquestionably male.

In the beginning, according to the people of Sermata in Indonesia, only the sky was inhabited. Then a sky woman climbed down a *rotan* palm to Earth, where she was impregnated by the south wind while she slept. The fossilised roots of that very tree are still shown to visitors on the island of Nolawana and, they say, all who live on Earth today are her descendants.

There are similar paternity suits outstanding against all the wayward winds. The Minahasser people of Sulawesi claim descent from an unwary girl impregnated by the west wind. The Algonquin hero Michabo, whom we know as Hiawatha, was born after his mother Wenonah received a visit from the same wily air. In a story taken from Finnish runes, the virgin Ilmater is said to have been inseminated by the east wind before giving birth to the wizard Vainamoinen. By some accounts, the goddess Hera conceived Hephaestus, who became craftsman to the gods, after inhaling a stray wind.

The Babylonian goddess Tiamat was born when her mother's belly was 'filled with the raging winds'. The Arunta women of Australia would hide in their huts whenever a storm blew out of the north, because it was believed to carry *ratapa*, the demon seeds that were responsible for producing twins. The early Arab traders told of an isle of women in the China Sea, where pregnancies could only be caused by the wind. The ancient Egyptians, who knew vultures well, insisted that there were no males amongst them, and that fertilisation was carried out entirely by the wind.[349] Virgil was certain that the same was true of the Lusitanian (Portuguese) mares. And Pliny, while recognising that there were male partridges, nevertheless believed that it was wind blowing from the male to the female that actually got the job done.[199]

The core of meaning, barely concealed in such beliefs, is that the highest being, the source of all life, is Earth herself, the eternal feminine, the creature Gaia. But – and this too is implicit – she lacks something, a seed, some kind of essence that wind provides. The nature of the interdependence is best described in Hindu mythology.

Vishnu, the preserver, takes on the form of water and a calm settles over the cosmic ocean, fathomless and subtle. But that is all that happens, until Vishnu reaches out again and a cleft appears in space, a cavern in the air which resounds with energy. Of this sound, the wind is born. Wind is not a divinity, but a spontaneous motor force with space to grow. It pervades space and expands relentlessly. Tumescent, violent, rushing, blowing, it invades and arouses the receptive waters. And from the commotion and the friction, a spark is struck, which flames into consciousness.[537]

The Toradja in Sulawesi, say that in the beginning the wind blew over the sea, raising great waves that drove spume upon the shore. It fell there in the shape of an egg, from which hatched the first human beings. The Earth Mother of the Zuni in New Mexico, is said to have started the human race by blowing the 'foam of life' off the waters of the Rio Grande.[305] In the Christian myth, 'the Spirit of God moved upon the face of the waters' and genesis began. And as soon as any divinity was identified with spirit, it was identified with wind, which 'bloweth where it listeth', and from this derive quite naturally the verbal symbols of omnipotence, omnipresence and inscrutability.[111]

There is an ecumenical understanding on this point at least, that the world, symbolised as liquid and feminine, is moved by the spirit, which is represented as airy and masculine – and that somehow both things are necessary for life. Because it is fleeting, wind is sometimes used on its own to represent life, but it cannot of itself produce life, it is simply an essence.

There is another Hindu image that expresses this difficult concept well. Brahma, the creator, soars through the air on the back of *hamsa*, a magnificent gander. The wild gander is like the wind, a free wanderer, flying north and south with the seasons, landing where it wishes. It is at ease in the air and on the water, bound to neither medium, able to live in either, but unable to live entirely on either. It is the divine essence which enlivens the lower sphere, but cannot in the end exist without it. The gander is the animal mask of the creative principle in the wind, touching us, even when we cannot actually see it, by the beauty and the mystery of its song.[249]

Bird symbols are common to many wind myths.

The world, say the Melanesians, began in chaos. And it stayed that way until the great bird god Tabuerik soared above the turmoil and beat it into shape with his wings.[105]

The Athapascan-speaking Indians of New Mexico tell of a raven who hovered over the primordial sea, stirring it into life with the wind from his wings. The Inuit on the shores of the Bering Sea, know him as Father Raven – Tulukauguk, creator of the world. While the Shoshone in California are convinced that he was actually a falcon.[156] In Norse myth, the wind bird is Hraesvelgr – the 'corpse eater', a giant in eagle's guise who sits on a hill at the edge of heaven and ventilates the world by the movement of his wings. In ancient Sumeria, he was Im Dugud, the winged demon of the storm. And in India his counterpart was Garuda, the eagle who carried Vishnu aloft, and now flies again as the symbol of Indonesia's national airline.[152]

There is everywhere a close association between creation and a wind force represented as a bird. Carl Jung found this totally appropriate, suggesting that the 'flight of a bird is reminiscent of intuition working its way up through the unconscious, making it the most fitting and accessible natural symbol of transcendence'.[245]

A bird was used as the symbol for the soul in ancient Egypt, and the same convention continues in Christian iconography where it represents the Holy Ghost. John the Baptist saw the spirit 'descending from heaven like a dove', and it became common amongst medieval artists to show it rising again from the pious at their death. Hence the countless painted nuns apparently in the process of regurgitating pigeons.

The wind serves a dual function. It brings awareness, but it also takes it away.

In some Eskimo languages, the place where the dead gather is known as Sillam Aipawe, the 'House of Winds'. The Bonda in central India say that the wind consists of ghosts of those who caught a fever and died in a single day.[126] Amongst American tribes of the Great Lakes area, it was believed that storms were battles between spirits of the recent dead, and it was those who won that hung out the bow of bright colours to let their living relatives know. On one occasion in Fiji, when a great chief died and a waterspout swept across his island lagoon, 'an old man who saw it, covered his mouth with his hand and said in an awestruck whisper – "There goes his spirit!"'[146]

It was inevitable that the more lively winds would be credited not only with spirits, but with personalities and names of their own.

Endowing something with a name, preferably one with some magic to it, provides a certain amount of power over that thing. It is a great deal more satisfying to curse something that has a good resounding name. A wind with a name changes from a nebulous entity into something more accessible, more personal, something that can be damned or praised, condemned or cajoled at will.

It is no coincidence that the *mistral*, *sirocco* and *föhn* have dozens of local names. There is one for almost every tribe or town that found it impossible to ignore these winds, and necessary to do, or at least say, something about their effects.

It is difficult, however, to understand why Anglo-Saxon people, alone in all the world, should have exercised such restraint in this respect. There are a hundred or more French names for various states or conditions of moving air, but across the Channel in England, there seems to be only one wind that has its own title, though there are plenty nasty enough to warrant one. This is the *helm*, a ruthless wind of the lakelands that is said to be 'strong enough to uproot turnips'. And this one owes its fame largely to the weather-sensitive art critic John Ruskin, who condemned it as one of the 'Plague Winds' of the world.

Things are not much better in the United States, where at least there are some Indian names such as *chinook* to enrich the meteorological scene. It is very sad however that the best that even Texans can do with a tidal wave of polar air that comes pouring down across the lone star state to paralyse the Panhandle, is to call it a *norther*. Across the border in Mexico things pick up a little when it becomes the *tehuantepecer*, and happy to relate, reach almost poetic heights in Costa Rica where this worthy wind receives its due in the splendid name of *papagayos*.

Australians seem to have managed better than most other English-speakers. Partly because they could draw on Aboriginal sources such as *koochee* for the whirlwind, but largely due to a determination to invent their own language anyway. Which leaves them with *cockeye*, for a summer squall in the northwest, and the wonderfully descriptive *buster* and *brickfielder* in New South Wales.

Names tend to sound more romantic in someone else's language. It comes as a great disappointment to discover that the sultry *haboob* in Arabic just means a 'blow', and that the dusty *karaburan* of the Gobi Desert, translates into a very dull 'black wind'. But there is more than adequate compensation in the Moroccan *arifi*, Arabic for 'the scorcher'; *bad-i-sad-o-bist-roz*, Afghanistan's 'wind of one hundred and twenty days'; *daibafu*, the 'hang horse', for a Japanese wind strong enough to lift both mount and rider; and *mezzer-ifoullousen*, the Berber name for a cold southeaster, which means 'the one that plucks the fowls'.

Many winds received their names from their areas of apparent origin. The *vallesaria* is a falling wind that descends on the eastern end of Lake Geneva from the direction of Valais, and the *vent d'espagne* of the French Pyrenees is self-explanatory. A few were named for their destination – *seguin* in Provençale means 'the one that follows the Sun'; and some for their timing – *prodromes*, the 'runners' of Greece, often preceded the rising of Sirius, the dog star.

Animal names crop up in *elephanta*, for a gale off the Indian coast; *biliku*, 'the spider', which wraps itself around the Andaman Islands; and *bai*, the cold northerly battering 'ram' of dynastic Egypt. Some of the best names, such as the gale known in Scotland as the *landlash*, simply describe what happens, or how the locals feel about it happening again. There is a miserable wet wind that descends at regular intervals on Shanghai, where they recognise it simply and laconically as *yuh*.

The Greeks had a more formal arrangement in which each of the eight winds that visited Athens was given divine status and personal, suitably godly, attributes. In the time of Aristotle and Theophrastus, the direction allocated to each wind was somewhat arbitrary, based partly on the situation of classical sites and partly on the position of the rising and setting sun during the seasons. But by the first century BC, when Andronikos built his 'Tower of Winds', each of the eight faces was aligned in a modern way, according to compass directions. It is an interesting feature of the Tower that all the winds are shown flying anticlockwise, from left to right, following precisely the pattern and the sequence in which they would naturally occur as a cyclonic low pressure centre passed through the Mediterranean.

The Polynesians exercised an even finer judgment. An anthropologist working in the Tokelau islands found that they used a wind compass of twelve points.[24] Another on Pukapuka island reported an awareness of sixteen different winds.[64] And a missionary living in the Southern Cook group in the nineteenth century, discovered that there are thirty-two traditional local wind names and directions, corresponding exactly to the thirty-two points on a western sailing compass.[157]

There is no good geographical reason why winds should blow

from 4, 8, 12, 16 or 32 different directions, rather than 3, 5, 7, 11 or 17. But there seem to be good human reasons, including our natural affinity for halving things and then halving them again, for all multiples of four.

The almost universal choice of four as the number of the winds, is a very basic one, common to much religious and conceptual thought. Hence the four evangelists and gospels of Christianity, the four great books of Buddhism, the four horsemen of the Apocalypse, the four cardinal points of the compass, the four corners of the world, the four elements, the four temperaments, the four seasons etc. As a symbol of completeness, there seems to be nothing as whole or as satisfying as something which stands four-square.

The sign for the number four, from a time before the use of numerals, was not a square but a cross. Usually a Greek cross with arms of equal length, something very common on Assyrian and Persian monuments and on Greek coins and statues. Even in those times it was already derivative of an older symbol, perhaps the most ancient of all – a cross with the ends of the arms bent at right angles, usually to the left to form a *swastika*.[527]

The name comes from the Sanskrit *su*, meaning 'good'; *asti*, meaning 'to be'; and the substantive suffix *ka* – running together to denote 'something of good fortune'. It seems to be of Aryan origin and was certainly in use as early as the Bronze Age. Apart from the Nazi symbol, which turns the other way, it is now common in Buddhist iconography and in India, China and Japan, where it is popular as a sign of good wishes and a long life. But a clue to its original meaning may be gained from the fact that amongst pre-literate people in the New World, everywhere from Ohio to the Andes, it was unquestionably a symbol that stood for the Four Winds.[306] The existence of such a symbol in portrayals of the Aztec god Quetzalcoatl, 'the serpent dressed with green feathers', shows that he was also a keeper of the winds, perhaps even a manifestation of the elusive west wind itself.

The similarity of wind myths from so many disparate sources, is a strong argument in favour of common ground. In some cases there are obvious cultural connections, but in others it seems that the people involved share little more than their basic biology – and an experience of wind.

The mythology they have created is the product of an intuitive understanding of that experience.

A message to life, from life, about itself.

Wind Wisdom

The weather is inherently chaotic.

Weathermen so often get it wrong, because beyond a certain point there are no useful rules. Long-range forecasts are impossible. Not even the best analysts, with all their sophisticated satellite surveys, can be certain what will happen next.

The main problem is that all future patterns are extremely dependent on present starting conditions. Which means that if a forecaster's assessment of the location of an existing air mass is only slightly off, then a prediction about next week's weather will be way off. After about ten days, prediction and reality inevitably diverge.

Weather is not, however, totally random. It is subject to the constraint of a force which mathematicians have begun to call the 'strange attractor'. This is a theoretical limit which can best be envisioned as a tightrope upon which the weather walker must perform. She can leap from point to point anywhere along the rope, but may not walk beside it. If she performs in the Amazon, she can rain or shine, but never snow. Which means that there is a higher order inherent in chaos, an arrangement which makes something chaotic radically different from something that is merely random.

Science is only just beginning to peer into the heart of chaos in an attempt to understand its rules, its harmonies – its strange attractors. And has already realised that, no matter how finely we analyse a system like the weather, its behaviour will remain beyond our ability to predict. We may be able to identify the strange attractors of the system, the nature of the tightrope that it walks, but we will never be able to forecast exactly where on the rope the walker will step from one moment to the next.

The action of the weather on living things is subject to the same constraints. As weather fronts roll by, bringing an ebb and flow of temperature and pressure, they influence our own internal tides. We may respond in synchrony with the atmospheric rhythms, flowing with them, or we may amplify or negate their influences. Our exact response depends on individual physique and temperament, but most of all it depends on a prior state of stimulation – precisely where we happen to be standing at that moment on our own personal tightrope.

Two people, or the same person at two different times, will react differently to the same weather change – or in the same way to different weather changes. Their future response depends on their immediate history. This is the Law of Initial Value.

What it means in terms of weather-sensitivity, is that there are no weather virgins. Every experience is a second or a third one, which must be related back to an earlier stimulus. All reaction depends on a prior level of arousal or excitation. According to the Law, the *lower* the initial level of arousal, the *higher* the response to an exciting stimulus. The colder you are when going into a warm room, the hotter you will feel after staying a while. Conversely, the Law also says that the *higher* the initial value of arousal, the *lower* the response to a further exciting stimulus. The third or fourth cup of coffee has nothing like the effect of the first, and may even produce a reverse or paradoxical response.

Our biological tides are driven by the involuntary nervous system. On the approach of a cold weather front – blood volume shrinks, shutting off the periphery of the body against the outside world; blood pressure rises; blood sugar increases; the pupils dilate; the upper eyelids are raised; sweat glands are stimulated; and there is an increase in breathing rate and in the flow of air to the lungs in all normal individuals. All these things happen unconsciously, arousing us, girding up our loins for the possible need to take avoiding action.

The usual reaction to a warm front – is that blood vessels dilate; blood pressure falls; blood sugar declines; the pulse rate drops; the pupils contract; the eyelids relax; sweating is controlled; and the flow of air to the lungs is reduced. All these are actions of the parasympathetic nervous system, the 'loyal opposition' which works through the inhibiting hormone noradrenalin to counteract the effects of adrenalin excitement.[408]

These responses are beyond our control, they take place unconsciously, automatically, keeping a delicate balance between stimulation and inhibition. They are responsible for maintaining our equilibrium in the face of a constantly-changing environment. Some people are more sensitive than others, showing more rapid changes in mood and well-being, but all healthy organisms respond at a level below conscious awareness. We are all weather walkers, poised on the tightrope, waiting for the subtle signals that will tell us where to place the next step.

We are not, however, totally at the mercy of chaos. We have our own 'strange attractor'. We are aware of wind.

Part of this awareness is purely somatic. A body that loses balance too badly becomes uneasy, even dis-eased. In the temperate zone of the northern hemisphere, there are well-documented statistical connections between warm westerly winds and exhaustion, industrial accidents, bleeding, embolism, fevers, some circulatory disturbances and suicide.

Warm easterly winds, which are relatively rare, tend to be depressive and may coincide with epileptic attacks. In the northern hemisphere, warm winds from the south are associated with appendicitis, bronchial asthma, diabetes, nephritis, spasms – including the onset of labour pains, and general excitability.

Polar fronts and cool northern air coincide with angina, apoplexy, gall bladder problems, meningitis, attacks of cramp, and an overall decrease in reaction time.

Wind sensitivity does not, of course, lead inevitably to illness. But there is something about infirmity which makes people more aware of being weather sensitive. As though a broken bone, a poor joint or a missing limb distorts the natural equilibrium in ways that change the Initial Value.

People with stomach complaints begin to feel uneasy even before the wind blows. Old wounds and scars start to ache when wind and rain are in the offing. A ringing in the ears of those who suffer from migraine is said to indicate a change of wind.

Jonathan Swift warned that:

> A coming storm your shooting corns presage,
> Old aches will throb your hollow tooth with rage.

And Samuel Butler, aware that the first to announce coming changes seem always to be old soldiers, arthritics and amputees, said:

> As old sinners have all points
> O' th' compass in their bones and joints –
> Can by their pangs and aches find
> All turns and changes of the wind.[237]

You do not, however, have to be injured or ill to be weather sensitive.

There is enough wind-shift in a healthy body, perhaps especially in a truly healthy body, for the message to be received loud and clear. Even if it only surfaces in recognition of secondary signs, such as the havoc that humidity plays with the hair when it is carried in on the first airs of an approaching front.

The English journalist A. P. Herbert could seldom resist a dig at authority and considered weathermen fair game:

> But when they hear the sibyl chant,
> 'All colourless, and feels like clay,
> All straight and horrible – I can't
> Do nothing with my hair today!'
> Then write they down, A deep depression runs
> South-west from Iceland – secondary ones
> Are busy in the Bay![280]

This is a frivolous example, but there is enough survival value in awareness of weather change for natural selection to insist, at some point in evolution, on bringing it to conscious attention. Perhaps those with some kind of infirmity notice it first, simply because they need the edge. It gives them a head start and the chance of being there at the finish amongst the fittest.

Before weather charts, the most useful and widely used of all weather indicators, was wind direction.

Each quadrant carried its own clues. In *The Shepherd of Banbury's Rules to Judge the Changes of the Weather*, published in England in 1744, we are warned that a 'northeast wind brings a long hard storm'; and 'when a summer wind's in the south, the rain's in its mouth', but 'a south wind in winter brings mild cloudy weather'; and 'when a wind shifts to the southwest, expect warm weather'.

Paul Marriott, a professional meteorologist and sceptical of easy answers, tested the rules in a long series of direct weather observations and found, to his surprise, that the good shepherd scored 66, 70, 73 and 92 per cent on these predictions alone. Which was at least as good as the forecasts provided by the British Meteorological Office.[321]

When wind signs are combined with all the other bits of local lore that accumulate around weather changes, they provide a formidable body of information.

Edward Jenner, who developed the first successful vaccine against smallpox, was fascinated by the sayings of his native Gloucestershire and two hundred years ago worked them all into a comprehensive rhyme:

> The hollow winds begin to blow,
> The clouds look blank, the grass is low;
> The soot falls down, the spaniels sleep,
> The spiders from their cobwebs peep.
> Last night the sun went pale to bed,
> The moon's halos hid her head;
> The boding shepherd heaves a sigh,
> For see, a rainbow spans the sky.
> The walls are damp, the ditches small,
> Closed is the pink-eyed pimpernel.
> Hark how the chairs and tables crack!
> Old Betty's nerves are on the rack;
> Loud quacks the duck, the peacocks cry,
> The distant hills are seeming nigh,
> How restless are the snorting swine,
> The busy flies disturb the kine,
> Low o'er the grass the swallow wings,
> The cricket too, how sharp he sings!
> Puss on the hearth, with velvet paws,
> Sits wiping o'er her whiskered jaws;
> Through the clear streams the fishes rise,
> And nimbly catch the incautious flies.
> The glow worms, numerous and light,
> Illumined the dewy dell last night;
> At dusk the squalid toad was seen,
> Hopping and crawling o'er the green;
> The whirling dust the wind obeys,
> And in the rapid eddy plays;
> The frog has changed his yellow vest,
> And in a russet coat is dressed.
> Though June, the air is cold and still,
> The mellow blackbirds voice is shrill;

315

My dog, so altered in his taste,
Quits mutton bones on grass to feast;
And see yon rooks, how odd their flight!
They imitate the gliding kite,
And seem precipitate to fall,
As if they felt the piercing ball.
'T will surely rain . . .[280]

It surely will. If even half these signs can be assembled in any northern temperate area, a cold front with rain is already on its way.

There is another aspect of wind awareness which has little or nothing to do with physical signs. It involves psychological change, and a philosophical flowering that seems to have come about as a direct result of exposure over millions of years to something as tangible and provocative, as numinous, as wind.

Evidence of our awareness at this level is manifest in the nature of wind myths that have grown in every culture in response to a need to come to terms with 'deeper meaning'. These are not the result of conscious attempts to analyse life, but the outcome of a direct instinctive unconscious knowledge that life, above a certain level, seems to have of itself. The awareness is coloured by individual and cultural experience, but there is enough similarity in the form of myth worldwide, to suggest that we are all being moved in the same way by the same catalytic force.

To understand its origin and development, we need the cultural equivalent of a number of key fossils. We need to look for an Olduvai Gorge of the mind, surviving evidence of the earliest patterns of religious belief.

In this century, cultural anthropology has made enormous strides. We have moved beyond a concern with what people look like or make, to an interest in what they say and believe to be useful and true. In many cases this is a rescue operation, picking up the last fragments of a dying culture amongst handfuls of San in the Kalahari or Vedda in Sri Lanka. But there is one very old system of belief that survives, almost unaltered, as part of a modern society. This living fossil is the creed that Japanese now call *shinto* – 'the way of the spirits' – a system that represents the oldest, purest and most refined distillation of early belief to be found anywhere in the world.

The lesson of Shinto is that everything is imbued with spirit and therefore necessarily divine. And as spirits ourselves, we resonate

316

in natural sympathy with all others. We are part of a vast symbiotic relationship with everything in nature, and recognise our mutual indebtedness. As a sign of this recognition, followers of Shinto pay respect with a bow, or a clap of the hands, or a simple offering, to shrines associated with a rock, a tree or a pool.

It is not true that someone visiting such a shrine claps their hands to attract the attention of the resident spirit. This would be rude and is anyway unnecessary. Clapping is simply a very ancient way of indicating both pleasure and respect, just as bowing is a traditional form of courtesy. And it is quite wrong to assume that offerings of food or drink are intended for the enjoyment of the spirit of the shrine. They are not meant to be used in any way. They are just tokens and symbols of respect.

There is nothing in the shrine ritual of a Shinto devotee that can be interpreted as worship in the usual sense, because there are no deities present.

So it is clear that in Shinto at least, which is the form of animism most accessible to us today, nobody worships a tree, and there is no god of the wind.

It begins to seem likely that early anthropologists and mythologists, many of whom were Christian missionaries to start with, have been too hasty in concluding that 'primitive' people anywhere see things in such a simplistic way. The evidence from all those who have succeeded in getting really close to a culture, suggests that they do not.

We share descent, it seems, from the same spiritual source. Somewhere, perhaps as much as three hundred thousand years ago, our joint ancestors began to think deeply about themselves and their relationship with the rest of the world.

They must have been impressed, right from the start, by the mysteries of birth and death. These are transitions reenacted in all the ceremonies of initiation and rites of passage that still punctuate an individual's progress through society. But the main stimulus to conceptual thought, the spur which even now starts the most worldly amongst us thinking in abstract, religious terms, is an experience of transcendence. The feeling, even if only for a moment, of belonging to something much bigger.

This comes to people in different ways. It can be sparked by a sunset, a piece of music, a mind-bending chemical, or the sensation of falling in love. But the most common and readily accessible experience of otherness and togetherness, of being involved in something very big and very strange, is that of being blown by the wind. If you miss it at first, if there happens to be something else on your mind at the time, it happens again. And again, until the message gets through.

You are touched by something invisible, something that bends trees and makes strange sounds. It throws the ocean into confusion, whips up fantastic froths of cloud and carries off a desert in its arms. And yet it communicates directly and intimately with you. You can feel its effects deep inside. You know, long before the first flicker of a leaf, when it is going to arrive. These sensations are subliminal, working on you unconsciously, resulting sometimes in feelings of paranoia. If the wind is an ill one, it seems intent on personal retribution, it points directly at you. But the net result of being in the wind, of being wind-sensitive, is that you end up feeling involved, as though you have some say in the workings of the world.

This is the essence of Shinto. The word *shin* is an old Chinese one, which means 'divine spirit'. The syllable *to* is derived from *tao*, 'the way', which was originally written as an ideograph of two separate characters meaning 'man at the crossroads'.

So when the Japanese combined *shin* and *to* into Shinto, they produced a composite that is simplified as the 'spirit way', but which actually reads 'humanity as divinity at the crossroads of the universe'.

And implicit in this reading is the assumption, born of experience with the wind, that we have the freedom and the right to choose our own direction.

Windfalls

There are, on a few Shinto shrines, some sacred curiosities. Stones that have fallen from the sky.

Nobody makes much fuss about them. They are simply there for people to take pleasure in, and as objects deserving of the respect accorded to everything that shares the spirit of divinity. The traditional explanation for their existence is very simple and matter-of-fact. 'There is a hole in the sky,' say the priests, 'and sometimes things just fall through it.'

That is the accepted myth. It is not really an explanation, and is not intended to be. You can believe it or not, as you choose. What it does, is to give the objects reality without moral judgment or censorship, and keep them available for later, possibly more enlightened, interpretation.

A stone fell out of the sky on to the shores of the Aegean in 468 BC. Several people saw it fall and the philosopher Anaxagoras examined it. He was a rationalist and concluded that the heavenly bodies must be made of the same stony substance as Earth. 'The stars,' he announced, 'are flaming rocks.' Athens, even then in the midst of its Golden Age, was not ready for such honesty. They put him on trial, accused of impiety and atheism, and he was forced to flee to the Hellespont, where he died.

Two thousand years later, in another Age of Reason, things were still falling from the sky. The French Academy of Sciences, in the late eighteenth century, appointed a special commission to investigate the matter and put the people's minds at rest. They examined the stones, including one weighing three kilograms that descended with a loud explosion near the Loire, and interviewed the witnesses. But they were unable to face the implications of the evidence. 'There are no stones in the sky,' the commission concluded rather smugly, 'therefore stones can not fall from the sky.' Their chairman was Antoine Lavoisier, a great scientist, considered by many to be the father of modern chemistry, but able nonetheless, even with a meteorite in his hand, to say, 'In spite of the belief of the ancients, true physicists have always been doubtful about the existence of these stones.'

This tendency to dismiss awkward facts, not by explaining them, but by explaining them away, still exists. And we are the poorer for it. It narrows our field of experience and robs the world of some of its more lively and revealing hues. It negates the very freedom of the wind.

Strange phenomena were once, quite rightly, held in much greater regard.

Provincial governors in China and ancient Babylon were expected to include in their annual reports to the central government everything strange which had taken place that year, peculiar objects in the sky, apparitions, monstrous births, irregularities in nature, popular delusions or unrest, every subtle symptom of psychological and thus of social disturbance . . . and the symptoms were treated or accommodated by appropriate changes in central orthodoxy.[334]

This is a very enlightened approach to both government and reality. The problem with most cosmologies is that they leave loose ends. There are always things which do not quite fit the current explanation of how the world works, and they are summarily dismissed as delusion, error or deliberate deceit. Which drives a wedge in between people and authority, relegating those who have had an unusual experience to the ranks of the credulous or crazy. 'The experience becomes illegitimate, no longer a source of wonder but of shame, a symptom of sickness, a portent of evil, a suitable case for treatment. This way of seeing things is excessively low-minded.'[334]

The truth is that odd things do happen. Some of them are merely statistically unlikely, stretching the limits of probability without bending the rules. Others are wildly improbable, one in a million happenstances, that seriously challenge the rules, but can still be accommodated as the exceptions that 'prove' them. The remaining few are the critical ones, which challenge all the assumptions on which the rules are based. These are the ones that get people poisoned, banned, burned or guillotined – and yet they refuse to go away.

The most durable of all such phenomena, are persistent reports throughout the course of history, of showers of things falling from the sky.

Pliny the elder, in an *Account of the World and the Elements* written in AD 77, says:

It rained milk and blood in the consulship of M. Acilius and C. Porcius. This was the case with respect to flesh, in the consulship of P. Voluminius and Servius Sulpicius, and it is said, that what was not devoured by birds did not become putrid. It also rained iron among the Lucanians, the year before Crassus was slain by the Parthians, and the substance which fell had the appearance of sponge. In the consulship of L. Paulus and C. Marcellus it rained wool, round the castle of Carissanum, near which place, a year after, T. Annius Milo was killed. It is recorded, among the transactions of that year, that when he was pleading his own cause, there was a shower of baked tiles.

Athenaeus in the fourth century AD, recorded a continuous three-day shower of fish in the Chersonesos district of Greece, so serious that 'the roads were blocked, people were unable to open their front doors and the town stank for weeks'. A rain of fish descended in Saxony during the reign of King Otto the Sixth in AD 689. The Archbishop of Uppsala in 1555 described and illustrated a fall of fish over Sweden. Another occurred in the Faroe Islands in 1673, when tons of fresh herring were found scattered on a mountain peak 'above two hundred fathoms high'. And there are literally dozens of other reports in the literature from Holland, England, South Africa, Argentina, Australia, the United States and the East Indies.[185]

These are not the wild-eyed, wish-fulfilling stories of those who expect miracles. They are for the most part, sober accounts by solid citizens who have nothing to gain, and everything to lose, by being associated with such mysteries. This report from a resident magistrate of the Thames Police, which tells of an incident that took place while he was an officer on military service in India, is typical:

> In a heavy shower of rain, while our army was on the march, a short distance from Pondicherry, a quantity of small fish fell with the rain, to the astonishment of all. Many of them lodged in the men's hats; when General Smith, who commanded, desired them to be collected, and afterwards, when we came to our ground, they were dressed, making a small dish that was served up and eaten at the general's table. These were not flying fish, they were dead, and falling from the well-known effect of gravity; but how they ascended, or where they existed, I do not pretend to account. I merely relate the simple fact.[197]

A lot of the best evidence gets eaten, but some does find its way into the hands of appropriate experts, who are seldom happy with it. In 1839 a wind storm accompanied by two short showers of rain hit the valley of Aberdare in Wales. A large number of live fish fell with both downpours, ten minutes apart, over the same area of about 750 square metres. This included a building whose gutters and waterspouts became blocked with the creatures. Some fell on a witness' hat and buckets full of others were swept up from the grass. Some were placed in fresh water, where they thrived and a few, still alive, were put on display at London Zoo a week later.[183]

The local vicar sent several specimens to the British Museum where John Gray, the Keeper of Zoology, identified them as common minnows *Phoxinus phoxinus* and three-spined stickle-back *Gasterosteus aculeatus*. He suggested, and his view was endorsed by the editor of the journal in which the article appeared, that they came from a local stream in a bucket of water thrown at the principal witness by a practical joker.[170]

And there, as is the way of things when experts have had their say, the matter rested. Until early this century, when a strange, walrus-like man called Charles Fort spent twenty-seven years working through the back files of newspapers and magazines in the libraries of New York City and the British Museum. He made notes of anything out of the ordinary until he had hundreds of thousands of little slips of paper filed in shoe boxes. Then he began to arrange them into patterns, and published these in a series of four delightful books, each of which contains an avalanche of data woven together loosely with his own whimsical interpretations.

Fort makes no attempt to explain anything, content to point to the abundance of strange phenomena and gently mock those who refuse to take them seriously. His comment on Gray's explanation of the Aberdare fish fall is simply to note that 'someone soused someone with a pailful of water in which were thousands of fishes four or five inches long, some of which covered roofs of houses, and some of which remained ten minutes in the air'.[141]

Charles Fort died in 1932, but the fish keep falling.

An ichthyologist at the American Museum of Natural History, after collecting seventy-eight reports from sixteen different countries, has become convinced that it happens, that 'fishes fall from the sky with rain'. One recent article lists fifty-three incidents in Australia alone.[334] The explanation offered is that whirlwinds and waterspouts are responsible.

A whirlwind starts in front of an approach storm, and as it gains in size the 'snout' elongates and approaches the water. This, caught by the whirling wind, rises up in a cone. The two unite, and the swirling column moves along, picking up water, fishes, and any other fairly light objects at or near the surface of the water . . . Furthermore, whirlwinds originating inland, will not only progress over land, picking up various objects, but over ponds and lakes – becoming fresh waterspouts . . . Sometimes the fishes are found in a long, narrow, fairly straight row over some distance, evidently having been dropped as the waterspout progressed over the country with lessening speed and carrying power.[186]

All of which sounds very reasonable, but the falls are seldom that linear and tend to be strangely exclusive. Fort quotes one incident in which a whirlwind was seen to travel over a pond. Things were not so much sucked up as rudely scattered about. He protests that it is easy to say that fish falling from the sky have been scooped up in a wind, but 'in the exclusionist imagination there is no regard for mud, debris from the bottom of the pond, floating vegetation, loose things from the shores . . . A pond going up would be quite as interesting as fishes coming down . . . It seems to me that someone who had lost a pond would be heard from.'[141]

There are indeed falls of things other than fish.

In 1870, the city of Sacramento in California was pelted with 'mudpuppies', a type of salamander with red gills, that lay two hundred to a puddle. In 1890, hundreds of freshwater crabs showered down on two separate districts of San Francisco. During a rainstorm in Arizona in 1941, a plummeting clam hit a boy on the shoulder. Hundreds of thousands of snails dropped all over Algiers in 1953. Live freshwater shrimps were deposited on Los Angeles in 1963.

Frogs and toads are perennial fallers, one of the best recent reports coming from Birmingham in England in 1954 when hundreds of little frogs scared a group of people in a park by bouncing off their umbrellas. And in that same year, a lady on a farm near Largo in Florida, reported a rain of crayfish. 'We gathered two ten-quart pails of them in the yard around the house. Mother cooked them and we ate them. They tasted just like any other crayfish.'[335]

The oddest feature of these assorted windfalls is the rarity of mixed showers. Most consist of one species only and are usually confined to individuals of the same size and age. Froglets fall all over the place, but tadpoles almost never do. Charles Fort wondered about such eccentricity and playfully proposed the existence of a 'Super Sargasso Sea' somewhere in the upper atmosphere, where windborne objects collected and sorted themselves out before being shaken loose again by the occasional storm. Nobody need take this suggestion seriously, but it becomes necessary in the face of mounting evidence, to account for such highly selective dispersal.

It is similar in many ways to poltergeist phenomena. Every police force, every newspaper in the world, has files full of stories about mysterious bombardments of stones which rain down on certain houses or locations, sometimes dropping on tables and floors inside closed rooms. The descent and fall of the objects is not random, but seems to be controlled to the extent that they move in unorthodox ways or land in carefully selected areas. All these goings-on seem however to be centred on a person rather than a place, if that individual moves, the focus of activity moves with them. Very often this person is a child, usually an adolescent girl, but always someone who is emotionally disturbed, going through a period of difficult adjustment. It is this instability which seems to attract the phenomena, just as weather signs first manifest in an arthritic or amputee.

Amongst Charles Fort's favourite oddities were reports of trees which in times of drought seemed to attract their own local rain, of people in the wilderness who had been showered with 'manna' in times of need, of newly dug ponds unaccountably found to be full of adult fish. He wondered whether there might not be an unknown force behind distribution which responded to inequities, a force that once piled up mountains on the plains and filled the empty quarters with elephants and tigers, but is restricted now in its dotage, to the occasional pebble or to juggling with little frogs.

If we are serious about Gaia, about the possibility of Earth being a giant organism, then this idea is not so absurd. Such a creature would have to control homeostatic mechanisms which automatically maintain equilibrium, in the same way that our sympathetic and parasympathetic nervous systems balance each other out. It seems that the atmosphere is a system of this kind and is maintained by life for its own ends; and it is not a large step from there to the assumption that part of this complex includes a feedback loop that makes it amenable to suggestion.

The Greek philosopher Plotinus, who founded the mystic discipline of Neoplatonism in the third century AD, said that free will included the ability to have whatever you wanted. All you had to do was create a suitable receptacle for it. Hang up a nesting box if you wish to attract a wild bird, build a shrine suitable for the spirit you seek, or simply make yourself receptive. All prayer is offered on the assumption that desire can be applied to nature in a way that produces fulfilment. Rain for the farmer, manna for the faithful, visions of the virgin or close encounters with extraterrestrials, all according to your need.

The fact that fish fall so often, when in theory, whirlwinds and waterspouts could equally easily provide hamburger or squid, is interesting. The fish is a potent symbol. It was the sign of the goddess Atargatis of Syria, one of the avatars or forms of Vishnu, and the symbol of immortality in Egypt. It was a very early symbol of baptism, representing the pious swimming in the waters of scripture, and was eagerly adopted by the first Christians. The letters of the Greek word for fish form an acrostic J.C. or, in Greek, the whole phrase 'Jesus Christ of God the Son Saviour'. Fish figured prominently in the feeding of the five thousand and took the place of wine in many early Eucharists.[213]

In short, there are very few motifs which move the unconscious mind more strongly or more easily, and none which have such a natural association with wind and water.

It is another interesting fact that paranormal events seem to need an audience. They seldom if ever take place except in the presence of people. Poltergeists never rattle around in empty houses. Strange falls, as far as we know, never take place in wild unpopulated areas. These things seem to be closely related to human personalities, to attitudes and states of consciousness; to energies or images released from, or inspired by, the unconscious; to powerful imaginations and strong wills. How else do you account for the fact that no less than eight species of fish – black bass *Micropterus salmoides*, goggle-eye *Chaenobryttus coronarius*, shad *Pomolobus mediocris*, two kinds of sunfish *Lepomis*, and several sorts of minnow – the largest and most varied collection ever to descend from the sky, fell in Marksville, Louisiana on October 23rd, 1947 in a great shower all around a visiting ichthyologist?[16]

And what is one to make of windfalls such as the flakes of fresh meat which drifted down over Kentucky in 1876; or the gelatinous 'bog butter' that landed on a field in Ireland in 1958; or the luminous jelly called *pwdre ser* that sometimes decorates hillsides in Wales; or the millions of seeds of mustard and cress that fell on the roof and garden of one suburban home in Southampton, England in 1979; or the golf balls, buck shot, nuts and bolts, foil and tinsel, bricks, tiles, nails, coins, banknotes, amulets and rosaries that keep dropping down on people all over the place?

Except perhaps that they represent the fallout from unconscious desires incompletely expressed, taking on odd forms with the sort of strange logic that one expects of objects or events in a dream.

Winds have the power to lift and carry any of these things, and in tornado and hurricane conditions they frequently do.

Storm burdens, however, tend to be rather briefly borne and haphazardly strewn. What we see in these non-random, highly selective, often appropriate, sometimes scary, windfalls – is another force at work.

We know that weather exercises considerable control over many of the processes of life. It is possible that, in this intricate symbiosis, there is room for a reciprocal influence.

Maybe we can and do get a little of what we ask for – and perhaps this is precisely what we deserve.

HEAVEN'S BREATH

A newborn child, mouth wide, reaches hungrily for air.

Eyes closed, arms flailing, it gags on the unfamiliar medium, holding the air, not yet quite sure what to do with it. Then suddenly, sometimes with the assistance of a friendly slap, there is a gasp, a release, a small puff followed almost instantly by a new and larger intake of air. The pump is primed, automatic processes take over, the body draws the third of six or seven hundred million breaths that constitute a lifetime.

It is a magic moment. Not the beginning of life, but such a clear transition, that it is easy to understand why the ideas of life and breath and spirit and creation should have become so intertwined.

In many languages they go by the same name. *Ruh* in Hebrew and Arabic means both 'breath' and 'spirit'. The Dakota and the Sioux called it *niya*, the Aztec *ehecatl*. Whatever the word, it is also one that grades easily into others from the same root that express concepts of wind and divinity. Quetzalcoatl, the god of sun and wind and air, is born when the lord of existence breathes on his mother. It is a common belief amongst some Mexicans still, that a man's breath is more responsible for conception than his semen.

The Melbourne tribes in Australia claimed that the creator Pundjel made humans of clay and then 'he lay upon them, and blew his breath into their mouths, into their noses, and into their navels; and breathing very hard, they stirred'.[105]

The Batek negrito people say that *nawa angin*, the 'wind-life-soul', is blown into their bodies through the soft fontanelle of the skull.[127] The Navaho agree and know the name of the god responsible. The Baiga in India called him Bhagavan and said, 'he blew also on the earth and made it hard and firm'. In Teutonic myth, Odin was the wind god, 'the wild huntsman amongst the raging host' who gave soul to logs of ash and alder that became human when he breathed upon them.[304] In Orissa, the god was Labosum and the creative wind just one of his stray belches. In other areas, the origin of vital air was more mundane, even scatological, but in every case life began with the inspiration of moving air, with wind as breath, bearing spirit.[125]

'Breath is life', say the yogis, because it contains *prana* – which they see as the source of all energy. *Pranayama* is the art of controlling breath in such a way that physical, mental, intellectual, sexual, spiritual and cosmic energy are maximised in an individual. 'When the breath is regular, so is the mind, it becomes attuned to the winds of the universe.'[239]

In China, the connection is clear. 'The breath of the universe,' said Tzu-Chhi, 'is called wind. At times it is inactive. But when it rises, then from a myriad apertures there issues its excited noise . . . Like swirling torrents or singing arrows, bellowing, sousing, trilling, wailing, roaring, purling, whistling in front and echoing behind, now soft with the cool breeze, now shrill with the whirlwind . . . Have you never listened to its deafening roar?'[356]

The primacy of wind is emphasised by the fact that the Sun symbol, hexagram fifty-seven in the *I Ching*, is made up of the trigrams Wind upon Wind.

There are differences in religion and belief which arise due to the different character and structure of the cultures in which they occur, but there are also awesome basic similarities.

Students of comparative religion once believed that the panoply of separate spirits required by animism, gradually combined into general gods – such as Aeolus for *all* the winds. And that further generalisation and abstraction transformed these departmental deities into a monotheistic God, who became the Lord of Heaven and Earth. But this is an unnecessarily ethnocentric attitude, which gives undue prominence to our own most-recent beliefs. The evidence suggests a reverse process, a growth of animism out of a very early monotheistic concept of a supreme force or being.[240]

There are a host of spirits and gods involved in world cultures. Most of these are attached to tangible or clearly substantive objects such as volcanoes, rivers, rocks, the soil, the sun and the moon. The myths and attributes, the patterns of ritual and worship connected with such deities, make it clear that they arise as part of a process of speculation about nature. They are animated theories about how things work. More a product of science than religion.

But there is evidence also of another process in action. Of an early and fundamental awareness of uniformity and identity, of something born of a certainty that we are all somehow intimately involved.

The Nuer in central Africa express it well with their concept of *Kwoth*, which they describe as 'the being of pure spirit', who is everywhere, like the wind. Who perhaps *is* the wind.

Of all natural forces, the wind has always been the most difficult to grasp. It touches us, moves us, but we cannot touch back. It was our first experience of the ineffable. Something of indescribable power, too remote to be seen, but near enough to be sensed in a very intimate, a very personal, way.

Religions set up sun gods, bird gods, beast gods, deities of water, earth and fire who could be named, drawn, sculpted, praised and blamed. But behind these convenient façades in almost every case is a sky god, a god of storms, of weather and the wind.

In some cultures he is male and has a name such as Enlil, Baal, Adad or Yahweh, but in even these there is something else involved. A power behind the throne, a less tangible, more universal, spirit. Something like the Fuegian Temaukel, a 'mist in the sky' whose name was never uttered; or Olorum of the Yoruba, the 'owner of the sky'; like Brahman, the World Soul; or the Christian Holy Ghost.

The rationales are not important. What matters is that we, as humans, feel instinctively that we are part of this power, and that it is part of us.

We are, or have the potential to be, more than human.

Our lives are woven into the wonderful fabric of Gaia, and we share her destiny of becoming something greater still.

And in all this, we are nourished and animated by the wind, brought into being by the touch of Heaven's Breath.

DICTIONARY OF WINDS

There are surprisingly few English language names for specific winds. This is a list of almost four hundred wind names, most of them in other languages, from all over the world – which occur in meteorological, mythological and anthropological works, and in contemporary and classic literature. It includes also a few simple definitions of the most common wind types and wind conditions:

AAJEJ	Whirlwind in southern Morocco. Attacked with knives by the fellahin.
ABROHOLOS	Easterly squall, between May and August on the southern coast of Brazil.
AFER	Southwesterly in ancient Rome. Also AFRICO, AFRICUS.
ALIZÉ	Trade wind from the northeast in southern France.
ALM	FALL wind in Yugoslavian Karst region.
ALTANUS	Southerly in ancient Rome.
ANABATIC	An upslope wind.
ANTI-TRADE	Reverse winds 2000 metres above the TRADES.
APELIOTES	Southeasterly with gentle rain, in ancient Greece. Portrayed as a young man bearing fruit.
AQUILO	Northwesterly in ancient Rome.
ARASHI	Storm wind in Japan.
ARGESTES	West-southwesterly in ancient Rome.

ARIFI	Strong southern of SIROCCO type in Morocco. Also AREF, IRIFI & RIFI. From Arabic 'to scorch'.
ASIFA-T	Circular tropical storm in the Arabian Gulf.
ASPRE	Warm breeze in southern France.
AUGER	DUST DEVIL, sometimes stationary, in California.
AURA	Breath of air at dawn in ancient Greece.
AUSTER	Southerly in ancient Rome.
AUSTRU	Dry, cold, winter westerly on Lower Danube.
AUTAN	Hot, dry, FALL wind in Corbières Mountains of southern France. Counteracts the cool MARIN.
AZIAB	Hot, wet southerly in the Red Sea.
BACKING	Wind changing direction counterclockwise.
BAD-I-SAD-O-BIST-ROZ	Hot, dry, dusty, north or northwester from June to September in Iran & Afghanistan. Means 'Wind of One Hundred and Twenty Days'.
BAGUIO	Circular tropical storm in Philippines. *See* HURRICANE.
BAI	Northerly in dynastic Egypt. 'The Ram'.
BALI	Strong easterly in Java.
BARAT	Strong, blustery northwester in Sulawesi, Indonesia.
BARBER	Freezing wind that ices up beards and hair in midwestern Canada and United States.
BARINE	Rare westerly gale from sea on Venezuelan coast.
BAYOMO	Violent, gusty northerly, associated with thunder, on slopes of Sierra Maestra in Cuba.
BELAT	Cool, winter northerly, hazy with sand, in Oman and on coast in Saudi Arabia.
BERG	Hot, dry FALL wind in South Africa.
BHOOT	Small DUST DEVIL in India.
BILIKU	Dry northeastern MONSOON in Andaman Islands.
BISE	Cold, dry northeaster, accompanied by heavy cloud and low pressure in Languedoc region of France.
BLAAST	Cold squall in Scotland.
BLACK ROLLER	Dust storm in western United States.
BLIZZARD	Cold, north or northwesterly gale with snow. From German *blitz*, 'lightning'. After 1820 in USA.
BOEKIFU	Northeasterly TRADE wind in Japan.
BOFU	Gale in Japan.
BOHOROK	Warm, dry FALL wind off Barison Range in Sumatra.

BORA	Northeaster, dry, cold, blustery KATABATIC wind. Off Karst and Dinaric Alps along Adriatic coast.
BORASCO	Gusty wind with thunderstorms in Mediterranean.
BOREAS	Northerly in ancient Greece. Cold and stormy. Portrayed as an old man blowing a conch shell.
BORNAN	Downdraft through Drause valley on to Lake Geneva.
BRAVES	Prevailing westerlies in temperate latitudes.
BRAW	FALL wind in the Schouten Islands.
BREEZE	Light wind, usually blowing along sea or lake shore.
BREVA	Late afternoon northerly on Lake Como in Italy.
BRICKFIELDER	Hot, dry, southwesterly in New South Wales & Victoria areas of Australia. Also BRICKLAYER.
BRISA	Northeastern TRADE wind on west coast of South America.
BRISOTE	Northeastern TRADE in Cuba.
BROBOE	Dry southeastern TRADE in Sulawesi, Indonesia. Also BRUBRU.
BRUSHCHA	Cold northwesterly in Besgell Valley of Switzerland.
BULL'S EYE	Sudden squall out of a clear sky in Nova Scotia and around Cape of Good Hope. Also WHITE SQUALL.
BURA	The BORA at Trieste.
BURAN	Cold, biting, black northeaster in Russia. In winter, also called PURGA.
BURGA	Gale with sleet or snow in Alaska.
BURSTER	Southerly, cold and wet, preceded by an eerie lull after BRICKFIELDER in south west Australia. Also BUSTER.
CACIMBO	Cool southwester off cold Benguela current along Angola coast between July and October.
CAECIAS	In Milton's *Paradise Lost*. Probably CIRCIAS.
CALM	See FORCE ZERO.
CANTALAISE	Violent wind with snow on Aubrac plateau in France.
CAPE DOCTOR	Strong, cool, southeaster between December and March, blows smog away from Cape Town.
CARABINERA	Squall in Spain.
CARBAS	East-northeasterly in ancient Rome.
CAT'S PAW	Enough air to ripple a pool in United States.

CAURUS	Northwesterly in ancient Rome.
CENCHRON	Breeze said by Hippocrates to blow in Phasis.
CHALLIHO	Strong southerly, precedes southwest MONSOON in India.
CHERGUI	Warm, dry southerly of SIROCCO type in Morocco.
CHI'ING FUNG	Gentle breeze in China.
CHILI	Hot, dry southerly of SIROCCO type in Tunisia. Also CHICHILI & CHIHILI in Algeria.
CHINOOK	Hot, dry FALL wind, producing rapid rise in temperature. In lee of Rockies, especially Montana.
CHOCOLATERO	Hot, sandy squall, coloured by dust, in Gulf of Mexico.
CHOM	Hot, dry southerly in Algeria.
CHUBASCO	Violent, short-lived gale in Gulf of California and Pacific coast of Mexico from May to November.
CHURADA	Fierce, rainy squall from January to March in the Marianas Islands.
CHWA	Breeze in Wales.
CIERCO	Cold, westerly of MISTRAL type in Ebro region, Spain.
CIRCIAS	Northwesterly in ancient Rome.
COCKEYE	Squall, with thunderstorm between December and March in Northwest Australia.
COLLA	Gale in the Philippines.
COLLADA	Strong, steady northerly in Gulf of California.
CONTRASTES	Winds which blow in pairs from opposite directions during the same day. In Mediterranean.
COROMELL	Nocturnal, land breeze between November and May at La Paz, Mexico.
CORONAZO	Southeasterly storm spawned by TYPHOON in Pacific, reaching western Mexico from May to November. Also CORONAZO DE SAN FRANCISCO & CORDONAZO.
CORUS	Northwesterly in ancient Rome.
CREITHLEAG	Gentle breeze in Ireland.
CRIADOR	Rain-bearing westerly in northern Spain.
CRIVETZ	Cold, northeasterly, bringing blizzards in Rumania.
CROSSWIND	One which blows in a perpendicular direction.
CYCLONE	Circular tropical disturbance around low pressure. Clockwise in southern hemisphere, anticlockwise in north. Also HURRICANE in Indian Ocean. From Greek *kyklon* 'coil of a snake'.

DATOO	Westerly sea breeze in Gibraltar.
DCHAOUI	Another name for HABOOB in Sudan.
DEMANI	Light airs in April in East Africa. From Swahili, 'fair winds'.
DOCTOR	Breeze, sometimes HARMATTAN in West Africa.
DOGODA	Gentle westerly wind in Slavic countries.
DOINIONN	'Wild weather' in Ireland.
DRINET	Cold mountain wind in Rumania.
DUST DEVIL	Wind vortex, produced by heating over land.
DUSTER	Dust-bearing wind in 'Dust Bowl' of United States.
DZHANI	Warm wind from south in North Africa.
EASTER	FALL wind in Oregon. Responsible for the great Tillamook fire in 1933.
EASTERLY	Wind of ill repute, usually cold in winter, hot in summer in Europe and Middle East.
ELEPHANTA	Southern gale in September and October, marking end of southwest MONSOON on Malabar coast, India.
ELISEOS	Northeast TRADE in Spain.
ELVEGAST	Cold, dry, easterly in winter on Norwegian coast.
ERH CHI CHIH FUNG	Northerly in China and Mongolia.
ETESIAE	Westerly in ancient Rome.
ETESIAN	Cool northwesterly on summer afternoons in Greece and Turkey. From Greek *etos* 'season'.
EUROCIRCIAS	Southeasterly in ancient Rome.
EUROCLYDON	Blustery northeasterly gale in eastern Mediterranean. Wrecked St. Paul on Malta on his way to Rome.
EUROS	Easterly in ancient Greece. Sultry and wet. Portrayed as an old man, well clad. Also EURUS.
FAKATIU	Northwesterly in Tikopia, Melanesia.
FALL WIND	Wind warmed and dried by descent.
FAVONIUS	Warm, westerly FALL wind in ancient Rome.
FEH	Dainty breeze in Shanghai, China.
FETCH	Distance blown by wind along sea surface.
FLAKT	Gentle breeze in Sweden.
FLAUWE	Gentle breeze in Holland.
FÖHN	Downhill, FALL wind, hot and dry, in Alps. From Latin FAVONIUS or Gothic *fôn*, 'fire'. Or from Phoenicias, indicating origin in Phoenicia. Also FOEHN, FÖ & FÜN.

FORCE ZERO	Wind with speed less than 4 kilometres per hour. Less than 2 knots.
FORCE ONE	Wind with speed between 5 and 10 kilometres per hour. From 3 to 5 knots.
FORCE TWO	Wind with speed between 10 and 15 kilometres per hour. From 6 to 8 knots.
FORCE THREE	Wind with speed between 15 and 23 kilometres per hour. From 9 to 12 knots.
FORCE FOUR	Wind with speed between 24 and 30 kilometres per hour. From 13 to 16 knots.
FORCE FIVE	Wind with speed between 30 and 40 kilometres per hour. From 17 to 21 knots.
FORCE SIX	Wind with speed between 40 and 50 kilometres per hour. From 22 to 26 knots.
FORCE SEVEN	Wind with speed between 50 and 60 kilometres per hour. From 27 to 31 knots.
FORCE EIGHT	Wind with speed between 60 and 70 kilometres per hour. From 32 to 37 knots.
FORCE NINE	Wind with speed between 70 and 80 kilometres per hour. From 38 to 43 knots.
FORCE TEN	Wind with speed between 80 and 90 kilometres per hour. From 44 to 50 knots.
FORCE ELEVEN	Wind with speed between 90 and 115 kilometres per hour. From 51 to 57 knots.
FORCE TWELVE	Wind with speed greater than 115 kilometres per hour. More than 58 knots. See HURRICANE.
FRESH BREEZE	See FORCE FIVE.
FRISK	Gale in Sweden.
FUGA	Strong wind in Crimea.
FUNG CHIAO HSUEH	Northeasterly in China.
FURICANO	Variation on HURRICANE in Caribbean.
GALE	See FORCE EIGHT.
GALERNA	Cold, squally northwesterly in Bay of Biscay.
GALLEGO	Cold northerly in Spain.
GALLICUS	Northerly in ancient Rome.
GARBAS	Easterly in ancient Rome.
GARBIN	Moist southwesterly in summer on Atlantic coast of Spain.
GARMSAL	Hot, dusty, summer westerly in Turkestan.
GARVI	Southerly of SIROCCO type in Algeria.
GENDING	FALL wind in Java.
GENEVA	Rain-bearing southwesterly in Swiss canton of Vaud.
GENTLE BREEZE	See FORCE THREE.

GEOSTROPHIC	Wind blowing parallel to straight isobars.
GERGUI	Hot, dry southeasterly with sandstorms in Algeria.
GERONA	Hot, dry SIROCCO type wind in Spain.
GHARBI	Strong, moist, dust-laden southwesterly in Morocco. Bringing 'red rain' to France, Italy & Greece.
GHIBLI	Dry, desert SIROCCO type wind in Tunisia. Produces nervous condition of 'ghiblitis'.
GIBA	The 'hang-horse' wind of Japan. Also DAIBAFU. Strong enough to smash down a horse and rider.
GRADIENT	Wind blowing parallel to curved isobars.
GREGALE	Strong northeasterly in autumn, coinciding with high pressure over central Europe. From Balkans. Also GRECALE & GRÉGAL – from 'Greek Wind'.
GRENOBLE	Rain-bearing southwesterly in southern France.
GUST	Transitory wind.
HABOOB	Dust storm in Sudan. From May to September. Bright yellow wall 1000 metres high, followed by rain. From Arabic *habb*, 'to blow'.
HALMIAK	FALL wind on Yugoslavian coast.
HALNE	FALL wind of the Tatra Mountains in Czechoslovakia.
HALNY	Warm wind in Poland.
HARMATTAN	Dry, dusty northeasterly TRADE out of Sahara. In December to February along West African coast. Blows dust hundreds of kilometres over Atlantic.
HAUR	Easterly wind in North Africa.
HAWA JANUBI	Southerly in Arabian peninsula.
HAWA SHIMALI	Northerly in Arabian peninsula.
HAYATE	Gale in Japan.
HEAD	Wind blowing in opposite direction.
HELM	Strong northeasterly in Cumberland, England. Accompanied by helmet-shaped cloud over Pennines.
HOKUTO	Northeasterly in Japan.
HOWLER	Strong westerlies in the 'fifties' latitudes.
HUPE	Cool land breeze in Tahiti.
HURLEBLAST	Variation on HURRICANE in old Caribbean.
HURRICANE	Circular tropical storm in Caribbean and North Atlantic. See FORCE TWELVE. From *Hunraken*; Mayan storm god; *Huracan* of Taino in Antilles; *Hyoracan* of Galibi in Guiana; and *Aracan* or *Huiranvucan* of the Caribs.

IBE	FALL wind in Caucasus.
IMBAT	Sea breeze in North Africa.
INVERNA	Late afternoon downdraft on Lake Maggiore, Italy.
ISERAN	Cold, gusty northerly in French Alps.
I TIEN TIEN FUNG	A 'Sigh in the Sky' of China.
JASNA BURA	The 'Bright Bora'. KATABATIC wind in Croatia.
JAUK	FALL wind in Karawanken region of Yugoslavia.
JET STREAM	Tubular high speed current of air near tropopause.
JIMMYCANE	Variation on HURRICANE in Caribbean.
JURA	Cold, gusty, late afternoon, mountain wind in the foothills of the Jura. Also JORAN.
KABEYUN	Westerly, 'Father of Winds' to the Algonquin.
KADANNEK	Southeasterly in Greenland.
KADJA	Sea breeze in Bali.
KAI	Balmy south wind of China.
KAIKIAS	Northeasterly in ancient Greece. Cold and wet. Portrayed as an old man with a shield.
KAMAKAZA	'Sickle wind' of Japan. Cuts like a knife. Also KAMAITACHI.
KAMIKAZE	The 'Spirit Wind' of Japan.
KAPALILUA	Sea breeze in Hawaii.
KARABURAN	East-northeasterly. The 'Black Wind' of the Gobi. Similar to HARMATTAN, but cold, with dust storms.
KASIKAZI	Northeast MONSOON in East Africa.
KATABATIC	Downslope wind.
KAUS	Sea breeze, bringing rain in winter to Persian Gulf.
KAWAIHAE	Squall in Hawaii.
KHAMSIN	Hot, dusty SIROCCO type wind in Egypt. From March to May. From Arabic, 'fifty'. Said to blow for fifty days. Ninth Plague of Egypt.
KHARIF	Sandstorm in Somalia & Yemen. Preceded by cumulus cloud and leaden sea. KHAMSIN type.
KIBIBONOKKA	Northerly, 'The Fierce One' of the Algonquin.
KLOD	Warm, FALL wind in Bali.
KNIK	Strong southeasterly in Alaska.
KOCHI	Easterly in Japan.
KOGARASHI	Cold, winter wind in Japan.
KOHALA	Gale in Hawaii.
KOHILO	Gentle breeze in Hawaii.
KOKAZE	'Little Wind' in Japan.

KOLAWAIK	Southerly of the Gran Chaco in Argentina.
KONA	Sultry southeastern in Hawaii.
KOOCHEE	Whirlwind in Australia.
KOSHAVA	Cold northeaster bringing snow from interior of Asia to Yugoslavia. Also KOSSAVA.
KUBANG	FALL wind in Java.
KUSI	Southwest MONSOON in East Africa.
KWAT	Downdraft in Amoy (Xiamen) region of China.
LAÂWAN	Westerly winnowing wind in Morocco. From Arabic, 'the helper'.
LABECH	Rain-bearing southwesterly in Marseilles, France.
LAKAWA	Easterly in the Gran Chaco of Argentina.
LANDLASH	Gale in Scotland.
LAXWAIK	Westerly in Argentina's Gran Chaco.
L'ESTE	Hot, dry, southeasterly, usually in January, in Madeira and nearby African coast.
LEUCONOTUS	Southerly in ancient Rome.
LEUNG	Cold northerly on China coast.
LEVANTER	Easterly from July to October in Balearic Islands.
LEVANTO	Hot, southeasterly in the Canary Islands.
LEVECHE	Hot, dry, southeaster, carrying dust and sand, visible as approaching brown cloud in southern Spain.
LIBECCIO	Southwesterly sea breeze in Italy. Mentioned in Milton's *Paradise Lost*.
LIBERATOR	Warm, moist westerly in Gibraltar.
LIBONOTUS	Southwesterly in ancient Rome.
LIGHT AIR	See FORCE ONE.
LIGHT BREEZE	See FORCE TWO.
LINER	Squall along a front in United States.
LIPS	Southwesterly in ancient Greece. Strong. Portrayed as a young man with ornament from the stern of a Greek ship – *aplustre*. Also LIBS.
LJUKA	Warm FALL wind in Yugoslavia.
LOMBARDE	Usually warm, but can be cold, in French Alps.
MAESTRAL	Cold, strong, northerly in Gulf of Genoa.
MAESTRO	Strong, summer northwester in western Adriatic.
MALEDETTO	FALL wind in northern Italy.
MALOJA	Warm, FALL wind of Maloja Pass in Switzerland.
MAMATELE	Hot northwester in Malta.
MAOI FUNG	Northeast TRADE in China.
MARIN	Moist, oppressive, southeaster with heavy rain in southern France.

MATINIERE	Mountain wind, violent in mornings in Alps.
MATO WAMNIYOMNI	Ogala Dakota name for the whirlwind.
MATSUBORI	'Secret Wind' of Japan. Also HOMACHI.
MATSUKAZE	Sound of light breeze in the pines, in Japan.
MBATIS	Southerly light breeze in Greek summer evening.
MEDINA	Land breeze in Cadiz in Spain.
MELAMBOREAS	Northerly, MISTRAL type, in Provence.
MELTEME	Northeasterly autumn wind in afternoons in Aegean. The 'bad tempered wind' of Greece and Turkey. Also MELTEMIA & MELTEMI.
MESETA	Warm, dry wind in Spain.
MEZZER-IFOULLOUSEN	Cold, violent southeasterly in Morocco. Known to Berbers as 'That Which Plucks the Fowls'.
MIDI	Rain-bearing southwesterly in Juras and Alps.
MINUANO	Cold wind in Brazil.
MISTRAL	Cold, dry, blustery northwesterly of Rhone Valley. When low pressure area lies over Gulf of Lyon. From Latin *magistralis*, 'masterly'.
MODERATE BREEZE	See FORCE FOUR.
MODERATE GALE	See FORCE SEVEN.
MOLAN	Downdraft in French Alps.
MONCÃO	Northeasterly TRADE in Portugal.
MONSOON	Seasonal wind characterised by diametrically opposite change of direction. From Arabic *mausin*, 'season'.
MONTAGNÈRE	Land breeze on southern slopes of Alps.
MORGET	Nocturnal land breeze over Lake Geneva.
MRAČNA BURA	'The Dark Bora', KATABATIC wind in Croatia.
MURWA	Southerly amongst the Bavenda in South Africa.
MYATEL	Northeasterly with storms in northern Russia.
NAALEHU	Dry land breeze in Hawaii.
NAF HAT	A blast off the Arabian desert.
NARAI	Cold northeasterly from Siberia, in Japan.
NASIM	Breeze in Saudi Arabia.
NEKRAYAK	Northeasterly in Greenland.
NEVERI	Strong, showery wind in Yugoslavia.
NOR'EASTER	Cold wind in New England.
NORTH CANADIAN	Cold, crusty wind that helped Benedict Arnold repel British fleet at Battle of Lake Champlain in 1776.
NORTHER	Tidal wave of polar air in Texas. Preceded by warm, moist wind from the south.

NOR'WESTER	FALL wind at Christchurch in New Zealand.
NOTUS	Sultry wet southerly in ancient Greece. Portrayed as young man with water jar. Also NOTOS.
NOWAKI	Hot, searing, autumn wind in Japan.
OE	Isolated whirlwind in the Faroe Islands.
OES	Waterspout in Holland.
OM	Squall in Canton.
ORA	Southerly in the mornings on Lago di Garda, Italy.
ORKAN	Gale in Norway.
ORLEANS	Wanton, wilful easterly – blew at Siege of Orléans until Joan of Arc prayed for a westerly to carry her troops across the Loire to attack the fort.
ORNITHIAE	Easterly in ancient Rome.
OROSHI	Cold, dry wind on Kanto plain in Japan.
OSTRO	Southerly. Italian sailor's name for AUSTER.
OUARI	Sandstorm in Somalian summer.
PAMPERO	Southwesterly squall bringing cool air and rain to Argentinian pampas.
PAPAGAYOS	Cool northerly wind in Costa Rica.
PEI FUNG	Northerly in China.
PITTARAK	Northwesterly in Greenland.
POLACK	Cold, dry wind in Sudetenland.
PONENTE	Westerly sea breeze on Italian coast.
PONTIA	Cold, dry night wind from Rhone Valley.
PORIAZ	Cold, summer northeaster off Black Sea in Bulgaria.
POUWAN-GUARTEK	Southwesterly in Greenland.
PREVAILING	Wind whose direction has highest frequency.
PRODROMES	Mild northeasters that preceded the rising of Sirius in ancient Greece. 'The Runners'.
PRUGA	Strong, cold wind in Alaska.
PUELCHE	FALL wind in Chilean Andes. Also PULCHE.
PURGA	Strong, cold northeaster, with snow in the Soviet Union. Also POORGA.
PYR	FALL wind of the Upper Danube. Also PYRN.
QUEXALCOATL	Westerly in Aztec times.
RAGHIEH	Cold easterly on coast of Syria.
RAIKIAS	Cold northeasterly on Turkish coast.
RAKI	Westerly in Tikopia, Melanesia.

REBAT	Southerly morning breeze on Lake Geneva.
REFFOLI	Powerful, gusty wind in the Adriatic.
REPPU	Circular tropical disturbance in Japan.
RESHABAR	Lusty, black, dry, northeaster out of the Caucasus in Kurdistan. *Rrashaba*, 'Black Wind'.
RIESENGEBIRG	FALL wind in East Germany.
ROARING FORTIES	Prevailing westerlies below 40° south.
ROETETURM	FALL wind in Rumania. Also ROTETUR.
ROK	Vindictive gale in Iceland. Also ROC.
SAMIEL	Hot, dry wind in Turkey. From Arabic *samm*, 'poison' + Turkish *yel*, 'wind'.
SANSAR	Icy, northwest wind in Iran. Also SARSAR.
SANTA ANA	Northeasterly FALL wind in southern California. Hot and dusty with an evil reputation.
SCHNEEFRESSER	'Snow Eater' of Switzerland. See FÖHN.
SÉCHARD	Land breeze over Lake Geneva.
SÉGUIN	Diurnal breeze in Provence. 'Following the Sun'.
SEISTAN	Cold, strong northeaster, blowing for four months in Seistan province of Iran.
SEPENTRIO	Northerly in ancient Rome.
SHAMAL	Mild, sometimes dusty, northwester in summer on the Mesopotamian plain.
SHAMSIR	Cold northerly wind in Iran.
SHARAV	Hot, dry, desert wind in Israel.
SHARKI	Southeaster along the Persian Gulf.
SHARKIYE	Cool, desert wind in Lebanon. Also SHARQIEH.
SHAWONDASEE	Southerly 'Lazy Wind' of the Algonquin.
SHIH LUNG	Northeast TRADE in China.
SHIMPU	'Divine Wind' of Japan. Routed the Mongols in 1281.
SIFFANTO	Hot wind from the heel of Italy.
SIMOOM	Hot, dry, desert wind of North Africa. 'Poison Wind'. From Arabic *samm*, 'poison'. Also SAMUM.
SIROCCO	Southeasterly, principally in spring, off Sahara. Hot, evil wind. Many local names. Also SCIROCCO.
SKIRON	Northwesterly in ancient Greece. Hot in summer. Portrayed as old man with a brazier.
SKYSWEEPER	Sea breeze in Majorca, named by English settlers.
SNO	Cold easterly in Norway. See ELVEGAST.
SOLANO	Southeasterly sea breeze in Spain. Laden with dust from Africa. Said to cause giddiness.
SOLANUS	Easterly in ancient Rome.

SONORA	Summer, desert wind in Arizona.
SOULÈDRE	Cold northeasterly in France.
SOUTHEASTER	Strong, cool wind of late summer in Cape Town. The CAPE DOCTOR.
SOVER	Northerly breeze of late afternoons on Lago di Garda in Italy.
SOYO KAZE	Gentle breeze in Japan.
SQUALL	Strong wind that rises suddenly and is short-lived. LINE SQUALL is associated with long arch of cloud.
STEPPENWIND	Cold northeaster in Germany.
STIKINE	Strong, gusty northerly on Alaskan coast.
STORM	See FORCE TEN.
STRONG BREEZE	See FORCE SIX.
STRONG GALE	See FORCE NINE.
SUBVESPERUS	Southwesterly in ancient Rome.
SUESTADA	Strong gale, bearing rain, in La Plata regions of Argentina and Uruguay.
SUHAILI	Cold, strong, wet, southwesterly on Persian Gulf.
SUKHOVEY	Warm easterly with dust storms on Gobi desert.
SUMATRA	Irregular squalls, MONSOON surges in Indonesia.
SUPERNAS	Northeaster in ancient Rome.
SURAZO	Cold, strong, mountain wind from the Peruvian Andes.
SURFACE	Wind blowing at 10 metres above the ground.
SVESZHEST	Gentle breeze in the Soviet Union.
SZ	First faint breeze of autumn in China.
TAIL	Wind blowing in same direction as movement.
TAKU	Strong northeaster in Alaska.
TANGA MBILI	The variable winds of September in East Africa. From Swahili, 'two ways'.
TAPAYAGUA	Squall in western Central America.
TARAI	Southwesterly MONSOON in the Andaman Islands.
TA TE KATA	Sioux name for CHINOOK.
TATSUMAKI	TORNADO in Japan. 'Dragon Whirl'.
TAUERN	FALL wind, FÖHN type at Salzburg.
TEBBAD	Hot, dusty wind in Turkestan.
TEGEN	Northeaster in Holland.
TEHUAN-TEPECER	Cold, strong northerly in Mexico. MISTRAL type. Sets in suddenly on Pacific coast of Panama.
TERRAL	Land breeze at Valparaiso in Chile.
TEZCATLIPOCA	'Divine Wind' of Aztecs. Northerly.
THALWIND	Valley breeze in Germany.

THAR	Hot dry wind in Rajasthan, India.
THERMAL	Local updraft above a relatively warm surface. Produced by a thermal gradient.
THRASCIAS	Northerly in ancient Rome. Also THRACIAS.
TIVANO	Southerly midday breeze on Lake Como in Italy.
TOKALAU	Northeasterly in Fiji.
TOKERAU	Northerly in Tikopia, Melanesia.
TONGA	Southeasterly in Tikopia, Melanesia.
TORNADO	Most violent storm of all, rotating at 300 kilometres per hour. Common in United States, particularly in May. From Latin *tornare*, 'to turn'.
TRADE	Prevailing wind blowing over the ocean within 30 degrees of equator. Northeasterly in north hemisphere, southeasterly in the south.
TRAMONTANA	Cold, strong Alpine wind. From south during afternoon on Lake Maggiore in Italy.
TRAUBEN-KOCHER	'The Grape Cooker'. Swiss name for the FÖHN.
TRAVERSE	Rain-bearing westerly in the Alps.
TROMBES GIRATOIRE	Whirlwind in France.
TSUJI	'Wind of the Crossroads' in Japan.
TSUMUJIKAZE	Whirlwind in Japan. Meaning 'whorl in the hair'.
TUAURA	Southerly wind in Tikopia, Melanesia.
TUNG SHANG	Northeast TRADE in China.
TY FUNG	'Great Wind' in China.
TYPHOON	Circular tropical storm in Pacific. HURRICANE.
UBERRE	Warm FALL wind over Neuchatel.
VALLESARIA	'Wind of Valais'. FÖHN type east of Lake Geneva.
VANOISE	Cold, northerly mountain wind off Vanoise massif.
VARDARAC	Strong, cold, dry, winter northwester in Yugoslavia.
VAUDAIRE	FÖHN on Lake Geneva.
VEERING	Wind that makes clockwise change in direction.
VENDAVALES	Hot, gentle, southwesterly, followed by heavy rain. In Morocco and Spain from September to March.
VENTANIA	Gale in Portugal.
VENT D'ESPAGNE	FÖHN from Pyrenees into southern France.
VENTO COADO	Playful breeze that whistles through the crannies of hillside homes in Portugal.

343

VENTO DE BAIXO	Sea breeze in Portugal.
VÉSINE	Updraft in Drôme Valley of Swiss Alps.
VIND-BLAER	Breeze in Icelandic sagas.
VINDS-GNYR	Blustery downdraft in ancient Ireland.
VIOLENT STORM	See FORCE ELEVEN.
VIRAZON	Persistent sea breeze at Valparaiso in Chile. And at Cadiz in Spain.
VIUGA	Cold, stormy northeaster in southern Siberia.
VOLTURNUS	Southeasterly in ancient Rome.
VYETEROK	Gentle breeze in Soviet Union.
WABUN	Easterly, 'The Morning Bringer' of the Algonquin.
WADDY	Puff of wind in Cornwall.
WAFF	Gentle breeze in Scotland.
WAIMEA	Misty, sea breeze in Hawaii.
WALTZING JINN	Name for DUST DEVIL in old Nubia.
WARM BRAW	Dry, warm southwesterly FALL wind in New Guinea.
WARYARAIK	Northerly in Gran Chaco of Argentina.
WATAKUSHI	The 'Private' or 'Me' wind of Japan.
WHIRLY	Small, violent storm in Antarctic.
WHITTLE	Any strong gust in Cheshire. Named after Captain Whittle of the 16th century, whose coffin was hurled to the ground just as his bearers approached the church.
WILLIWAW	Brief, local downdraft or bluster in both Alaska and the Straits of Magellan.
WILLY-WILLY	Circular tropical storm in the Timor Sea.
WISPER	Cool, evening, bluster of great strength in narrow parts of the Rhine Valley.
WITCH	Another name for SANTA ANA in California.
XLOKK	Hot, dry wind in Malta.
YAMA OROSHI	Fall wind in Japan.
YAMO	Whirlwind in Uganda. From Lango, 'Wind in the Body'.
YUH	Wet gale in Shanghai.
ZEPHYR	Mild breeze bringing pleasant warm weather in Italy.
ZEPHYRUS	Westerly in ancient Greece, warm and pleasant. Portrayed as a young man bearing flowers.
ZONDA	Hot, dry westerly FALL wind. From Argentinian Andes.

BIBLIOGRAPHY

All of the works listed here are referred to in the text by their appropriate numbers. Authors may be located in the text by reference to the Index.

This is a comprehensive list, designed to make it possible to refer to the original sources quoted, and to find further reading matter on each of the many subjects covered.

1 ABDULLAH, A. J. 'Some aspects of the dynamics of tornadoes', *Monthly Weather Review* 83: 83–94.
2 ADAMS, J. B. et al. 'The aerial aphid plankton over the research station . . .' *Canadian Entomologist* 108: 1069–78, 1976.
3 ALERSTAM, T. 'Wind as selective agent in bird migration', *Ornis Scandanavica* 10: 76–93, 1979.
4 ALLEN, A. *The Story of Clothes*. Faber & Faber: London, 1967.
5 ANESAKI, M. *Japanese Mythology*. Marshall Jones: Boston, 1916.
6 ANON. *Yorkshire Observer*: 25th February, 1911.
7 ARAKAWA, H. 'Twelve centuries of blooming dates of the cherry blossoms . . .' *Geofisica Pura e Applicata* 30: 36–50, 1955.
8 ARATA, G. F. 'A note on the flying behaviour of certain squids', *The Nautilus* 68: 1–3, 1954.
9 ARNOLD, Matthew. '*Empedocles on Etna*'. Macmillan: London, 1852.
10 ARRHENIUS, S.A. *Worlds in the Making*. Harper: New York, 1908.

11   ASSAEL, M. et al. 'Influence of artificial air ionisation on the human electroencephalogram', *International Journal of Biometeorology* 18: 306–12, 1974.

12   ASTRUC, J. *De Morbis Veneris Libri Sex*. Paris, 1684.

13   BADT, K. *John Constable's Clouds*. Routledge & Kegan Paul: London, 1950.

14   BAGLEY, D. *Wyatt's Hurricane*. Collins: London, 1966.

15   BAGNOLD, R. A. *The Physics of Blown Sand and Desert Dunes*. Methuen: London, 1951.

16   BAJKOV, A.D. 'Do fish fall from the sky?' *Science* 109: 402, 1949.

17   BAKER, R. R. *Human Navigation*. Hodder & Stoughton: London, 1981.

18   BARNES, H. (ed.) *Some Contemporary Studies in Marine Science*. Allen & Unwin: London, 1966.

19   BASU, S. (ed.) *Proceedings of the Symposium on Monsoons of the World*. Hind Union Press: New Delhi, 1958.

20   BATEMAN, A. J. 'Is gene dispersal normal?' *Heredity* 4: 353–63, 1950.

21   BATHURST, G. B. 'The earliest recorded tornado'. *Weather* 19: 202–4, 1964.

22   BAULE, G. M. & McFEE, R. 'Detection of the magnetic field of the heart'. *American Heart Journal* 66: 95, 1963.

23   BEAGLEHOLE, J. C. (ed.) *The Journals of Captain James Cook*. University Press: Cambridge, 1955.

24   BEAGLEHOLE, E. & BEAGLEHOLE, P. 'Ethnology of Pukapuka', *Bulletin of the Bishop Museum* No. 150, 1938.

25   BECKER, R. O. & MARINO, A. A. *Electromagnetism and Life*. New York State University: Albany, 1982.

26   BECKWITH, M. *Hawaiian Mythology*. University of Hawaii Press: Honolulu, 1970.

27   BEEBE, W. 'Insect migration at Rancho Grande in north central Venezuela', *Zoologica* 34: 107–10, 1949.

28   BEEDELL, S. *Windmills*. Scribners: New York, 1975.

29   BELLROSE, F. C. 'Radar in orientation research', *Proceedings of the XIV International Ornithological Congress*: 281–309, 1967.

30   BELLWOOD, P. S. *Man's Conquest of the Pacific*. Oxford University Press: New York, 1979.

31   BENVENISTE, P. E. & TODARO, G. J. 'Evolution of C-type viral genes', *Nature* 252: 456–8, 1974.

32   BERESFORD, J. (ed.) *The Diary of a Country Parson*. Oxford University Press: London, 1949.

33   BERLAND, L. 'Recherches en avion sur la faune de l'atmosphère', *Nature* (Paris) 62: 341–5, 1934.

34   BERNARD, H. W. *The Greenhouse Effect*. Ballinger: Cambridge, Massachusetts, 1980.

35   BEVERIDGE, W. I. B. *Influenza*. Heinemann: London, 1977.

36 BIGELOW, F. H. 'The waterspout seen off Cottage City . . .'
    Monthly Weather Review 34: 307–15, 1906.
37 BLEEK, W. H. I. & LLOYD, L. C. *Specimens of Bushman
    Folklore.* Allen Lane: London, 1911.
38 BLOOMFIELD, M. *Hymns of the Atharva-Veda.* Clarendon:
    Oxford, 1897.
39 BLUMENSTOCK, D. I. *The Ocean of Air.* Rutgers University
    Press: New Brunswick, 1959.
40 BOAS, F. 'The central Eskimo', *Annual Report of the Bureau of
    Ethnology* 6: 593, 1888.
41 BOAS, F. 'Reports on the north western tribes', *Report of the
    British Association for the Advancement of Science*, 1895.
42 BOOKER, C. A. 'Tower damage provides key to Worcester
    tornado data', *Electrical World* 170: 22–4, 1953.
43 BOSHIER, A. & BEAUMONT, P. B. 'Beyond the mists of
    mining', *Nuclear Antiquity* 2: 21–6, 1974.
44 BOUCHER, K. *Global Climate.* English University Press:
    London, 1975.
45 BOURDILLON, R. B. & LIDWELL, O. M. 'Sneezing and the
    spread of infection', *The Lancet* 1946: 365, 1941.
46 BOWDEN, J. et al. 'Possible wind transport of coffee leaf rust
    across the Atlantic', *Nature* 229: 500–1, 1971.
47 BOWEN, E. G. 'Lunar and planetary tails in the solar wind',
    Journal of Geophysical Research 69: 4969, 1964.
48 BOYCO, A. D. et al. 'Aeroionization in industrial hygiene', *All
    Union Conference in Leningrad*, November 1963.
49 BOYLE, T. E. *Symbol and Meaning in the Fiction of Joseph
    Conrad*, Mouton: London, 1965.
50 BRADSHAW, A. D. & McNEILLY, T. *Evolution and Pollution.*
    Edward Arnold: London, 1981.
51 BRAMWELL, C. D. 'Aerodynamics of *Pteranodon*', *Journal of
    the Linnean Society* 3: 313–28, 1971.
52 BRIDGES, ROBERT. *The South Wind.*
53 BRIMBLECOMBE, P. & OGDEN, C. 'Air pollution in art and
    literature', *Weather* 32: 285–91, 1977.
54 BRONTË, Emily. *The Night Wind*, 1846.
55 BROOKS, C. E. P. *Climate Through the Ages.* Benn: London,
    1949.
56 BROWN, S. *World of the Wind.* Alvin Redman: London, 1962.
57 BROWNING, I. In ROSEN (Ref 408).
58 BRUCH, C. W. 'Microbes in the upper atmosphere and beyond',
    in GREGORY & MONTEITH (Ref 177).
59 BRYDONE, P. *A Tour Through Sicily and Malta.* Strahan &
    Cadell: London, 1776.
60 BRYSON, R. A. & PADOCH, C. 'On the climates of history', in
    ROTBERG & RABB (Ref 411).

61 BULLER, A. H. R. *Researches on Fungi.* Longmans Green: New York, 1909.

62 BURKE, J. *Musical Landscapes.* Webb & Bower: London, 1983.

63 BURROUGHS, J. *'The Heart of Burroughs Journals'*, 1920.

64 BURROWS, W. 'Some notes and legends on a south sea island', *Journal of the Polynesian Society* 32: 143–73, 1923.

65 BYRON, Lord. *'Childe Harold'*, 1812.

66 CAIRNS-SMITH, A. *The Life Puzzle.* Oliver & Boyd: Edinburgh, 1971.

67 CAMPBELL, J. G. *Superstitions of the Highlands and Islands of Scotland.* Glasgow, 1900.

68 CARSON, S. L. 'Human energy under varying weather conditions', Doctoral dissertation: University of Washington, 1947.

69 CHANDLER, Raymond. *The Midnight Raymond Chandler.* Houghton Mifflin: Boston, 1971.

70 CHENEY, M. 'G. 'The decretal of Pope Celestine III on tithes of windmills', *Bulletin of Medieval Canon Law* 1: 63–6, 1971.

71 CHEREMISINOFF, N. P. *Fundamentals of Wind Energy.* Ann Arbor Science: Michigan, 1978.

72 CLAIBORNE, R. *Climate, Man and History.* Norton: New York, 1970.

73 CLARIDGE, J. *Shepherd of Banbury's Rules to Judge the Changes of the Weather.* London, 1744.

74 CLARKE, W. E. *'Vanessa cardui* and other insects at the Kentish Knock lightship', *Entomologists Monthly Magazine* 39: 289–90, 1903.

75 COHEN, D. 'Magnetoencephalography', *Science* 175: 664, 1972.

76 COLLINS, C. W. 'Dispersion of gipsy moth larvae by the wind', *Bureau of Entomology Bulletin* 273: 22, 1915.

77 COMSTOCK, J. H. *The Spider Book*, Comstock: Ithaca, 1948.

78 CONE, C. D. 'Control of cell division by the electrical voltage of the surface membrane', *Proceedings of the Twelfth Annual Session of the American Cancer Society*, 1970.

79 CONRAD, Joseph. *Youth.* McClure Phillips: New York, 1903.

80 CONRAD, Joseph. *The Nigger of the Narcissus.* Dent: London, 1897.

81 CONRAD, Joseph. *Typhoon.* Dent: London, 1903.

82 CONRAD, Joseph. *The Mirror of the Sea.* Doubleday: New York, 1916.

83 COON, C. S. *Climate and Race.* Harvard University Press: Cambridge, 1953.

84 CORRINGTON, R. H. *The Melanesians.* Clarendon: Oxford, 1891.

85 COURT, A. 'Windchill', *Bulletin of the American Meteorological Society* 29: 487–93, 1948.

86 COUSTEAU, J-Y. & COUSTEAU, P. *The Shark*. Doubleday: New York, 1970.

87 CROSBY, C. R. & BISHOP, S. C. 'Aeronautic spiders with a description of a new species', *Journal of the New York Entomological Society* 44: 43–9, 1936.

88 CRUTCHER, H. L. 'Winds, numbers and Beaufort', *Weatherwise* 28: 260–71, 1975.

89 CUMMING, C. F. G. *In the Hebrides*. London, 1883.

90 DANA, Richard Henry. *Two years Before The Mast*. New York, 1840.

91 DANON, A. & SULMAN, F. G. 'Ionizing effects of winds of ill repute on serotonin metabolism', In TROMP & WIEHE (Ref 492).

92 DARLING, C. A. & SIPLE, P. A. 'Bacteria of the Antarctic', *Journal of Bacteriology* 42: 83–98, 1941.

93 DARLINGTON, P. J. 'The origin of the fauna of the Greater Antilles', *Quarterly Review of Biology* 13: 274–300, 1938.

94 DARWIN, C. 'An account of the fine dust which often falls on vessels in the Atlantic Ocean', *Quarterly Journal of the Geological Society of London* 2: 26–30, 1846.

95 DARWIN, C. *The Voyage of the Beagle*. Dent: London, 1959.

96 DAVENPORT, W. H. 'Marshallese folklore types', *Journal of American Folklore* 66: 219–37, 1953.

97 DAVIS, R. B. & WEBB, T. 'The contemporary distribution of pollen in eastern North America', *Quaternary Research* 5: 395–434, 1975.

98 DAVIS, R. B. et al. 'Mexican free-tailed bats in Texas', *Ecological Monograph* 32: 311–46, 1962.

99 DAWSON, G. M. 'On the Haida Indians . . .' *Geological Survey of Canada*: 124, 1879.

100 DE LA RUE, A. *Man and the Winds*. Hutchinson: London, 1955.

101 DELISLE, L. 'On the origin of windmills in Normandy and England', *Journal of the British Archaeological Association* 6: 403–6, 1851.

102 DESSENS, J. 'Man-made tornadoes'. *Nature* 193: 13–14, 1962.

103 DICKSON, R. C. 'Aphis dispersal over southern California deserts', *Annals of the Entomological Society of America* 52: 368–72, 1959.

104 DIMMICK, R. L. 'Delayed recovery of airborne *Serratia marcescens* . . .' *Nature* 187: 251–2, 1960.

105 DIXON, R. B. *Oceanic Mythology*. Marshall Jones: Boston, 1916.

106 DOUGHERTY, E. C. (ed.) *The Lower Metazoe*. University of California Press: Berkeley, 1963.

107 DOUGLAS, G. N. *South Wind*. John Lane: London, 1917.

108 DRABBLE, M. *A Writer's Britain*. Knopf: New York, 1979.

109 DRAPER, L. 'Freak ocean waves', *Weather* 21: 2–4, 1966.

110 DUGUID, J. P. 'The numbers and sites of origins of the droplets

expelled during expiratory activities', *Edinburgh Medical Journal* 52: 385, 1945.

111 DUNBAR, B. H. F. *Symbolism in Medieval Thought.* Yale University Press: New Haven, 1929.

112 DUNN, G. E. & MILLER, B. I. *Atlantic Hurricanes.* Louisiana State University Press: Baton Rouge, 1960.

113 DURAND, A. L. 'Landbirds over the north Atlantic', *British Birds* 65: 428–42, 1972.

114 DUTTON, J. A. *The Ceaseless Wind.* McGraw Hill: New York, 1976.

115 EATON, G. P. 'Volcanic ash deposits as a guide to atmospheric circulation in the geological past', *Journal of Geophysical Research* 68: 521–8, 1963.

116 EDWARDS, C. '*Velella velella*: the distribution of its dimorphic forms', in BARNES (Ref 18).

117 EDWARDS, J. A. 'Arthropod fallout on Alaskan snow', *Arctic and Alpine Research* 4: 167–76, 1972.

118 EDWARDS, M. B. *Through Spain to the Sahara.* Hurst and Blackell: London, 1868.

119 EGEDE, H. *A Description of Greenland.* T. & J. Allman: London, 1818.

120 EICHENWALD, H. F. et al. 'The cloud-baby', *American Journal of Diseases of Children* 100: 161, 1960.

121 EISELEY, Loren. *The Immense Journey.* Knopf: New York, 1957.

122 EISELEY, Loren. *The Unexpected Universe.* Harcourt, Brace Jovanovich: New York, 1964.

123 ELDRIDGE, F. R. *Wind Machines.* Van Nostrand: New York, 1980.

124 ELTON, C. S. 'The dispersal of insects to Spitzbergen', *Transactions of the Entomological Society of London*: 289–99, 1925.

125 ELWIN, V. *Myths of Middle India.* Oxford University Press: London, 1949.

126 ELWIN, V. *Tribal Myths of Orissa.* Oxford University Press: London, 1954.

127 ENDICOTT, K. *Batek Negrito Religion.* Clarendon Press: Oxford, 1979.

128 EVANS, B. *Dictionary of Mythology.* Centennial: Lincoln, Nebraska, 1970.

129 EVANS, I. H. N. *The Religion of the Tempasuk Dusuus of North Borneo.* University Press: Cambridge, 1953.

130 FABRE, J. H. C. *The Life of the Spider.* Dodd, Mead: New York, 1919.

131 FELT, E. P. 'Dispersal of insects by air currents', *Bulletin of the New York State Museum* No. 274: 59–129, 1928.

132 FERREL, W. *A Popular Treatise on the Winds . . .* Macmillan: London, 1889.

133 FICKER, H. V. & DE RUDDER, B. *Föhn und Föhnwirkunger.* Leipzig, 1948.

134 FIELDS, W. G. & MACKIE, G. O. 'Evolution of the Chondrophora', *Journal of the Fisheries Research Board of Canada* 28: 1595–1602, 1971.

135 FIRTH, R. *We, the Tikopia.* Allen & Unwin: London, 1936.

136 FISHER, F. R. (ed.) *Man Living in the Arctic.* National Academy of Sciences: Washington, DC., 1961.

137 FLAVIN, C. 'Harnessing the wind', *Weatherwise* 35: 211–17, 1982.

138 FLETTNER, A. *The Story of the Rotor.* Willhoft: New York, 1926.

139 FLORA, S. D. *Tornadoes of the United States.* University of Oklahoma Press: Norman, 1954.

140 FLOWER, D. (ed.) *Voltaire's England.* Folio Society: London, 1950.

141 FORT, C. *The Book of the Damned.* Boni & Liveright: New York, 1919.

142 FORT, C. *New Lands.* Boris & Liveright: New York, 1923.

143 FORT, C. *Lo!* Kendall: New York, 1931.

144 FORT, C. *Wild Talents.* Kendall: New York, 1932.

145 FRANK, N. L. & HUSAIN, S. A. 'The deadliest tropical cyclone in history?' *Bulletin of the American Meteorological Society* 52: 438–44, 1971.

146 FRAZER, J. G. *The Golden Bough.* Macmillan: London, 1920.

147 FRAZER, J. G. *The Worship of Nature.* Macmillan: London, 1926.

148 FREUD, S. *Moses and Monotheism.* Knopf: New York, 1939.

149 FUJITA, T. T. 'Proposed characterization of tornadoes and hurricanes by area and intensity', *Satellite and Mesometeorology Research Project*: University of Chicago, 1971.

150 FULTON, J. D. 'Micro-organisms of the upper atmosphere', *Applied Microbiology* 14: 237–50, 1966.

151 FURNEAUX, R. *Krakatoa.* Secker & Warburg: London, 1965.

152 GASTER, T. H. *Myth, Legend and Custom in the Old Testament.* Duckworth: London, 1969.

153 GEIGER, R. *The Climate Near the Ground.* Harvard University Press: Cambridge, 1950.

154 GENTRY, R. C. 'Project Stormfury', *Bulletin of the American Meteorological Society* 50: 404–9, 1969.

155 GENTRY, R. C. 'Hurricane Debbie modification experiments', *Science* 168: 473–5, 1970.

156 GIFFORD, E. W. 'Western Mono myths', *Journal of American Folklore* 36: 301–67, 1923.

157 GILL, W. W. *Myths and Songs of the South Pacific.* King: London, 1876.

158 GISLEN, T. 'Aerial plankton and its conditions of life', *Biological Reviews of the Cambridge Philosophical Society* 23: 109–27, 1947.

159   GIVONI, B. *Man, Climate and Architecture*. Applied Science: London, 1976.

160   GLADWIN, T. *East is a Big Bird*. Harvard University Press: Cambridge, 1970.

161   GLICK, P. A. 'The distribution of insects, spiders, and mites in the air', *US Department of Agriculture Technical Bulletin* No. 673, 1939.

162   GLICK, P. A. 'Collecting insects by airplane in southern Texas', *US Department of Agriculture Technical Bulletin* No. 1158, 1957.

163   GLICK, P. A. 'Collecting insects by airplane . . .' *US Department of Agriculture Technical Bulletin* No. 1222, 1960.

164   GOLDING, E. W. *The Generation of Electricity by Wind Power*, Spon: London, 1955.

165   GOMME, G. L. (ed.) *Popular Superstitions*. Elliot Stock: London, 1884.

166   GÖTZ, F. W. P. 'Staubfälle in Arosa in Spätwinter 1936', *Meteorologische Zeitung* 6: 227, 1936.

167   GRACE, J. 'Some effects of wind on plants', in GRACE et al (Ref 168).

168   GRACE, J. et al. *Plants and Their Atmospheric Environment*. Blackwell: Oxford, 1981.

169   GRAHAM, D. C. 'Songs and stories of the Ch'uan Miao', *Smithsonian Miscellaneous Collection* 123: 1–336, 1954.

170   GRAY, J. E. 'The shower of fishes', *Zoologist* 17: 6540–1, 1859.

171   GRAY, Thomas. *Ode on a Distant Prospect of Eton College*, 1742.

172   GREELEY, R. et al (eds.) *Aeolian Features of Southern California*. Arizona State University: Tempe, 1978.

173   GREENHILL, B. *The Life and Death of the Merchant Sailing Ship*. HMSO: London, 1980.

174   GREGORY, P. H. 'The operation of the puffball mechanism . . .' *Transactions of the British Mycological Society* 32: 11–15, 1949.

175   GREGORY, P. H. *The Microbiology of the Atmosphere*. Leonard Hill: London, 1973.

176   GREGORY, P. H. & HIRST, J. M. 'The summer air-spora at Rothamstead in 1952'. *Journal of General Microbiology* 17: 135–52, 1957.

177   GREGORY, P. H. & MONTEITH, J. L. (eds.) *Airborne Microbes*. University Press: Cambridge, 1967.

178   GRESSIT, J. L. 'Problems in the zoogeography of Pacific and Antarctic insects', *Pacific Insects Monograph* 2: 1–94, 1961.

179   GRESSIT, J. L. & LEECH, L. E. 'Insect habitats in Antarctica', *Polar Record* 10: 501–5, 1961.

180   GRIAULE, M. *Conversations with Ogotemmêli*. Oxford University Press: London, 1970.

181   GRIBBIN, J. R. *Future Weather*. Delacorte: New York, 1982.

182 GRIBBIN, J. R. 'El Chichon and Britain's weather', *New Scientist* 88–9, April 14th, 1983.

183 GRIFFITH, J. 'The shower of fishes', *Zoologist* 17: 6493, 1859.

184 GUALTIEROTTI. et al. *Aerionotherapy*. Carlo Erba Foundation: Milan, 1968.

185 GUDGER, E. W. 'More rains of fishes', *Annals and Magazine of Natural History* 3: 1–26, 1929.

186 GUDGER, E. W. 'Rains of fishes – myth or fact?' *Science* 103: 693–4, 1929.

187 GURNEY, A. B. 'Grasshopper glacier of Montana . . .' *Reports of the Smithsonian Institution*: 305–26, 1952.

188 HADDON, A. C. 'Head-hunters', *Report on the Cambridge Anthropological Expedition to the Torres Straits* 6:201, 1908.

189 HADFIELD, A. M. *Time to Finish the Game*. Phoenix: London, 1964.

190 HAFSTEN, U. 'A pollen analytic investigation of two peat deposits from Tristan da Cunha', *Reports of the Norwegian Expedition of 1937–1938* 22: 1–42, 1951.

191 HAGEN, B. *Unter der Papuas*. Wiesbaden, 1899.

192 HAMMEL, H. T. 'The cold climate man', in FISHER (Ref 136).

193 HANKIN, E. H. *Animal Flight*. Iliffe: London, 1913.

194 HARDING, A. F. (ed.) *Climatic Change in Later Prehistory*. University Press: Edinburgh, 1982.

195 HARDY, A. C. & MILNE, P. S. 'Studies in the distribution of insects by aerial currents', *Journal of Animal Ecology* 7: 199–229, 1938.

196 HARLFINGER, O. & JENDRITZKY, G. In SOYKA & EDMONDS (Ref 456).

197 HARRIOTT, J. *Struggles Through Life*. Longman: London, 1815.

198 HARRIS, B. 'Volcanic particles in the stratosphere', *Australian Journal of Physics* 17: 472–9, 1964.

199 HARTLAND, E. S. *Primitive Paternity*. Harrap: London, 1909.

200 HEDIN, S. A. *Across the Gobi Desert*. Greenwood: New York, 1931.

201 HEININGER, S. K. *A Handbook of Renaissance Meteorology*. Duke University Press: Durham, 1960.

202 HEMENWAY, C. L. & SOBERMAN, R. K. 'Studies of micro-meteorites obtained from a recoverable sounding rocket', *Astronomical Journal* 67; 256–66, 1962.

203 HENRY, W. A. 'Bijdrage tot de kennis der Bataklanden', *Tijdskrif voor Indische Taal, Land en Volkenkunde* 17: 23, 1878.

204 HERBERT, G. *Jacula Prudentum*. London, 1633.

205 HERRING, J. L. 'Evidence for hurricane transport and dispersal . . .' *Pan-Pacific Entomologist* 34: 174–5, 1958.

206 HEYERDAHL, T. *The Kontiki Expedition*. Rand McNally: Chicago, 1950.

207 HEYERDAHL, T. *Early Man and the Ocean*. Doubleday: New York, 1979.

208 HIDE, R. 'Motions in planetary atmospheres', *Meteorological Magazine* 100: 268–76, 1971.

209 HIPPOCRATES. In LLOYD (Ref 288).

210 HIPPOCRATES. *Airs, Waters, Places*. Heineman: London, 1923 (412 BC).

211 HOLFORD, I. (ed.) *The Guinness Book of Weather Facts and Feats*. Guinness Superlatives: London, 1977.

212 HOLZAPFEL, E. P. & HARRELL, J. C. 'Transoceanic dispersal studies of insects', *Pacific Insects* 10: 115–53, 1968.

213 HOOKE, S. H. 'Fish Symbolism', *Folklore* 72: 535–8, 1961.

214 HOTCHIN, J. et al. 'Survival of micro-organisms in space', *Nature* 206: 442–5, 1965.

215 HOUSE, D. C. 'Forecasting tornadoes and severe thunderstorms', *Meteorological Monographs* 5: 141–55, 1963.

216 HOWARD, L. 'Essay on the modifications of clouds', *Tillochs Philosophical Magazine* 16 and 17, 1803.

217 HOWARD, L. *The Climate of London*. London, 1818.

218 HOWARD, L. O. 'On gossamer spiders web', *Proceedings of the Entomological Society of Washington* 3: 191–2, 1893.

219 HOWE, G. M. & LORAINE, J. A. (eds.) *Environmental Medicine*. Heineman: London, 1973.

220 HOWITT, A. W. 'On some Australian ceremonies', *Journal of the Anthropological Institute* 13: 442, 1884.

221 HOYLE, F. & WICKRAMASINGHE, N. C. 'Influenza from space', *New Scientist* 79: 946–8, 1978.

222 HOYLE, F. & WICKRAMASINGHE, N. C. *Life Cloud*. Dent: London, 1978.

223 HOYLE, F. & WICKRAMASINGHE, N. C. *Diseases from Space*. Dent: London, 1979.

224 HUDSON, W. H. *A Hind in Richmond Park*. Dent: London, 1923.

225 HUMPHREYS, W. J. *Weather Rambles*. Bailliere: London, 1937.

226 HUNTINGFORD, G. W. B. (ed.) *The Periplus of the Erythraean Sea*. Hakluyt Society: London, 1980.

227 HUNTINGTON, E. *Civilization and Climate*. Yale University Press: New Haven, 1915.

228 HUNTINGTON, E. *Climate Change*. Yale University Press: New Haven, 1922.

229 HUNTINGTON, E. *Main Springs of Civilization*. Harper & Row: New York, 1945.

230 HURD, W. E. 'Influence of the wind on the movements of insects', *Monthly Weather Review* for February: 94–7, 1920.

231 HURD, W. E. 'Waterspouts', *Monthly Weather Review* 56: 207–11, 1928.

232 IMSHENETSKY, A. A. et al. 'Upper boundary of the biosphere', *Applied and Environmental Microbiology* 35: 1–5, 1978.

233 INGER, R. F. 'Systematics and zoogeography of Philippine amphibia', *Fieldiana* 33: 181–531, 1954.

234 INGOLD, C. T. *Spore Discharge in Land Plants*. Clarendon Press: Oxford, 1939.

235 INGRAM, M. J. et al. 'Past climates and their impact on man', in WIGLEY et al. (Ref 842).

236 INOUE, E. 'Studies of the phenomena of waving plants caused by wind', *Journal of Agricultural Meteorology* (Tokyo), 11: 87–90, 1955.

237 INWARDS, R. *Weather Lore*. Stock: London, 1898.

238 IRVING, W. *Lives of the Successors of Mahomet*. Bohn: London, 1850.

239 IYENGAR, B. K. S. *Light on Pranayama*. Allen & Unwin: London, 1981.

240 JAMES, E. O. *The Worship of the Sky God*. Athlone Press: London, 1963.

241 JANKOWIAK, J. 'Effects of wind on man,' in LIGHT (Ref 453).

242 JENSEN, M. *Shelter Effect*. Danish Technical Press: Copenhagen, 1954.

243 JOHNSON, C. G. 'The changing numbers of *Aphis fabae* . . .' *Annals of Applied Biology* 37: 441–50, 1952.

244 JOHNSON, C. G. *Migration and Dispersal of Insects by Flight*. Methuen: London, 1969.

245 JUNG, C. G. *Man and His Symbols*. Aldus: London, 1964.

246 KALS, W. A. *The Riddle of the Winds*. Doubleday: New York, 1977.

247 KAPLAN, M. M. & WEBSTER, R. G. 'The epidemiology of influenza', *Scientific American*: 88–106, December 1977.

248 KEATS, John. *I Stood Tip-Toe Upon a Little Hill*, 1820.

249 KEITH, A. B. *Indian Mythology*. Marshall Jones: Boston, 1917.

250 KELLER, F. *The Amazon and Madeira Rivers*. London, 1874.

251 KENDREW, W. G. *The Climates of the Continents*. Oxford University Press: London, 1955.

252 KETCHEN, E. E. et al. 'The biological effects of magnetic fields on man', *Journal of the American Industrial Hygiene Association* 39: 1–11, 1978.

253 KING, J. W. 'Weather and the earth's magnetic field', *Nature* 247: 131, 1974.

254 KINGSLEY, Charles. *'Prose Idylls'*, 1873.

255 KINSMAN, B. 'Historical notes on the original Beaufort scale', *The Marine Observer* 39: 116–24, 1969.

256 KINSMAN, B. 'Who put the wind speeds in Admiral Beaufort's scale?' *Oceans* 2: 18–25, 1969.

257   KIRKPATRICK, T. W. 'Records of migrating insects . . .' *Bulletin of the Entomological Society of Egypt* 10: 224–56, 1927.

258   KOEPPL, G. W. *Putnam's Power from the Wind.* Van Nostrand: New York, 1982.

259   KÖHLER, C. *A History of Costume.* Harrap: London, 1928.

260   KRUEGER, A. P. 'Are air ions biologically significant?' *International Journal of Biometeorology* 16: 313–22, 1972.

261   KRUEGER, A. P. 'Are negative ions good for you?' *New Scientist* 58: 668–70, 1973.

262   KRUEGER, A. P. & REED, E. J. 'Effect of the air ion environment on influenza in the mouse', *International Journal of Biometeorology* 16: 25–48, 1972.

263   KRUEGER, A. P. & SMITH, R. F. 'Effects of air ions on the living mammalian trachea'. *Journal of General Physiology* 42: 69–82, 1958.

264   KRUEGER, A. P. et al. 'The action of air ions on bacteria', *Journal of General Physiology* 41: 359–81, 1957.

265   KRUEGER, A. P. et al. 'Studies on the effects of gaseous ions on plant growth', *Journal of General Physiology* 45: 879–95, 1962.

266   KRUEGER, A. P. et al. 'The effect of abnormally low concentration of air ions on the growth of *Hordeum vulgaris*', *International Journal of Biometeorology* 9: 201–9, 1965.

267   KUTSCHENREUTER, P. H. 'A study of the effect of weather on mortality', New York Academy of Sciences 22: 126–38, 1959.

268   LAMB, H. H. 'The southern westerlies: a preliminary survey', *Quarterly Journal of the Royal Meteorological Society* 85: 1–23, 1959.

269   LAMB, H. H. 'Britain's changing climate', *Geographical Journal* 133: 445–68, 1967.

270   LAMB, H. H. 'Volcanic dust in the atmosphere', *Philosophical Transactions of the Royal Society* (A) 266: 425–99, 1970.

271   LAMB, H. H. *Climate: Present, Past and Future.* Methuen: London, 1977.

272   LAMB, H. H. *Climate, History and the Modern World.* Methuen: London, 1982.

273   LAMB, H. H. 'Reconstruction of the course of postglacial climate over the world', in HARDING (Ref 194).

274   LAMB, H. H. & WOODROFFE, A. 'Atmospheric circulation during the last ice age', *Quaternary Research* 1: 29–58, 1970.

275   LANE, F. W. *The Elements Rage.* David & Charles: Newton Abbot, 1966.

276   LANG, A. *Custom and Myth.* Longmans: London, 1910.

277   LAOUST, E. *Mots et Choses Berbères.* Paris, 1920.

278   LAWRENCE, D. H. *Song of a Man Who Has Come Through,* 1920.

279 LE BLANC, J. *Man in the Cold*. Thomas: Springfield, Illinois, 1975.
280 LEE, A. *Weather Wisdom*. Doubleday: New York, 1976.
281 LESLIE, C. R. *Memoirs of the Life of John Constable*. Phaidon: London, 1951.
282 LEVISON, M. et al. *The Settlement of Polynesia*. University of Minnesota Press: Minneapolis, 1973.
283 LEWIS, D. *We, the Navigators*. The University Press of Hawaii: Honolulu, 1973.
284 LEWIS, H. E. et al. 'Aerodynamics of the human microenvironment', *The Lancet* 1273: 7, 1969.
285 LEWIS, T. 'The species, aerial density and sexual maturity of Thysanoptera . . .' *Annals of Applied Biology* 55: 219–25, 1965.
286 LICHT, S. *Medical Climatology*. Licht: New Haven, 1964.
287 LINDSAY, J. F. 'Sound-producing dune and beach sands', *Bulletin of the Geological Society of America* 87: 463–73, 1976.
288 LLOYD, G. E. R. (ed.) *Hippocratic Writings*. Pelican Classics: London, 1978.
289 LÖBSACK, T. *Earth's Envelope*. Collins: London, 1959.
290 LOCKHART, J. G. *Memoirs of the Life of Sir Walter Scott*. Edinburgh, 1839.
291 LONGFELLOW, Henry, *A Day of Sunshine*, 1842.
292 LORCA, Federico. *Tree, Tree*, 1921.
293 LOVELOCK, J. E. 'Gaia as seen through the atmosphere', *Atmospheric Environment* 6: 579–80, 1972.
294 LOVELOCK, J. E. *Gaia*. Oxford University Press: London, 1979.
295 LOVELOCK, J. E. & MARGULIS, L. 'Atmospheric homeostasis by and for the biosphere: the Gaia hypothesis', *Tellus* 26: 2–10, 1974.
296 LUOMALA, K. *Voices on the Wind*. Bishop Museum Press: Honolulu, 1955.
297 McCLELLAND, D. C. *The Achieving Society*. Free Press: New York, 1961.
298 MacDONALD, J. D. *Murder in the Wind*. Fawcett: New York, 1956.
299 MacDONALD, J. D. *Condominium*. Lippincott: New York, 1977.
300 McGOWAN, A. *The Century Before Steam*. HMSO: London, 1980.
301 McGOWAN, A. *Tiller and Whipstaff*. HMSO: London, 1981.
302 McGRAIL, S. *Rafts, Boats and Ships*. HMSO: London, 1981.
303 McGREGOR, E. A. 'Painted lady butterfly', *Insect Pest Survey Bulletin* 4: 70–1, 1924.
304 MacKENZIE, D. A. *Teutonic Myth and Legend*. Gresham: London, 1912.
305 MacKENZIE, D. A. *Myths of Pre Columbian America*. Gresham: London, 1926.

306 MacKENZIE, D. A. *The Migration of Symbols*. Kegan Paul: London, 1926.

307 MACKIE, G. O. 'Studies on *Physalia physalis*: behaviour', *Discovery Reports* 30: 369–408, 1960.

308 MACKIE, G. O. 'Locomotion, flotation and dispersal', in MUSCATINE & LENHOFF (Ref 354).

309 McLEAN, R. C. 'Microbiology of the air', *Nature* 152: 258–9, 1943.

310 MACLEISH, A. *'Cook County'*, 1928.

311 McNEILLY, T. 'Evolution in closely adjacent plant populations', *Heredity* 23: 99–108, 1968.

312 McPHERSON, J. M. *Primitive Beliefs in the Northeast of Scotland*. Longmans: London, 1929.

313 MAGUIRE, B. 'The passive dispersal of small aquatic organisms . . .' *Ecological Monographs* 33: 161–85, 1963.

314 MAITANI, T. 'An observational study of wind-induced waving in plants', *Boundary-Layer Meteorology* 16: 49–65, 1979.

315 MANI, M. S. *Ecology and Biogeography of High Altitude Insects*. Junk: Hague, 1968.

316 MANN, Thomas. *Death in Venice*. Knopf: New York, 1925.

317 MANSFIELD, K. *'The Journal of Katherine Mansfield'*, 1927.

318 MARGULIS, L. & LOVELOCK, J. E. 'Biological modulation of the earth's atmosphere', *Icarus* 21: 471, 1974.

319 MARINO, A. A. & BECKER, R. O. 'High voltage lines', *Environment* 20: 6, 1978.

320 MARKHAM, S. F. *Climate and the Energy of Nations*. Oxford University Press: London, 1942.

321 MARRIOTT, P. J. *Red Sky at Night, Shepherd's Delight?* Sheba: Oxford, 1981.

322 MARSH, G. P. *Man and Nature*. New York, 1864.

323 MASEFIELD, John. *The West Wind*, 1910.

324 MATTHEWS, W. 'The mountain chant', *Annual Report of the Bureau of Ethnology* 5: 435, 1887.

325 MAURY, M. F. *The Physical Geography of the Sea*. Sampson Low: London, 1857.

326 MAYNARD, N. G. 'Significance of air-borne algae', *Zeitschrift fur Allgemeine Mikrobiologie* 8: 225–6, 1968.

327 MAYR, E. *Systematics and the Origin of Species*. Columbia University Press: New York, 1949.

328 MEADEN, G. T. 'Tornadoes in Britain', *Journal of Meteorology* 1: 242–51, 1976.

329 MEIER, F. C. 'Collecting micro-organisms from winds above the Caribbean sea', *Phytopathology* 26: 102, 1936.

330 MELARAGNO, M. G. *Wind in Architectural and Environmental Design*. Van Nostrand: New York, 1982.

331 MELLANBY, K. *The Biology of Pollution.* Edward Arnold: London, 1980.

332 MENEN, A. *Room for Ourselves.* McGraw Hill: New York, 1960.

333 MERWIN, William. *Night Wind,* 1971.

334 MICHELL, J. & RICKARD, R. J. M. *Phenomena.* Thames & Hudson: London, 1977.

335 MICHELL, J. & RICKARD, R. J. M. *Living Wonders.* Thames & Hudson: London, 1982.

336 MILLER, S. L. 'A production of amino acids under possible primitive earth conditions', *Science* 117: 528–9, 1953.

337 MILLER, W. H. 'Santa Ana winds and crime', *The Professional Geographer* 20: 23–7, 1968.

338 MILTON, John. *Hymn on the Morning of Christ's Nativity,* 1667.

339 MIQUEL, P. *Les Organismes Vivants de l'Atmosphère.* Gauthier-Villars: Paris, 1883.

340 MOAR, N. T. 'Possible long-distance transport of pollen to New Zealand', *New Zealand Journal of Botany* 7: 424–6, 1969.

341 MOFFIT, John. 'To Look at Any Thing', 1962.

342 MONCKTON, W. 'Some recollections of New Guinea customs', *Journal of the Polynesia Society* 5: 186, 1896.

343 MOONEY, J. 'Sacred formulas of the Cherokees', *Annual Report of the Bureau of Ethnology* 7: 387, 1891.

344 MOORE, W. L. *The New Air World.* Little, Brown: Boston, 1922.

345 MORRISON, J. *Long Ships and Roundships.* HMSO: London, 1980.

346 MOSSOP, S. C. 'Volcanic dust collected at an altitude of 20 kilometres', *Nature* 203: 824–7, 1964.

347 MOULTON, S. (ed.) *Aerobiology.* American Association for the Advancement of Science Publication No. 17: Washington, 1942.

348 MUKHERJEE, R. N. 'Background to the discovery of the symbol for zero', *Indian Journal of the History of Science* 12 225–31, 1977.

349 MÜLLER, W. M. *Egyptian Mythology.* Marshall Jones: Boston, 1918.

350 MUNN, R. E. *Biometeorological Methods.* Academic Press: New York, 1970.

351 MURCHIE, G. *Song of the Sky.* Secker & Warburg: London, 1955.

352 MURDOCH, J. 'Ethnological results of the Point Barrow Expedition', *Annual Report of the Bureau of Ethnology* 9: 432, 1892.

353 MURPHY, J. J. 'Meteorological features and history of tornado at Norfolk, Virginia', *Bulletin of the American Meteorological Society* 16: 252–5, 1935.

354 MUSCATINE, L & LENHOFF, H. M. *Collenterate Biology.* Academic Press: New York.

355 NAKAJIMA, K. et al. 'Recent human influenza A viruses . . .' *Nature* 274: 334–9.

356 NEEDHAM, J. *Science and Civilization in China*. University Press: Cambridge, 1956.

357 NELSON, E. W. 'The Eskimo about Bering Strait', *Annual Report of the American Bureau of Ethnology* 18: 19–651, 1899.

358 NEUBERGER, H. 'Climate in art', *Weather* 25: 46–56, 1970.

359 NEUHAUSS, R. *Deutsch Neue Guinea*. Berlin, 1911.

360 NEVINS, R-G. et al. 'How to be comfortable . . .' *ASHRAE Journal* 16: 41–3, 1974.

361 NIELSEN, E. T. 'On the habits of the migratory butterfly *Ascia monuste*', *Biol. Meddr.* 23: 1–81, 1961.

362 NIETZSCHE, F. W. *Also Sprach Zarathustra*. Leipzig, 1902.

363 NISHIKI, S. 'On the aerial migration of spiders', *Acta Arachnologica* 20: 24–34, 1966.

364 NOYES, Alfred. *The Highwayman*, 1912.

365 NOYES, Alfred. *Forty Singing Seamen*, 1912.

366 OKE, T. R. *Boundary Layer Climates*. Methuen: London, 1978.

367 OLIVE, L. 'Spore discharge mechanism in Basidiomycetes', *Science* 146: 542–3, 1964.

368 OLIVER, J. E. *Climate and Man's Environment*. Wiley: New York, 1973.

369 PADY, S. M. & KAPICA, L. 'Fungi in air over the Atlantic Ocean', *Mycologia* 47: 34–50, 1955.

370 PALMER, B. *Body Weather*. Stackpole: Harrisburg, 1976.

371 PAPP, R. P. 'A nival ecosystem in California', *Arctic and Alpine Research* 10: 117–31, 1978.

372 PARK, J. *The Wind Power Book*. Cheshire: Palo Alto, 1981.

373 PARKER, B. C. 'Rain as a source of vitamin B 12', *Nature* 219: 617–18, 1968.

374 PARKIN, D. W. 'Trade winds through the glacial cycles', *Proceedings of the Royal Society of London* A 337: 73–100, 1974.

375 PARRY, W. E. *Narrative of an Attempt to Reach the North Pole* . . . John Murray: London, 1828.

376 PEACOCK, M. 'Wind and weather holes', *Folklore* 10: 249–50, 1899.

377 PEARCE, F. 'The menace of acid rain', *New Scientist*: 419–24, August 12th, 1982.

378 PELHAM, D. *Kites*. Penguin: Harmondsworth, 1976.

379 PERHAM, J. 'Sea Dyak religion', *Journal of the Straits Branch of the Royal Asiatic Society* 10: 241, 1882.

380 PERRY, T. *The Butcher's Boy*. Scribners: New York, 1982.

381 PETERSEN, W. F. *The Patient and the Weather*. Edwards: Ann Arbor, 1934.

382 PETERSEN, W. F. *Man-Weather-Sun*. Thomas: Springfield, Illinois, 1947.

383 PLEYTE, C. M. 'Ethnographische Beschrijving der Kei Eilanden'.

Tijdschrift van het Nederlandsh Aardrijkskundig Genootschap 10: 827, 1893.

384   POLO, M. *The Description of the World*. Routledge: London, 1938.

385   POLUNIN, N. 'Arctic aeropalynology', *Canadian Journal of Botany* 33: 401–14, 1955.

386   POPE, Alexander '*An Essay on Criticism*', 1711.

387   PORTNER, D. M. et al. 'Effect of ultra-high vacuum on the viability of micro-organisms', *Science* 134: 2047, 1961.

388   POTTER, G. L. et al. 'Possible climatic impact of tropical deforestation', *Nature* 258: 697–8, 1975.

389   POTTER, L. D. & ROWLEY, J. 'Pollen rain and vegetation . . .' *Botanical Gazette* 122: 1–25, 1960.

390   PROCTOR, B. E. & PARKER, B. W. 'Micro-organisms in the upper air', in MOULTON (Ref 347).

391   RAPPOPORT, A. S. *Superstitions of Sailors*. Stanley Paul: London, 1928.

392   READ, H. *The Meaning of Art*. Faber & Faber: 691, London, 1931.

393   REITER, R. 'Atmospheric electricity and natural radioactivity', in LICHT (Ref 286).

394   REYNOLDS, G. *Turner*. Thames & Hudson: London, 1969.

395   REYNOLDS, G. W. 'Venting as practical means of reducing damage from tornado low pressures', *Bulletin of the American Meteorological Society* 39: 14–20, 1958.

396   REYNOLDS, J. *Windmills and Watermills*. Praeger: New York, 1972.

397   RICHARDSON, J. *A Dictionary of Persian, Arabic and English*. London, 1829.

398   RICHARDSON, W. J. 'Timing and amount of bird migration in relation to weather: a review', *Oikos* 30: 224–72, 1978.

399   RIDLEY, H. N. *The Dispersal of Plants Throughout the World*. Reeve: Ashford, Kent, 1930.

400   RIEHL, H. *Tropical Meteorology*. McGraw Hill: New York, 1954.

401   ROBINSON, Edwin. *New England*, 1920.

402   ROBINSON, N. & DIRNFELD, F. S. 'The ionization state of the atmosphere . . .' *International Journal of Biometeorology* 4: 101–50, 1963.

403   ROBOCK, A. 'The Little Ice Age', *Science* 206: 1402, 1979.

404   ROGERS, C. *Social Life in Scotland*. Edinburgh, 1884.

405   ROGOT, E. 'Association between coronary mortality and the weather', *Public Health Report* 89: 330, 1974.

406   ROSCOE, J. *The Baganda*. London, 1911.

407   ROSCOE, J. *The Bakitara*. University Press: Cambridge, 1923.

408   ROSEN, S. *Weathering*. Evans: New York, 1979.

409   ROSSETTI, Christina. *Sing-Song*, 1872.

410 ROSSMAN, F. O. 'Differences in the physical behaviour of tornadoes and waterspouts', *Weather* 13: 259–63, 1958.

411 ROTBERG, R. I. & RABB, T. K. *Climate and History*. University Press: Princeton, 1981.

412 RUDOFSKY, B. *Architecture Without Architects*. Doubleday: New York, 1964.

413 RUSKIN, John. *Modern Painters*. Dent: London, 1856.

414 RUSSELL, S. P. *Art in the World*. Rinehart: San Francisco, 1975.

415 SAAR, J. 'Japanese divers discover wreck of Mongol fleet', *Smithsonian* 12 (9): 118–28, 1981.

416 SAAR, J. 'Space-age wind tunnels . . .' *Smithsonian* 12 (10): 76–85, 1982.

417 SAGAN, C. 'Interstellar panspermia', *Symposium on Extraterrestrial Biochemistry and Biology*: Denver, 1961.

418 SALMON, J. T. & HORNER, N. V. 'Aerial dispersion of spiders in northcentral Texas', *Journal of Arachnology* 5: 153–7, 1977.

419 SALOMONSEN, F. 'The immigration and breeding of the fieldfare in Greenland', *Proceedings of the X International Ornithological Congress*: 515–26, 1951.

420 SANDBURG, Carl. *Wind Song*, 1926.

421 SASSOON, Siegfried. *South Wind*, 1971.

422 SAVONIUS, S. J. 'The S-rotor and its applications', *Mechanical Engineering* 53: 333–8, 1931.

423 SAVORY, T. H. *The Biology of Spiders*. Wiley: New York, 1928.

424 SCHISLER, L. C. et al. 'Transmission of a virus disease of mushrooms by infected spores', *Phytopathology* 53: 888, 1963.

425 SCHLICHTING, H. E. 'The importance of airborne algae and protozoa', *Journal of Air Pollution and Control* 19: 946–51, 1969.

426 SCHOLANDER, P. F. et al. 'Cold adaptation in Australian aborigines', *Journal of Applied Physiology* 13: 211, 1958.

427 SCORER, R. S. & LUDLAM, F. H. 'Bubble theory of penetrative convection', *Quarterly Journal of the Royal Meteorological Society* 79: 70, 1953.

428 SCOTT, Duncan. *Prairie Wind*, 1927.

429 SCOTT, Walter. *The Lady of the Lake*, 1810.

430 SEFERIS, George. *On Stage*, 1966.

431 SELBY, E. & SELBY, M. 'Beware the witch's wind'. *National Wildlife Magazine*: 34–5, August 1972.

432 SEZNEC, J. *The Survival of the Pagan Gods*. University Press: Princeton, 1972.

433 SHAKESPEARE, William. *Measure for Measure*.

434 SHAKESPEARE, William. *Midsummer Night's Dream*.

435 SHAND, A. 'The Moriri people of the Chatham Islands', *Journal of the Polynesian Society* 7: 73–88, 1898.

436 SHARP, E. L. 'Atmospheric ions and germination . . .' *Science* 156: 1359–60, 1967.

437 SHARP, R. P. & CAREY, D. L. *Bulletin of the Geological Society of America* 87: 1704–17, 1976.

438 SHAW, N. *Manual of Meteorology*. Cambridge University Press: London, 1926.

439 SHAW, N. *The Drama of Weather*. University Press: Cambridge, 1933.

440 SHELLEY, Percy. *Ode to the West Wind*, 1820.

441 SHEPARD, F. P. & YOUNG, R. 'Distinguishing between beach and sand dunes', *Journal of Sedimentary Petrology* 31: 196–214, 1961.

442 SIGURDSSON, H. 'Volcanic pollution and climate', *Eos*: 601–2, August 10th, 1982.

443 SILVERBERG, R. *The Challenge of Climate*. Meredith: New York, 1969.

444 SILVERMAN, G. J. et al. 'Exposure of micro-organisms to simulated extraterrestrial space ecology', *Life Sciences and Space Research* 2: 372, 1964.

445 SIMMONS, D. M. *Wind Power*. Noyes Data Corp: London, 1975.

446 SIMPSON, R. H. & MALKUS, J. S. 'Experiments in hurricane modification', *Scientific American* 211: 27–37, 1964.

447 SINCLAIR, J. *Statistical Account of Scotland*. Edinburgh, 1791–1799.

448 SIPLE, P. A. & PASSEL, C. F. 'Measurements of dry atmospheric cooling . . .' *Proceedings of the American Philosophical Society* 89: 117–99, 1945.

449 SLOCUM, J. *Sailing Alone Around the World*. Appleton Century: New York, 1935.

450 SMALLEY, I. J. (ed.) *Loess*. Dowden, Hutchinson & Ross: Stroudsburg, Pennsylvania, 1975.

451 SMITH, Alexander. *A Life Drama*, 1863.

452 SMITH, R. L. & HOLMES, D. W. 'Uses of doppler radar in meteorological observations', *Monthly Weather Review* 89: 1–7, 1961.

453 SMYTH, A. H. (ed.) *The Writings of Benjamin Franklin*. Haskell House: New York, 1970.

454 SOBERMAN, R. K. 'Noctilucent clouds', *Scientific American* 208: 51–9, 1963.

455 SORLEY, Charles *Barbary Lamp*, 1913.

456 SOYKA, F. & EDMONDS, A. *The Ion Effect*. Dutton: New York, 1977.

457 SPENSER, Edmund. *Prothalamion*, 1585.

458 STAIR, J. S. 'Jottings on the mythology and spirit lore of old Samoa', *Journal of the Polynesian Society* 5: 33–58, 1896.

459 STEIN, Gertrude. *An American and France*, 1936.

460 STEPHENS, James. *The Wind*, 1916.

461 STEVENSON, Robert Louis. *A Child's Garden of Verses*, 1885.

462 STEWART, G. *Storm*. Random House: New York, 1941.

463 STOMMEL, H. & STOMMEL, E. 'The year without a summer', *Scientific American* 240: 176–86, 1979.

464 STUART, J. & REVETT, N. *The Antiquities of Athens*. London, 1762.

465 STUMPER, R. & WILLEMS, A. 'La pluie de boue du 29 mars 1947', *Mathematical Archives* 17: 105–6, 1947.

466 SULMAN, F. G. *Health, Weater and Climate*. Karger: Basel, 1976.

467 SULMAN, F. G. *The Effect of Air Ionization, Electric Fields, Atmospherics and Other Electrical Phenomena on Man and Animal*. Thomas: Springfield, Illinois, 1980.

468 SULMAN, F. G. et al. 'Urinalysis of patients suffering from climatic heat stress', *International Journal of Biometeorology* 14: 45–53, 1970.

469 SULMAN, F. G. et al. 'Air ionometry of hot dry desert winds . . .' *International Journal of Biometeorology* 18: 313–18, 1974.

470 SUN, S. H. et al. 'Identification of some fungi from soil and air of Antarctica', *Terrestrial Biology*: Antarctic Research Series 30: 1–26, 1978.

471 SUTTON, O. G. 'Micrometeorology', *Scientific American* 211: 62–76, 1964.

472 SWAN, L. W. 'The ecology of the high Himalayas', *Scientific American* 205: 68–78, 1961.

473 SWAN, L. W. 'Aeolian zone', *Science* 140: 77–8, 1963.

474 SWAN, L. W. 'Alpine and aeolian regions of the world', in WRIGHT & OSBURN (Ref 535).

475 SYMONS, G. J. (ed.) 'The eruption of Krakatoa and subsequent phenomena', *Report of the Krakatoa Committee of the Royal Society*, 1888.

476 TANNEHILL, I. R. *Hurricanes*. Oxford University Press: London, 1956.

477 TCHIJEVSKY, A. L. *Air Ionization*. State Planning Commission Moscow, 1960.

478 TELEMANN, G. *The Compleat Harmony*, 1760.

479 THOM, E. C. 'The discomfort index', *Weatherwise* 12: 57–60, 1974.

480 THOMAS, Lewis. *The Lives of a Cell*. Viking Press: New York.

481 THOMPSON, James. *Castle of Indolence*, 1748.

482 THOMSON, W. A. R. *A Change of Air*. Scribners: New York, 1979.

483 THORNES, J. 'Constable's clouds', *Burlington Magazine* 141: 697–704, 1979.

484 TOLBERT, W. W. 'Aerial dispersal behaviour of two orb weaving spiders', *Psyche* 84: 13–27, 1977.

485 TOOMER, J. *Cane*, 1923.

486 TOTTON, A. K. 'Studies on *Physalia physalis*: natural history', *Discovery Reports* 30: 301–68, 1960.

487 TOYNBEE, A. *A Study of History*. Oxford University Press: London, 1934.

488 TREGEAR, R. T. 'Hair density, wind speed and heat loss in mammals', *Journal of Applied Physiology* 20: 796–801, 1965.

489 TROMP, S. W. *Medical Biometeorology*. Elsevier: Amsterdam, 1963.

490 TROMP, S. W. 'The relationship of weather and climate to health and disease', in HOWE & LORAINE (Ref 219).

491 TROMP, S. W. *Biometeorology*. Heyden: London, 1980.

492 TROMP, S. W. & WEIHE, W. H. 'Biometeorology', *Supplement to International Journal of Biometeorology* No. 13, 1969.

493 TWAIN, Mark. *New England Weather*, 1876.

494 TYLOR, E. B. *Primitive Culture*. John Murray: London, 1929.

495 VAN DER PIJL, L. *Dispersal in Higher Plants*. Springer: Berlin, 1982.

496 VAN TASSEL, E. L. 'The North Platte Valley tornado outbreak . . .' *Monthly Weather Review* 83: 255–64, 1955.

497 VEN, N. van de. *Construction Manual for a Cretan Windmill*. University of Technology: Eindhoven, Holland, 1977.

498 VILLIERS, A. *Wild Ocean*. McGraw Hill: New York, 1957.

499 VILLIERS, A. *Voyaging With the Wind*. HMSO: London, 1975.

500 VONNEGUT, B. 'Electrical theory of tornadoes', *Journal of Geographical Research* 65: 203–12, 1960.

501 VONNEGUT, B. 'Chicken plucking as measure of tornado wind speed', *Weatherwise* 28: 217, 1975.

502 WAILES, R. *The English Windmill*. Routledge & Kegan Paul: London, 1967.

503 WALOFF, Z. 'Orientation of flying locusts in migrating swarms', *Bulletin of Entomological Research* 62: 1–72, 1972.

504 WALTON, Izaak, *The Compleat Angler*. London, 1653.

505 WARD, M. *Mountain Medicine*. St. Albans: New York, 1975.

506 WARD, R. de C. 'The tornadoes of the United States as a climatic phenomenon', *Quarterly Journal of the Royal Meteorological Society* 43: 317–29, 1917.

507 WARD, R. de C. *The Climates of the United States*. Ginn: Boston, 1925.

508 WARMINGTON, E. H. *The Commerce Between the Roman Empire and India*. University Press: Cambridge, 1928.

509 WATANABE, K. 'Bora and man', in YOSHINO (Ref 536).

510 WATTS, A. *Wind and Sailing Boats*. Adlard Coles: London, 1965.

511 WAUGH, N. *The Cut of Women's Clothes*. Theatre Arts: New York, 1968.

512 WEBSTER, F. M. 'Winds and storms as agents in the diffusion of insects', *American Naturalist* 36: 799, 1902.

513 WEBSTER, P. J. 'Monsoons', *Scientific American*: 71–80, 1982.
514 WEINBERG, S. *The First Three Minutes*. Deutsch: London, 1977.
515 WERNER, E. T. C. *Myths and Legends of China*. Harrap: London, 1922.
516 WESTERMARCK, E. *Ritual and Belief in Morocco*. Macmillan: London, 1926.
517 WHEELER, R. H. 'The effects of climate on human behaviour in history', *Cycles*: 342–52, December 1962.
518 WHITE, J. *The Ancient History of the Maori*. Wellington: New Zealand, 1886.
519 WHITE, L. *Medieval Religion and Technology*. University of California Press: Berkeley, 1978.
520 WHITMAN, Walt. *I Sing the Body Electric*, 1881.
521 WIGLEY, T. M. L. et al (eds.) *Climate and History*. University Press: Cambridge, 1981.
522 WILKINS, E. T. 'Air pollution aspects of the London fog of December 1952', *Quarterly Journal of the Royal Meteorological Society* 80: 267–71, 1954.
523 WILLIAMS, C. B. et al. 'Studies in the migration of Lepidoptera', *Transactions of the Royal Entomological Society of London* 92: 101–283, 1942.
524 WILLIAMSON, K. 'Recent climatic influences on the status and distribution of some British birds', *Weather* 31: 362–84, 1976.
525 WILSON, A. T. 'Organic nitrogen in New Zealand snows', *Nature* 183: 318, 1959.
526 WILSON, A. T. 'Surface of the ocean as a source of airborne nitrogenous material . . .' *Nature* 184: 99–101, 1959.
527 WILSON, T. *The Swastika*. Annual Report of the United States National Museum, 1894.
528 WINKLER, Captain. 'On sea charts formerly used in the Marshall Islands', *Annual Report of the Smithsonian Institution*: 487–509, 1899.
529 WINSOR, T. & BECKETT, J. C. 'Biological effects of ionized air in man', *American Journal of Physiological Medicine* 37: 83–9, 1958.
530 WOLFORD, L. 'Tornado occurrences in the United States', *United States Weather Bureau Technical Paper* No. 20, 1960.
531 WOODCOCK, A. H. 'A theory of surface water motion deduced from the wind-induced motion of *Physalia*', *Journal of Marine Research* 5: 196–205, 1944.
532 WOODCOCK, A. H. 'Dimorphism in the Portuguese man-of-war', *Nature* 178: 253–5, 1956.
533 WOODCOCK, A. H. 'Note concerning *Physalia* behaviour at sea', *Limnology and Oceanography* 16: 551–2, 1971.
534 WORLD METEOROLOGICAL ASSOCIATION. *International Cloud Atlas*. Geneva, 1969.

535 WRIGHT, H. E. & OSBURN, W. H. *Arctic and Alpine Environments*. Indiana University Press: Indianapolis, 1967.
536 YOSHINO, M. M. (ed.) *Local Wind Bora*. University of Tokyo Press: Tokyo, 1976.
537 YOUNG, L. B. *Earth's Aura*. Knopf: New York, 1977.
536 ZIMMER, H. *Myths and Symbols in Indian Art and Civilization*. University Press: Princeton, 1972.
539 ZIMMERMAN, E. C. *Insects of Hawaii*. University of Hawaii Press: Honolulu, 1948.

CONVERSION FROM METRIC TO CUSTOMARY UNITS:

For convenience, some measurements are approximate.

LENGTH

25 millimetres	=	1 inch		
300 millimetres	=	30 centimetres	=	1 foot
100 centimetres	=	1 metre	=	40 inches
100 metres	=	110 yards		
1000 metres	=	1 kilometre	=	0.62 miles
1.6 kilometres	=	1 mile		
100 kilometres	=	62 miles		

SPEED

1 kilometre per hour		=	0.62 miles per hour	
1.6 kilometres per hour		=	1 mile per hour	
100 kilometres per hour		=	62 miles per hour	
160 kilometres per hour		=	100 miles per hour	
1 knot	=	1.85 kilometres per hour	=	1.15 miles per hour
10 knots	=	5.4 metres per second	=	12 miles per hour

AREA

1 square centimetre		=	0.15 square inches	
7 square centimetres		=	1 square inch	
929 square centimetres		=	1 square foot	
1 square metre		=	10.76 square feet	
4047 square metres		=	1 acre	
10,000 square metres	=	1 hectare	=	2.47 acres

PRESSURE

1 atmosphere	=	1.03 kilograms per square centimetre
1 atmosphere	=	14.69 pounds per square inch

VOLUME

1 cubic centimetre	=	0.061 cubic inches		
16.3 cubic centimetres	=	1 cubic inch		
196 cubic centimetres	=	1 cubic foot		
1000 cubic centimetres	=	1 litre	=	61 cubic inches
1000 litres	=	1 cubic metre	=	35.3 cubic feet

WEIGHT

1 gram	=	0.03 ounces		
28.3 grams	=	1 ounce		
453 grams	=	1 pound		
1000 grams	=	1 kilogram	=	2.2 pounds
1000 kilograms	=	1 metric ton	=	1.1 short tons or 2200 pounds

TEMPERATURE

1 degree Centigrade	=	1.8 degrees Fahrenheit
0.56 degrees C.	=	1 degree F.
10 degrees C.	=	50 degrees F.
20 degrees C.	=	68 degrees F.
30 degrees C.	=	86 degrees F.
100 degrees C.	=	212 degrees F.

It is a useful rule of thumb to remember that degrees Fahrenheit for most ordinary air temperatures can be obtained by doubling the degrees Centigrade and adding 30.

This works precise at 10°C – at higher temperatures, it is necessary to subtract a degree or two, at lower temperatures to add.

INDEX

Main references are in **bold**. A page number preceded by 'q' (e.g. q151) denotes a reference to, or a quotation from, an author named in the bibliography (which is itself alphabetically arranged) but not named with the quotation on the page.

TITLES IN SERIES

For a complete list of titles, visit www.nyrb.com or write to:
Catalog Requests, NYRB, 435 Hudson Street, New York, NY 10014

J.R. ACKERLEY My Dog Tulip*
J.R. ACKERLEY My Father and Myself*
J.R. ACKERLEY We Think the World of You*
HENRY ADAMS The Jeffersonian Transformation
RENATA ADLER Pitch Dark*
RENATA ADLER Speedboat*
AESCHYLUS Prometheus Bound; translated by Joel Agee*
ROBERT AICKMAN Compulsory Games*
LEOPOLDO ALAS His Only Son *with* Doña Berta*
CÉLESTE ALBARET Monsieur Proust
DANTE ALIGHIERI The Inferno
KINGSLEY AMIS The Alteration*
KINGSLEY AMIS Dear Illusion: Collected Stories*
KINGSLEY AMIS The Green Man*
KINGSLEY AMIS Lucky Jim*
KINGSLEY AMIS The Old Devils*
KINGSLEY AMIS Take a Girl Like You*
ROBERTO ARLT The Seven Madmen*
U.R. ANANTHAMURTHY Samskara: A Rite for a Dead Man*
IVO ANDRIĆ Omer Pasha Latas*
WILLIAM ATTAWAY Blood on the Forge
W.H. AUDEN (EDITOR) The Living Thoughts of Kierkegaard
W.H. AUDEN W.H. Auden's Book of Light Verse
ERICH AUERBACH Dante: Poet of the Secular World
EVE BABITZ Eve's Hollywood*
EVE BABITZ Slow Days, Fast Company: The World, the Flesh, and L.A.*
DOROTHY BAKER Cassandra at the Wedding*
DOROTHY BAKER Young Man with a Horn*
J.A. BAKER The Peregrine
S. JOSEPHINE BAKER Fighting for Life*
HONORÉ DE BALZAC The Human Comedy: Selected Stories*
HONORÉ DE BALZAC The Memoirs of Two Young Wives*
HONORÉ DE BALZAC The Unknown Masterpiece *and* Gambara*
VICKI BAUM Grand Hotel*
SYBILLE BEDFORD Jigsaw*
SYBILLE BEDFORD A Legacy*
SYBILLE BEDFORD A Visit to Don Otavio: A Mexican Journey*
MAX BEERBOHM The Prince of Minor Writers: The Selected Essays of Max Beerbohm*
STEPHEN BENATAR Wish Her Safe at Home*
FRANS G. BENGTSSON The Long Ships*
WALTER BENJAMIN The Storyteller Essays*
ALEXANDER BERKMAN Prison Memoirs of an Anarchist
GEORGES BERNANOS Mouchette
MIRON BIAŁOSZEWSKI A Memoir of the Warsaw Uprising*
ADOLFO BIOY CASARES The Invention of Morel
PAUL BLACKBURN (TRANSLATOR) Proensa*
CAROLINE BLACKWOOD Great Granny Webster*

* *Also available as an electronic book.*

G.B. EDWARDS The Book of Ebenezer Le Page*

JOHN EHLE The Land Breakers*

MARCELLUS EMANTS A Posthumous Confession

EURIPIDES Grief Lessons: Four Plays; translated by Anne Carson

J.G. FARRELL The Siege of Krishnapur*

J.G. FARRELL The Singapore Grip*

ELIZA FAY Original Letters from India

KENNETH FEARING The Big Clock

FÉLIX FÉNÉON Novels in Three Lines*

M.I. FINLEY The World of Odysseus

THOMAS FLANAGAN The Year of the French*

BENJAMIN FONDANE Existential Monday: Philosophical Essays*

SANFORD FRIEDMAN Conversations with Beethoven*

MARC FUMAROLI When the World Spoke French

CARLO EMILIO GADDA That Awful Mess on the Via Merulana

BENITO PÉREZ GÁLDOS Tristana*

MAVIS GALLANT The Cost of Living: Early and Uncollected Stories*

MAVIS GALLANT Paris Stories*

MAVIS GALLANT Varieties of Exile*

GABRIEL GARCÍA MÁRQUEZ Clandestine in Chile: The Adventures of Miguel Littín

LEONARD GARDNER Fat City*

WILLIAM H. GASS On Being Blue: A Philosophical Inquiry*

GE FEI The Invisibility Cloak

JEAN GENET Prisoner of Love

ÉLISABETH GILLE The Mirador: Dreamed Memories of Irène Némirovsky by Her Daughter*

NATALIA GINZBURG Family Lexicon*

FRANÇOISE GILOT Life with Picasso*

JEAN GIONO Hill*

JEAN GIONO A King Alone*

JOHN GLASSCO Memoirs of Montparnasse*

P.V. GLOB The Bog People: Iron-Age Man Preserved

NIKOLAI GOGOL Dead Souls*

EDMOND AND JULES DE GONCOURT Pages from the Goncourt Journals

ALICE GOODMAN History Is Our Mother: Three Libretti*

EDWARD GOREY (EDITOR) The Haunted Looking Glass

JEREMIAS GOTTHELF The Black Spider*

A.C. GRAHAM Poems of the Late T'ang

JULIEN GRACQ Balcony in the Forest*

HENRY GREEN Caught*

HENRY GREEN Loving*

HENRY GREEN Nothing*

WILLIAM LINDSAY GRESHAM Nightmare Alley*

HANS HERBERT GRIMM Schlump*

VASILY GROSSMAN An Armenian Sketchbook*

VASILY GROSSMAN Everything Flows*

VASILY GROSSMAN Life and Fate*

VASILY GROSSMAN Stalingrad*

LOUIS GUILLOUX Blood Dark*

OAKLEY HALL Warlock

PATRICK HAMILTON The Slaves of Solitude*

PATRICK HAMILTON Twenty Thousand Streets Under the Sky*

PETER HANDKE Short Letter, Long Farewell

PETER HANDKE Slow Homecoming